Problems for General Chemistry and Qualitative Analysis

Problems for General Chemistry and Qualitative Analysis

Fourth Edition

C. J. Nyman
Washington State University, Pullman

G. B. King
Washington State University, Pullman

J. A. Weyh
Western Washington University, Bellingham

John Wiley & Sons
New York Chichester Brisbane Toronto

Cover photo by FUNDAMENTAL PHOTOGRAPHS, New York

Library of Congress Cataloging in Publication Data

Nyman, Carl John, 1924–
  Problems for general chemistry and quantitative
analysis.

  Includes index.
  1. Chemistry—Problems, exercises, etc. 2. Chemistry,
Analytic—Qualitative. I. King, George Brooks,
1905– joint author. II. Weyh, J. A., joint
author. III. Title.
QD42.N98 1980 540′.76 79-24489
ISBN 0-471-05299-X

Printed in the United States of America
10 9 8 7 6 5 4 3 2 1

Preface to the Fourth Edition

The fourth edition represents a thorough revision of this problem book. All chapters have been extensively rewritten and updated to accommodate the changes in general chemistry that have occurred since the appearance of our previous editions.

We still retain, however, the basic objectives of the first edition: (1) to furnish students with simple and direct methods for solving numerical problems that illustrate chemical principles and to provide for them many examples of the types of problems encountered in a general chemistry course, and (2) to give instructors supplementary problems and exercises for use in tutorial and recitation sections and as homework assignments.

The arrangement is designed to be self-teaching. The subject matter has been divided into chapters in which the principles needed for solving a given type of problem are presented briefly at the start of each major section. These introductions are followed by examples of problems embodying the principles discussed, for which detailed solutions are provided. At the ends of the example sections are sets of problems involving the principles illustrated. Then, at the conclusion of each chapter, a set of general problems covering all the principles presented in the chapter is given. Answers to all numerical problems are found in Appendix XI.

This book represents a strong commitment to the dimensional analysis method of problem solving and strict adherence to significant figures is observed in numerical solutions. The SI system of units is introduced and certain of these units are used throughout. Most of the problems of the previous edition have been changed, and many new examples and problems have been added.

We owe special thanks to the Western Washington University Bureau for Faculty Research and especially Florence Preder for support in the typing of this manuscript. Acknowledgment is made to the following reviewers for their work in helping to refine the final manuscript: Jefferson Davis, University of South Florida; Barbara Drescher, Middlesex County College; Joseph R. Crook, Western Washington University; and Alice S. Corey, Pasadena City College. We also thank Richard W. Cohen for his help in obtaining and checking solutions to all the problems.

C. J. Nyman
G. B. King
J. A. Weyh

Contents

Problems for General Chemistry and Qualitative Analysis

Chapter 1
Introduction

Every student in beginning chemistry will be expected to solve many numerical problems since most of the principles of chemistry are best illustrated by the use of mathematical concepts. The student is apt to find that the most effective way to learn these principles is to work many, many problems. To do this one must be familiar with the elementary operations of mathematics; hence one should carefully and thoroughly review these methods before attempting chemical problems. A brief mathematical review is found in the Appendixes.

To solve any problem, certain logical steps should be followed:

1. Read the problem carefully to get a clear understanding of just what is desired.

2. Tabulate the data that are given, **including the units** in which the data are expressed. At this point you will know what is given and what is unknown.

3. Develop a plan for solving the problem. To do this you must determine the chemical principle which applies.

4. If a chemical reaction is involved, write the balanced equation.

5. If there is a mathematical relationship between the items of data given, set up the proper mathematical formulation.

6. Work out the problem by direct reasoning from the chemical and mathematical principles that apply. After obtaining an answer, consider it carefully to determine if it is **possible** and **reasonable**.

7 Some problems involve several steps and in reality are two or three consecutive problems. Always keep in mind the objective of the problem and the units in which the answer is to be expressed.

8. When arriving at an answer, make certain you obtain the proper number of significant figures. The necessary rules to be followed are outlined in Appendix IV.

We suggest that the student make the utmost use of pencil and paper in solving problems. Following the procedure outlined above, write down in sequence the details of the steps involved. The solution should be recorded neatly on paper in such a manner that the procedure followed is obvious to anyone examining it. Recording unconnected bits of information helter-skelter on a sheet of paper and then expecting someone else to follow your line of reasoning is unrealistic and often unacceptable. Remember that a clearly and concisely presented solution to a problem makes a good impression on the instructor, while a poorly presented one makes a bad impression. Great effort should be made to cultivate proper work habits; they will reap benefits not only in presenting the solutions to homework and examination questions in chemistry, but also in presenting solutions to all types of problems encountered by students throughout their entire careers.

Chapter 2

Units of Measurement

2-A. Metric System

The metric system of measurement is used universally in scientific work. This is a decimal system in which changes are made from one unit to another merely by moving the decimal point an appropriate number of places.

In 1960 an international group, the General Conference of Weights and Measures, reached agreement on the preferred usage of certain metric units and units derived from them. The **International System of Units** (abbreviated SI) is based on seven units, six of which are important to the chemist. These six units are listed in Table 2-1. Also included is the **derived** unit for volume.

Table 2-1

Measurement	SI Unit	Abbreviation
length	meter	m
mass	kilogram	kg
time	second	s
temperature	kelvin	K
amount of substance	mole	mol
electric current	ampere	A
volume*	cubic meter	m^3

* Derived from the basic linear measurement unit, meter.

Table 2-2

Name	Abbreviation	Factor	Name	Abbreviation	Factor
pico	p	10^{-12}	deci	d	10^{-1}
nano	n	10^{-9}	kilo	k	10^{3}
micro	μ	10^{-6}	mega	M	10^{6}
milli	m	10^{-3}	giga	G	10^{9}
centi	c	10^{-2}	tera	T	10^{12}

Common prefixes to the names of the various metric units, their abbreviations, and the factors relating each to a standard quantity are shown in Table 2-2.

In introductory chemistry it is not convenient to use all the SI base units as our reference point. For instance, a chemistry student seldom works with kilograms of material. A much more convenient reference point is the gram.

Table 2-3 Units of Measurement

Linear Measurement

The basic unit of length is the **meter** (m) (about 39.37 inches).

1 kilometer (km) = 10^3 meters	1 micrometer (μm) = 10^{-6} meter
1 decimeter (dm) = 10^{-1} meter	1 nanometer (nm) = 10^{-9} meter
1 centimeter (cm) = 10^{-2} meter	1 angstrom (Å) = 10^{-10} meter
1 millimeter (mm) = 10^{-3} meter	1 picometer (pm) = 10^{-12} meter

Mass Measurement

The reference unit of mass is the **gram** (g) (about 1/454 pound).

1 kilogram (kg) = 10^3 grams	1 nanogram (ng) = 10^{-9} gram
1 milligram (mg) = 10^{-3} gram	1 picogram (pg) = 10^{-12} gram
1 microgram (μg) = 10^{-6} gram	

Volume Measurement

The reference unit of volume is the **liter** (l) (about 1.06 quarts).

1 liter = exactly 10^3 milliliters (ml)	1 liter = 1 cubic decimeter (dm^3)
1 liter = exactly 10^3 cubic centimeters (cm^3)	1 liter = 10^{-3} cubic meter (m^3)
Hence 1 ml = 1 cm^3 = 10^{-3} liter	

Similarly, chemists normally work with milliliters of liquids and not cubic meters. For example, a common volume of 25 milliliters would translate into 2.5×10^{-5} m^3. In this book we use as our reference point for mass and volume the units gram and liter, respectively. We should point out, however, that, although they are not preferred, the gram and liter are acceptable SI units.

The more commonly encountered units of length, mass, and volume in chemistry and other sciences are tabulated in Table 2-3. Frequently used prefixes for these units are also included. The SI units second, mole, and ampere are introduced later in our study of chemistry. The unit K is discussed in Section 2-C.

2-B. Numbers, Units, and Dimensional Analysis

Most numbers that appear in problems are measurements on some scale and as such must include the units in which the quantities are expressed. Only occasionally do we encounter numbers that are unitless. Students should acquire the habit of always writing down the units following the numerical values when appropriate. These units should be carried along with the numbers through all the mathematical operations in multiplication, division, and so on. In a formulation involving several terms, cancellations should be made wherever possible; this applies to units as well as numbers. This procedure or process is called **dimensional analysis**. Throughout this book, we have tried to follow this practice in the solutions of problems worked as examples. To illustrate, let us analyze the following very simple problem.

How many pennies are equivalent to two nickels?

The answer of course is 10 pennies. But, more important, what logical thought process did we use to get to the answer?

First, we know that 5 pennies = 1 nickel. An equivalent statement would be 5 pennies per 1 nickel or 5 pennies/1 nickel.

Hence we see that the answer to our problem is obtained by multiplying the given quantity by our determined **conversion factor** between pennies and nickels. Multiply and divide units just as we do numbers.

$$2\ \cancel{\text{nickels}} \times \frac{5\ \text{pennies}}{1\ \cancel{\text{nickel}}} = 10\ \text{pennies}$$

Note that we could also express 5 pennies = 1 nickel as 1 nickel per 5 pennies or 1 nickel/5 pennies.

We do **not** use this equivalent statement to solve our problem, however, because the answer must have the unit pennies. Using the factor 1 nickel/5 pennies we obtain

$$2 \text{ nickels} \times \frac{1 \text{ nickel}}{5 \text{ pennies}} = 0.4 \frac{\text{nickel}^2}{\text{penny}}$$

This, of course, bears no resemblance to the answer required and hence, because of dimensional analysis, we know this factor does not correctly solve the problem. In effect we have been told by the units that we must devise a conversion factor that will yield, as an answer, the proper unit "pennies," that is, the factor must be 5 pennies/1 nickel.

Illustrations more closely aligned with our purpose in chemistry now follow.

Suppose we are asked to convert 100 g to kilograms. Using the prefixes for the metric system, we determine that 1 kg = 1000 g or 1 kg/1000 g. Hence

$$\text{Number of kilograms} = 100 \cancel{\text{g}} \times \frac{1 \text{ kg}}{1000 \cancel{\text{g}}} = 0.1 \text{ kg}$$

Perhaps we want to convert 50 mg to kilograms. Since we know that 1 mg = 0.001 g and 1 g = 0.001 kg, we can write

$$\text{Number of kg} = 50 \cancel{\text{mg}} \times \frac{0.001 \cancel{\text{g}}}{1 \cancel{\text{mg}}} \times \frac{0.001 \text{ kg}}{1 \cancel{\text{g}}} = 0.00005 \text{ kg}$$

These represent good illustrations of conversion from one unit to another by the process of dimensional analysis.

Example 2-a. Add the following masses and express the answer decimally in grams: 0.00200 kg, 450 mg, 30.0 cg, 0.11 dg.

Solution. Each unit must be converted to grams or some other convenient common unit prior to addition.

$$0.00200 \cancel{\text{kg}} \times \frac{1000 \text{ g}}{1 \cancel{\text{kg}}} = 2.00 \text{ g}$$

$$450 \cancel{\text{mg}} \times \frac{1 \text{ g}}{1000 \cancel{\text{mg}}} = 0.45 \text{ g}$$

$$30.0 \cancel{\text{cg}} \times \frac{1 \text{ g}}{100 \cancel{\text{cg}}} = 0.300 \text{ g}$$

$$0.11 \cancel{\text{dg}} \times \frac{1 \text{ g}}{10 \cancel{\text{dg}}} = 0.011 \text{ g}$$

$$\text{Total} = 2.761 \text{ g}$$

When it is rounded to the proper number of significant figures (see Appendix IV), the number becomes 2.76 g.

Example 2-b. Determine the number of liters in a box with the dimensions: 3.02 m long, 25 cm deep, and 41 mm wide.

Solution. Since 1 liter = 1000 cm^3, convert each dimension to centimeters. The volume is then obtained in cubic centimeters by multiplying length × width × depth. Finally, multiply by 1 liter/1000 cm^3 to obtain the number of liters.

$$\text{Volume} = 302\ \text{cm} \times 25\ \text{cm} \times 4.1\ \text{cm} = 30{,}955\ \text{cm}^3$$

This number must be rounded to 31,000 cm^3 to obtain the proper number of significant figures.

$$\text{Volume} = 31{,}000\ \cancel{\text{cm}^3} \times \frac{1\ \text{liter}}{1000\ \cancel{\text{cm}^3}} = 31\ \text{liters}$$

Example 2-c. The surface area of a crystal is found to be 1.9×10^2 mm^2. Calculate the area in m^2.

Solution. We know that 1 meter = 1000 mm. Note that the area is given in **square** millimeters or mm^2. The square term means mm × mm and hence we **must** cancel mm **twice** in our conversion.

$$1.9 \times 10^2\ \text{mm}^2 = 1.9 \times 10^2\ \cancel{\text{mm}^2} \times \frac{1\ \text{m}}{1000\ \cancel{\text{mm}}} \times \frac{1\ \text{m}}{1000\ \cancel{\text{mm}}}$$

$$= 1.9 \times 10^{-4}\ \text{m}^2$$

The following solution is **incorrect** as affirmed by attention to the units:

$$1.9 \times 10^2\ \text{mm}^2 = 1.9 \times 10^2\ \text{mm}^{\cancel{2}\,1} \times \frac{1\ \text{m}}{1000\ \cancel{\text{mm}}} = 1.9 \times 10^{-1}\ (\text{mm})\,(\text{m})$$

Note here that the final unit is **not** the desired unit m^2.

Example 2-d. A drop of oleic acid having a volume of 0.054 cm^3 is spread out on a surface to a uniform thickness of exactly 10 Å. What surface area in square meters is covered?

Solution.

$$\text{Area} = \frac{\text{Volume}}{\text{Thickness}}$$

Since the area is wanted in square meters, the volume should be expressed in m^3 and thickness in m.

$$\text{Volume in m}^3 = 0.054\ \cancel{\text{cm}^3} \times \frac{0.010\ \text{m}}{1\ \cancel{\text{cm}}} \times \frac{0.010\ \text{m}}{1\ \cancel{\text{cm}}} \times \frac{0.010\ \text{m}}{1\ \cancel{\text{cm}}}$$

$$= 5.4 \times 10^{-8}\ \text{m}^3$$

$$\text{Thickness in meters} = 10\ \cancel{\text{Å}} \times \frac{10^{-10}\ \text{m}}{1\ \cancel{\text{Å}}} = 1.0 \times 10^{-9}\ \text{m}$$

$$\text{Area} = \frac{\text{Volume (m}^3)}{\text{Thickness (m)}} = \frac{5.4 \times 10^{-8}\ \text{m}^{\cancel{3}2}}{1.0 \times 10^{-9}\ \cancel{\text{m}}}$$

$$= 5.4 \times 10^{1}\ \text{m}^2 = 54\ \text{m}^2$$

Problems (Metric System)

2-1. Add the following masses and express the answer decimally in grams: 5.000000 kg, 491 mg, 61.9 cg, 221.000 g, 8.14 dg.

2-2. Add the following masses and express the sum decimally in grams: 0.8 cg, 91 mg, 0.35 dg, 0.003050 kg, 0.127 g.

2-3. Add the following units of linear measurement and express the sum decimally in meters: 360.0 cm, 7.20000 km, 490 mm, 6.900 dm, 82.31 m.

2-4. Assuming 3.50 liters of paint are spread uniformly on a wood surface to a thickness of 100.0 nm, what surface area in square meters would be covered?

2-5. Suppose 3.0 liters of gasoline are spilled on lake water. If a gasoline film is formed that covers an area of $1.0 \times 10^6\ \text{m}^2$ what is the average thickness of the film in Å units?

2-6. Calculate the number of liters in 1 m^3.

2-7. Make the calculations indicated:

(a) $6.00 \times 10^4\ \mu\text{m}$ = ________ m
(f) 6.7 kg = ________ g
(b) 57.0 mg = ________ kg
(g) 0.92 cm = ________ m
(c) 27.6 ml = ________ liter
(h) 7.53 liters = ________ cm^3
(d) 10.0 mm^3 = ________ m^3
(i) 531 mg = ________ kg
(e) 3.5 dm^3 = ________ liter
(j) 2.2 km = ________ cm

2-8. Yellow light in the middle range of the visible spectrum has a wavelength of about 5.8×10^{-5} cm. Express this wavelength in micrometers, nanometers, and angstrom units.

2-C. Temperature Scales

The relationships between the Fahrenheit, Celsius (centigrade), and Kelvin (absolute) scales of temperature are shown in Figure 2-1. Note that between

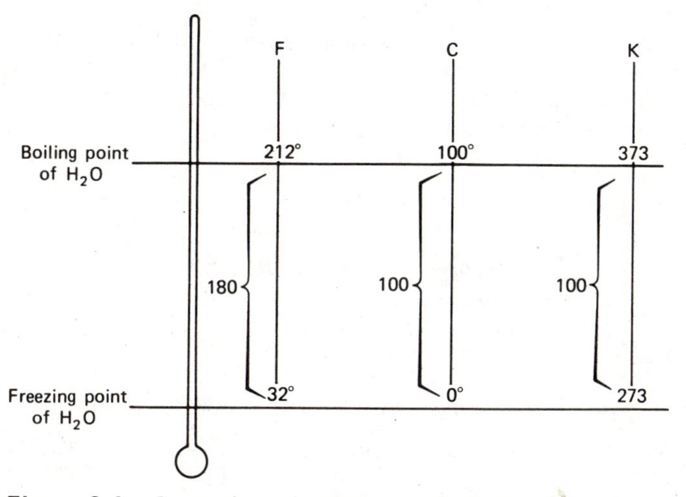

Figure 2-1 Comparison of temperature scales: F = Fahrenheit, C = Celsius or Centigrade, K = Kelvin or absolute.

the fixed points—the freezing and boiling points of water—there are 180 divisions or degrees on the Fahrenheit scale and 100 divisions or degrees on the Celsius and Kelvin scales. Hence 1 Celsius degree = 180/100 = 9/5 = 1.8 Fahrenheit degrees* or 1 Fahrenheit degree = 5/9 Celsius degree. To convert Celsius temperature to Fahrenheit temperature, multiply the Celsius reading (°C) by 9/5, then add 32 since the reference point for the °F scale is 32 and not 0:

$$°F = \tfrac{9}{5}°C + 32 \qquad (1)$$

To convert from Fahrenheit to Celsius, first subtract 32 from the Fahrenheit reading, then multiply by 5/9:

$$°C = (°F - 32)\tfrac{5}{9} \qquad (2)$$

* 1 Celsius degree $= 1\ \text{\sout{Celsius degree}} \times \dfrac{180 \text{ Fahrenheit degrees}}{100\ \text{\sout{Celsius degrees}}} = 1.8$ Fahrenheit degrees

On the Celsius and Kelvin scales, the degrees are the same size, but the Kelvin reading (K) will be 273 higher than the Celsius reading:

$$K = {}^\circ C + 273 \tag{3}$$

Example 2-e. What is the Fahrenheit equivalent of 20°C?

Solution. Each Celsius degree is equivalent to 9/5 Fahrenheit degrees. Therefore, 20 degrees or divisions on the Celsius scale correspond to $20 \times 9/5 = 36$ divisions or degrees on the Fahrenheit scale. However, since these degrees or divisions are measured above the freezing point of water taken as 32, this point must be added; therefore,

$$^\circ F = \tfrac{9}{5}(20) + 32 = 36 + 32 = 68$$

Example 2-f. Iron begins to glow red at about 900°F. What is this temperature on the Celsius scale?

Solution.

$$^\circ C = (^\circ F - 32) \times \tfrac{5}{9} = (900 - 32)\tfrac{5}{9} = 482$$

Example 2-g. What is the Fahrenheit temperature corresponding to absolute zero on the Kelvin scale?

Solution. We must change the Kelvin temperature first to Celsius temperature, then to Fahrenheit.

$$K = {}^\circ C + 273$$

If $K = 0$, then $^\circ C = -273$

$$^\circ F = \tfrac{9}{5}(-273) + 32 = -459$$

Problems (Temperature Scales)

2-9. Make the following temperature conversions:

(a) 65°C to °F (f) 0°F to °C
(b) 98.6°F to °C (g) −77°C to K
(c) −273°C to K (h) −319°F to K
(d) 158 K to °F (i) 212°F to K
(e) −95°F to °C (j) 323 K to °F

2-10. Make the temperature conversions as indicated by the blanks in the following table.

	Fahrenheit	Celsius	Kelvin (absolute)
(a)	———	———	273
(b)	———	125°	———
(c)	203°	———	———
(d)	23°	———	———
(e)	———	−20°	———
(f)	113°	———	———
(g)	———	———	173
(h)	———	35°	———
(i)	392°	———	———
(j)	11.4°	———	———
(k)	———	———	50

2-11. At what temperature will a Celsius and a Fahrenheit thermometer have the same reading?

2-D. Density and Specific Gravity

Density is a measure of the mass of a substance in relation to its volume and is defined

$$\text{Density} = \frac{\text{Mass}}{\text{Volume}}$$

Any units of mass and volume may be used, but they must always be specified. For example, if 10 ml of a liquid has a mass of 8 g, its density is 8 g/10 ml = 0.8 g/ml. In chemical work, densities are usually expressed as grams per milliliter or grams per cubic centimeter.

Specific gravity is a **number** that is the ratio of the masses of equal volumes of the substance in question and some standard substance. In other words, it is a ratio of the density of the substance in question and the density of a standard substance, with both densities expressed in the same units. For liquids and solids, the standard chosen is usually water with a density of 1.00 g/ml, and when the comparison is to water in this unit, the specific gravity of a substance is **numerically** equal to its density. For example, if 10.0 ml of a metal has a mass of 195 g, its density is 195 g/10.0 ml = 19.5 g/ml. Its specific gravity is 19.5 since specific gravity equals

$$\frac{\text{Density of metal}}{\text{Density of water}} = \frac{19.5 \not{g/ml}}{1.00 \not{g/ml}} = 19.5$$

Example 2-h. A cube of platinum (Pt) metal 10.0 cm along each edge has a mass of 21.5 kg. What is the density of Pt in g/cm^3?

Solution.

$$\text{Volume of cube} = (10.0\ \text{cm})^3 = 1.00 \times 10^3\ \text{cm}^3$$

$$\text{Mass in grams} = 21.5\ \not{kg} \times \frac{1000\ \text{g}}{\not{kg}} = 21{,}500\ \text{g}$$

$$\text{Density} = \frac{\text{Mass}}{\text{Volume}} = \frac{21{,}500\ \text{g}}{1.00 \times 10^3\ \text{cm}^3} = 21.5\ \text{g/cm}^3$$

Example 2-i. The specific gravity of concentrated sulfuric acid, H_2SO_4, is 1.85. What volume of H_2SO_4 would have a mass of 925 g?

Solution.

$$\text{Specific gravity} = \frac{\text{Density of } H_2SO_4}{\text{Density of } H_2O}$$

The density of H_2O is 1.00 g/ml, hence

$$\text{Density of } H_2SO_4 = (1.85)(1.00\ \text{g/ml}) = 1.85\ \text{g/ml}$$

$$\text{Volume} = \frac{\text{Mass}}{\text{Density}} = \frac{925\ \text{g}}{1.85\ \text{g/ml}}$$

$$= 925\ \not{g} \times \frac{1\ \text{ml}}{1.85\ \not{g}} = 5.00 \times 10^2\ \text{ml}$$

Example 2-j. A rectangular piece of metal measuring 20.0 cm × 70.0 mm × 3.00 dm has a mass of 46.2 kg. What is the density of the metal in grams per cubic centimeter?

Solution.

$$\text{Volume of metal} = 20.0\ \text{cm} \times 7.00\ \text{cm} \times 30.0\ \text{cm} = 4.20 \times 10^3\ \text{cm}^3$$

$$\text{Density} = \frac{\text{Mass}}{\text{Volume}}$$

$$= \frac{46.2\ \not{kg}}{4.20 \times 10^3\ \text{cm}^3} \times \frac{1000\ \text{g}}{1\ \not{kg}} = 11.0\ \text{g/cm}^3$$

Example 2-k. Assuming that atoms of mercury, Hg, are spherical in the liquid state, calculate the percentage of unoccupied space in liquid mercury if the radius of a mercury atom is 1.56 Å. The density of liquid mercury is 13.6 g/ml. Volume of a sphere = $\frac{4}{3}\pi r^3$ where r is the radius.

Solution.

$$\text{Volume of 1 atom} = \tfrac{4}{3}\pi(1.56 \times 10^{-8}\ \text{cm})^3 = 1.59 \times 10^{-23}\ \text{cm}^3$$

Volume of 1 mole* (atomic volume)

$$= \frac{1.59 \times 10^{-23}\ \text{cm}^3}{1\ \cancel{\text{atom}}} \times \frac{6.02 \times 10^{23}\ \cancel{\text{atoms}}}{1\ \text{mole}}$$

$$= 9.57\text{cm}^3/\text{mole}$$

$$\text{Theoretical density} = \frac{\text{Atomic weight}}{\text{Atomic volume}}$$

$$= \frac{200.6\ \text{g}/\cancel{\text{mole}}}{9.57\ \text{cm}^3/\cancel{\text{mole}}} = 21.0\ \text{g/cm}^3$$

By comparing the theoretical density with the actual density the percent of unoccupied space may be calculated thus:

$$\%\ \text{Unoccupied space} = \frac{\text{Theoretical density} - \text{Actual density}}{\text{Theoretical density}} \times 100$$

$$\%\ \text{Unoccupied space} = \frac{21.0 - 13.6}{21.0} \times 100 = 35.2\%$$

Problems (Density and Specific Gravity)

2-12. A piece of galena (PbS) with a mass of 364.5 g displaces 48.6 ml of water. What is the density of galena (PbS)?

2-13. Lead has a density of 11.4 g/cm^3. What is the mass of a lead brick measuring 20.0 cm long × 13.3 cm wide × 6.50 cm high?

2-14. The specific gravity of mercury, Hg, is 13.60 relative to water. What volume of Hg would have the same mass as 476 ml of H_2O?

* 1 mole = 6.02×10^{23} particles; in this case, 6.02×10^{23} atoms.

2-15. Assuming that bromine, Br_2, molecules are spherical, determine the percentage of vacant space in liquid bromine that has a density of 2.93 g/ml. Assume the radius of a Br_2 molecule to be 2.28 Å.

General Problems

2-16. A metal tank has the dimensions 30.0 cm × 75.0 mm × 0.800 m. What is the volume in liters?

2-17. A brick of silver with a mass of 31.5 kg measures 0.100 m long, 40.0 cm wide, and 75.0 mm thick. Calculate the density of silver in grams per cubic centimeter.

2-18. What is the density in grams per milliliter of a block of wood that has a mass of 648 g and a volume of 0.800 dm^3?

2-19. A chunk of sulfur, mass 50.9 g, displaces 24.6 ml of water. What is the density of sulfur?

2-20. Water has a density of 1.0 g/cm^3. What is the density of water in kilograms per cubic meter?

2-21. What mass of alcohol can be contained in a tank whose volume is 345 liters if the density of alcohol is 0.79 g/ml?

2-22. What mass of mercury, density 13.6 g/ml, will occupy the same volume as 45.0 g of chloroform, density 1.50 g/ml?

2-23. What volume, in liters, would be occupied by 772 g of gold? The density of gold is 19.3 g/cm^3.

2-24. Concentrated (100%) sulfuric acid has a specific gravity of 1.85 or a density of 1.85 g/ml. What volume of 100% H_2SO_4 should be measured out for an experiment requiring 0.370 kg of the acid?

2-25. Exactly one cubic decimeter of platinum has a mass of 21.5 kg. Calculate the specific gravity of platinum (relative to water).

2-26. The volume of a cylinder is given by the formula $\pi r^2 h$, where r is the radius and h is the height. Calculate the capacity of the cylinder in liters if the radius is 2.0 dm and the height is 0.40 m.

2-27. How many grams of alcohol would a spherical container with a diameter of 34 cm inside measurement hold if the density of the alcohol is 0.79 g/ml? Volume of sphere $= \frac{4}{3}\pi r^3$, where $r =$ radius.

2-28. Copper metal has a density of 8.92 g/cm^3. What is the mass of a cube of copper that measures 0.250 m along each edge?

2-29. Four liters of paint are used to cover an area of 2×10^2 m^2. Assuming the coat of paint has uniform thickness, what is the thickness in micrometers?

2-30. Gold may be hammered to a thickness of 2.5×10^{-5} cm. Starting with 5.0 g of gold, what area in square meters of gold foil 2.5×10^{-5} cm thick could be obtained? The density of gold is 19 g/cm^3.

2-31. What is the temperature difference in °F between −60°C and 163 K?

2-32. Determine the temperature at which (a) a Celsius and a Fahrenheit thermometer will read numerically the same but opposite in sign, (b) a Fahrenheit thermometer will read 20 degrees higher than a Celsius thermometer, and (c) a Kelvin thermometer has the same reading as a Fahrenheit thermometer.

2-33. An iceberg of mass 9.2×10^{10} kg, with a density of 0.92 g/ml, will have what volume above the waterline? *Hint*: A floating object displaces its own mass of water.

2-34. The spherical atomic nucleus of calcium has a diameter of 1.00×10^{-12} cm, and 6.02×10^{23} of these nuclei have a mass of 40.1 g. Calculate the approximate density of this nuclear matter in metric tons per cubic centimeter. The volume of a sphere is $\frac{4}{3}\pi r^3$, where r is the radius; 1 metric ton = 1000 kg.

2-35. From the beginning of time to 1980 the world production of gold, density 19.3 g/ml, has been estimated to be approximately 64 million kilograms. Assuming that all this gold is packed together in a cube, what would be the approximate length of a side of the cube in meters?

2-36. 0.0010 ml of human blood is diluted to 1.0 liter. On microscopic examination, the diluted sample is found to contain seven red corpuscles per cubic millimeter. What is the red corpuscle count per cubic centimeter of the original blood?

2-37. Assuming that sea water contains 0.040 mg of gold per cubic meter, what mass of sea water in metric tons (1000 kg) would contain 8.0 g of gold? Assume the density of sea water is 1.03 g/ml.

2-38. Bromine is recovered commercially from sea water that contains 65 parts by weight of this element per 1.00 million parts of water. Assuming 100% recovery and that sea water has a density of 1.03 g/cm^3, calculate the volume in liters of sea water required to produce 8.0×10^1 g of bromine.

2-39. One mole of atoms of any element contains 6.02×10^{23} atoms. Assuming that every individual on the earth's surface (population 4 billion) is engaged in counting the atoms in one mole of an element at the rate of 2 per second and working 24 hours per day and 365 days a year, how long would it take to complete the job?

2-40. The density of liquid cesium is 1.87 g/ml and its atomic radius is 2.62 Å. Calculate the percentage of vacant space in liquid Cs, assuming spherical atoms.

2-41. A certain hypothesis proposes that the interior of the earth is composed of three principal layers. The outer layer is primarily a crust of rock that has a density of about 3.0 g/ml and is 16 km thick. The next layer is composed of molten silicates of density about 4.4 g/ml and is approximately 3.0×10^3 km thick. The inner core, thought to be composed of molten iron with some cobalt and nickel, has a radius of 3600 km and a density that averages about 11.0 g/ml. From this information (a) estimate the mass of the earth in kilograms, and (b) the average density of the earth in grams per milliliter.

Chapter 3

Atoms, Molecules, and Formulas: The Mole Concept

3-A. Atomic, Molecular, and Formula Weights

Atomic Weight (At Wt)

The atomic weight of an element is a number that expresses the mass of one atom of the element relative to an atom of carbon, ^{12}C, which has been chosen as the standard and assigned a mass of exactly 12. Fluorine, for example, has an atomic weight of 19. This means that one atom of fluorine is 19/12 as heavy as one atom of ^{12}C; aluminum has an atomic weight of 27, which means that one atom of aluminum is 27/12 as heavy as one atom of ^{12}C.

Molecular Weight (MW)

The molecular weight refers to the mass of one molecule of a substance relative to the mass of one atom of ^{12}C. A molecule is represented by a chemical formula that shows the number and types of atoms that compose the substance. The molecular weight of the molecule is obtained by adding the atomic weights of all atoms in the formula for the substance as shown in Example 3-a.

Example 3-a. Sulfuric acid, H_2SO_4, is composed of two H atoms, one S atom, and four O atoms with atomic weights 1, 32 and 16, respectively. What is the molecular weight of H_2SO_4?

Solution.

Two atoms H have mass 2 × the atomic weight of H = 2 × 1 = 2
One atom S has mass 1 × the atomic weight of S = 1 × 32 = 32
Four atoms O have mass 4 × the atomic weight of O = 4 × 16 = 64
MW H_2SO_4 = 98

Therefore, one molecule of H_2SO_4 is 98/12 as heavy as one atom of ^{12}C.

Formula Weight (FW)

In dealing with ionic compounds and others where discrete molecules do not exist, it is customary to use the term formula weight instead of molecular weight. It is obtained in the same way as a molecular weight—that is, the sum of the masses of all atoms in the formula for the substance. Formula weight may be used for all substances whether ionic or molecular. Thus the formula weight of H_2SO_4 is 98; the formula weight of NaCl is 58.5, and so on. The distinction between formula weight and molecular weight lies entirely in the desirability of using precise terminology. The term molecular weight should be reserved for those substances that do in fact exist in the form of molecules as represented by the chemical formula. Insofar as chemical calculations are concerned, formula weights and molecular weights are handled by identical procedures.

Example 3-b. The ionic compound zinc nitrate has the formula $Zn(NO_3)_2$. Determine the formula weight of $Zn(NO_3)_2$.

Solution. One formula unit of $Zn(NO_3)_2$ is composed of 1 zinc, 2 × 1 = 2 nitrogens, and 2 × 3 = 6 oxygens. Hence,

1 Zn has mass 1 × atomic weight of Zn = 1 × 65.4 = 65.4
2 N have mass 2 × atomic weight of N = 2 × 14.0 = 28.0
6 O have mass 6 × atomic weight of O = 6 × 16.0 = 96.0
FW $Zn(NO_3)_2$ = 189.4

Problems (Atomic, Molecular, and Formula Weights)

3-1. A certain atom is 186/31.0 as heavy as a single ^{12}C atom. What is its atomic weight?

3-2. Determine the molecular or formula weight for each of the following substances. See the inside front cover for a table of atomic weights. Express the answer to three significant figures.

(a) SbH_3 (d) Rb_2O (g) $NaClO_4$
(b) CdO (e) $POBr_3$ (h) $K_2Cr_2O_7$
(c) $LiNH_2$ (f) H_3PO_4 (i) $Na_2S_2O_3$

3-3. A particular molecule is 37.0/3.00 as heavy as a single ^{12}C atom. What is its molecular weight?

3-4. Determine the molecular or formula weight for each of the following. Express the answer to three significant figures.

(a) $NaHCO_3$	(d) $Al(NO_3)_3$	(g) $Co(C_2H_3O_2)_2$
(b) $HC_2H_3O_2$	(e) $(NH_4)_3AsO_4$	(h) $Ba_3(PO_4)_2$
(c) $Mg(NO_2)_2$	(f) $C_{12}H_{22}O_{11}$	(i) $Cr_2(SO_4)_3$

3-5. The formula weight of the compound Na_3MF_6 is 2.10×10^2. What is the atomic weight of M?

3-6. A molecule has the formula X_4 and its molecular weight is 124. Determine the atomic weight of X.

3-B. Moles and the Mole Concept

Because of the extremely small size and mass of atoms and molecules it is unrealistic to attempt to work with one or even a dozen such units in the laboratory. A convenient number of particles for the chemist to work with is 6.02×10^{23}.* This number is sometimes called Avogadro's number. Whenever we are dealing with 6.02×10^{23} particles of a substance we are working with one **mole** of the substance. Hence we see that the term mole defines a certain number of units just as the term dozen implies twelve of those units.

Why is Avogadro's number convenient for the chemist to use? At first glance (and even later) it appears that 6.02×10^{23} is cumbersome and far from convenient. The answer to the convenience question is as follows: To obtain 6.02×10^{23} **atoms** of any element one simply weighs out an amount of the element equal to its atomic weight expressed in grams. Hence one mole of carbon atoms is obtained by weighing out 12.0 g of carbon. 23.0 g of sodium contain 6.02×10^{23} atoms of sodium (one mole of sodium). Similarly, we can say that 1.00 mole of sulfur atoms contains 6.02×10^{23} sulfur atoms and has a mass of 32.1 g.

A similar relationship exists when considering molecules. Hence, the molecular weight of a compound, expressed in grams, contains 6.02×10^{23} **molecules** of that compound. Consequently, we see that 1.00 mole of carbon dioxide (CO_2; MW = 44.0) has a mass of 44.0 g and contains 6.02×10^{23} molecules of CO_2; 1.00 mole of table sugar ($C_{12}H_{22}O_{11}$; MW = 342) contains 6.02×10^{23} sugar molecules and has a mass of 342 g.

The same relationship pertains to ionic substances. When we say 1.00 mole of KCl we are specifying a sample that contains 6.02×10^{23} **formula units**

* To four significant figures the number is 6.022×10^{23}.

of KCl. This sample must have a mass of 74.6 g since the atomic weight of potassium is 39.1 and the atomic weight of chlorine is 35.5. One must remember, however, that in reality 6.02×10^{23} KCl formula units (1.00 mole KCl) contain 1.00 mole of K^+ ions (6.02×10^{23} K^+ ions) and 1.00 mole of Cl^- ions (6.02×10^{23} Cl^- ions).

It is imperative to remember that whenever the term mole is used, the type of particle being considered must be specified. For example, one mole of oxygen **atoms** (6.02×10^{23} atoms O; mass = 16.0 g) is quite different from one mole of oxygen **molecules** (6.02×10^{23} molecules O_2; mass = 32.0 g). It should be clear from this example that to state simply "one mole of oxygen" is confusing and in fact offers no clear direction to the student or chemist. The nature of species being considered can be defined either by word or formula. For example, the formula O_2 specifies molecules of oxygen and not oxygen atoms. The symbol O, on the other hand, would clearly specify oxygen atoms.

Occasionally, when dealing with masses of individual atoms, a different unit of mass, the **atomic mass unit** (abbreviated amu) is employed. One atomic mass unit is defined as 1/12 the mass of the ^{12}C atom, or in other words, one ^{12}C atom has a mass exactly equal to 12 amu. Notice that by this definition the mass of one atom in amu's is numerically equal to its atomic weight. A similar relationship exists for the mass of a single molecule in amu's and its molecular weight. It should also be clear that the atomic mass unit is related to our basic mass unit, the gram. Since there are 12.00 g for 6.022×10^{23} ^{12}C atoms and since there is one ^{12}C atom per 12.00 amu, the mass in grams of one atomic mass unit is given by

$$\text{Mass in grams of 1 amu} = \frac{12.00\ \text{g}}{6.022 \times 10^{23}\ \cancel{^{12}\text{C atoms}}} \times \frac{1\ \cancel{^{12}\text{C atom}}}{12.00\ \text{amu}}$$

$$= 1.661 \times 10^{-24}\ \text{g/amu}$$

All the above terms are perhaps best understood by considering illustrative examples.

Example 3-c. How many grams of $BaCl_2 \cdot 2H_2O$ are contained in 1.50 moles of the compound?

Solution. First we calculate the formula weight of $BaCl_2 \cdot 2H_2O$. This compound is a "hydrate." The "dot" indicates that two molecules of H_2O are intimately associated with each formula unit of $BaCl_2$. Hence the formula weight of the compound must include the contribution from the two H_2O molecules. The formula unit contains 1 Ba, 2 Cl, 4 H, and 2 O. Atomic weights are: Ba, 137.3; Cl, 35.5; H, 1.0; and O, 16.0. Hence:

$$\begin{aligned}
\text{Mass contribution of Ba} &= 1 \times 137.3 = 137.3 \\
\text{Mass contribution of Cl} &= 2 \times 35.5 = 71.0 \\
\text{Mass contribution of H} &= 4 \times 1.0 = 4.0 \\
\text{Mass contribution of O} &= 2 \times 16.0 = \underline{32.0} \\
\text{FW} &= 244.3
\end{aligned}$$

Now, using dimensional analysis, we convert from moles to grams of compound:

$$1.50\ \cancel{\text{mole BaCl}_2 \cdot 2\text{H}_2\text{O}} \times \frac{244.3\ \text{g BaCl}_2 \cdot 2\text{H}_2\text{O}}{1\ \cancel{\text{mole BaCl}_2 \cdot 2\text{H}_2\text{O}}} = 366\ \text{g BaCl}_2 \cdot 2\text{H}_2\text{O}$$

Example 3-d. Contained in 284 g of Cl_2 are (a) how many moles of chlorine molecules, Cl_2, (b) how many molecules of Cl_2, (c) how many moles of chlorine atoms, Cl, and (d) how many atoms of chlorine?

Solution. (a) First determine the MW of Cl_2. The atomic weight of Cl is 35.5. Since there are two atoms of Cl per molecule of Cl_2, the MW of $Cl_2 = 2 \times 35.5 = 71.0$. Next we let dimensional analysis (i.e., the units) do the work.

$$284\ \cancel{\text{g Cl}_2} \times \frac{1\ \text{mole Cl}_2}{71.0\ \cancel{\text{g Cl}_2}} = 4.00\ \text{mole Cl}_2$$

(b) The number of molecules of Cl_2 is obtained by multiplying the number of moles of Cl_2 by the number of Cl_2 molecules per mole.

$$4.00\ \cancel{\text{mole Cl}_2} \times \frac{6.02 \times 10^{23}\ \text{molecules Cl}_2}{1\ \cancel{\text{mole Cl}_2}} = 24.1 \times 10^{23}\ \text{molecules Cl}_2$$

(c) Moles of Cl atoms obtainable are calculated as follows. Since the formula Cl_2 tells us that there are 2 moles Cl for each mole of Cl_2 we have

$$4.00\ \cancel{\text{mole Cl}_2} \times \frac{2\ \text{mole Cl}}{1\ \cancel{\text{mole Cl}_2}} = 8.00\ \text{mole Cl}$$

(d) Finally, the number of Cl atoms is obtained from the definition of mole:

$$8.00\ \cancel{\text{mole Cl}} \times \frac{6.02 \times 10^{23}\ \text{atoms Cl}}{1\ \cancel{\text{mole Cl}}} = 48.2 \times 10^{23}\ \text{atoms Cl}$$

Example 3-e. 0.250 mole of $Al_2(SO_4)_3$ contains (a) how many grams of $Al_2(SO_4)_3$, (b) how many formula units of $Al_2(SO_4)_3$, and (c) how many moles of SO_4^{2-} ions? (FW of $Al_2(SO_4)_3 = 342$).

Solution. (a) Using the definition of the mole and dimensional analysis, we have

$$0.250\ \cancel{\text{mole } Al_2(SO_4)_3} \times \frac{342\ \text{g } Al_2(SO_4)_3}{1\ \cancel{\text{mole } Al_2(SO_4)_3}} = 85.5\ \text{g } Al_2(SO_4)_3$$

(b) Since there is an Avogadro's number of formula units per mole of $Al_2(SO_4)_3$ we have

$$0.250\ \cancel{\text{mole } Al_2(SO_4)_3} \times \frac{6.02 \times 10^{23}\ \text{formula units } Al_2(SO_4)_3}{1\ \cancel{\text{mole } Al_2(SO_4)_3}}$$

$$= 1.50 \times 10^{23}\ \text{formula units } Al_2(SO_4)_3$$

(c) Since the formula for $Al_2(SO_4)_3$ tells us that there are 3 moles of SO_4^{2-} ion per mole of $Al_2(SO_4)_3$ we have

$$0.250\ \cancel{\text{mole } Al_2(SO_4)_3} \times \frac{3\ \text{mole } SO_4^{2-}}{1\ \cancel{\text{mole } Al_2(SO_4)_3}} = 0.750\ \text{mole } SO_4^{2-}$$

Example 3-f. Determine the mass in grams of (a) one molecule of H_2O, and (b) 5.08×10^{24} molecules of H_2O.

Solution. (a) Since we know that 6.02×10^{23} H_2O molecules have a mass of 18.0 g (MW of H_2O = 18.0) we have

$$1\ \cancel{\text{molecule } H_2O} \times \frac{18.0\ \text{g } H_2O}{6.02 \times 10^{23}\ \cancel{\text{molecule } H_2O}} = 2.99 \times 10^{-23}\ \text{g } H_2O$$

or 2.99×10^{-23} g for each molecule of H_2O.

(b) Similar to (a) we have

$$5.08 \times 10^{24}\ \cancel{\text{molecules } H_2O} \times \frac{18.0\ \text{g } H_2O}{6.02 \times 10^{23}\ \cancel{\text{molecules } H_2O}}$$

$$= 1.52 \times 10^{2}\ \text{g } H_2O$$

Example 3-g. Ferrous ammonium sulfate, $Fe(NH_4)_2(SO_4)_2$, is a common fertilizer and moss killer for lawns. How many $Fe(NH_4)_2(SO_4)_2$ formula units are present in 71.0 g of ferrous ammonium sulfate? (FW of $Fe(NH_4)_2(SO_4)_2$ = 284).

Solution. This is a two-step problem but quite straightforward using dimensional analysis. First convert from grams $Fe(NH_4)_2(SO_4)_2$ to moles $Fe(NH_4)_2(SO_4)_2$, and then convert to formula units:

$$71.0\ \cancel{\text{g } Fe(NH_4)_2(SO_4)_2} \times \frac{\cancel{1\ \text{mole } Fe(NH_4)_2(SO_4)_2}}{284\ \cancel{\text{g } Fe(NH_4)_2(SO_4)_2}}$$

$$\times \frac{6.02 \times 10^{23}\ Fe(NH_4)_2(SO_4)_2\ \text{formula units}}{\cancel{1\ \text{mole } Fe(NH_4)_2(SO_4)_2}}$$

$$= 1.50 \times 10^{23}\ Fe(NH_4)_2(SO_4)_2\ \text{formula units}$$

Example 3-h. Five molecules of NH_3 have a mass of 85.0 amu. What is the corresponding mass of these five molecules in grams?

Solution. We have shown that 1.00 amu is equivalent to 1.66×10^{-24} g. Hence

$$85.0 \ \cancel{\text{amu}} \times \frac{1.66 \times 10^{-24} \text{ g}}{1 \ \cancel{\text{amu}}} = 1.41 \times 10^{-22} \text{ g}$$

Problems (Moles and the Mole Concept)

3-7. How many moles of K_2SO_4 are represented by 87.1 g of the compound?

3-8. How many moles are represented by 3.00×10^2 g of each of the following substances?
(a) AgCl (c) SiH_4 (e) CdS
(b) $Ca(CN)_2$ (d) RbOH (f) Na_2MnO_4

3-9. What mass in grams is represented by
(a) 0.91 mole of CO_2? (c) 3.62 moles of H_5IO_6?
(b) 0.32 mole of K_2SO_4? (d) 2.17 moles of NH_3?

3-10. How many atoms of silver, Ag, are contained in 3.26 moles of Ag?

3-11. A 10 lb sack of table sugar, $C_{12}H_{22}O_{11}$, contains (a) how many moles of sugar, and (b) how many molecules of sugar? (One pound = 4.5×10^2 g).

3-12. How many moles of phosphorus are present in 2.00×10^2 g of each of the following substances?
(a) PH_3 (c) $H_4P_2O_7$ (e) P_4O_6
(b) H_3PO_2 (d) P_4O_{10} (f) $(HPO_3)_3$

3-13. How many atoms of each element are present in 2.00 moles of pyrophosphoric acid, $H_4P_2O_7$?

3-14. (a) How many moles of sodium ion are contained in 328 g of Na_3PO_4? (b) What number of sodium ions are contained in the sample in (a)? (c) What number of phosphate ions are contained in the same sample?

3-15. What is the mass in grams of a single
(a) H atom of mass 1.008 amu?
(b) F atom of mass 19.00 amu?
(c) U atom of mass 238.0 amu?

3-16. A molecule X_8 has a mass of 256 amu. (a) What is the molecular weight of X_8? (b) What is the atomic weight of X? (c) What is the mass in grams of one molecule of X_8?

3-17. A particular vitamin C tablet contains 0.250 g of vitamin C (ascorbic acid, $C_6H_8O_6$). (a) How many molecules of vitamin C does this represent? (b) How many atoms of C are contained in 0.250 g of vitamin C?

3-18. Calculate the mass, in grams, of a one million (1.00×10^6) atom sample of the element gold, Au.

3-19. The mass of a single atom of element M is 2.86×10^{-22} g. What is its atomic weight?

3-20. A pure sample of Li_3AlF_6 contains 4.50×10^{24} lithium ions. Calculate the mass, in grams, of the entire sample.

3-C. Isotopes and Determination of Chemical Atomic Weights

Most elements in their natural state are composed of mixtures of **isotopes** of that element. Atoms of the same element that have **different** numbers of **neutrons** are called **isotopes**. Note that isotopic atoms have the **same** number of **protons** since they are atoms of the same element.

The composition of the isotope mixture is usually expressed in terms of atom percentage. This means a percentage comparison of the atoms of a particular isotope to the total number of atoms present.

$$\text{Atom \% (for a given isotope)} = \frac{\text{Number of atoms of a given isotope}}{\text{Total number of isotopic atoms}} \times 100$$

The chemical atomic weight is the "weighted" average of the masses of all isotopes present in the sample of the element. The word "weighted" in the previous sentence indicates that we must take into account the atom percentage of a given isotope.

A mass spectrometer is an instrument that measures the exact mass of each isotope and its relative abundance. From such data the chemical atomic weight may be calculated.

Example 3-i. Strontium, as it exists naturally, is composed of four isotopes. These are listed below with their relative abundances expressed in atom percentage. From this information, calculate the atomic weight of naturally occurring strontium.

Sr Isotope	Exact Mass	% Abundance
84	83.913	0.560
86	85.909	9.86
87	86.909	7.02
88	87.906	82.56
		100.00

Solution. The contribution of each isotope to the atomic weight will be in proportion to its relative abundance. Therefore,

$$
\begin{aligned}
{}^{84}\text{Sr contributes } 83.913 \times 0.00560 &= 0.47 \\
{}^{86}\text{Sr contributes } 85.909 \times 0.0986 &= 8.47 \\
{}^{87}\text{Sr contributes } 86.909 \times 0.0702 &= 6.10 \\
{}^{88}\text{Sr contributes } 87.906 \times 0.8256 &= \underline{72.58} \\
\text{Chemical atomic weight} &= 87.62
\end{aligned}
$$

The calculated atomic weight based on $^{12}C = 12.0000$ is 87.62, the number listed inside the front cover of this book.

Example 3-j. Chlorine, which exhibits an atomic weight of 35.453, is composed of two isotopes, ^{35}Cl and ^{37}Cl. The exact mass of ^{35}Cl is 34.969 and the exact mass of ^{37}Cl is 36.966. With this information, calculate the percentage of each isotope in the naturally occurring mixture.

Solution. Let X = the fraction of ^{35}Cl atoms. Since only two isotopes of Cl are found naturally, the quantity $(1 - X)$ equals the fraction of ^{37}Cl atoms. The sum of the contributions of each species to the atomic weight must equal the chemical atomic weight. Consequently,

$$
\begin{aligned}
(X)(34.969) + (1 - X)(36.966) &= 35.453 \\
34.969X + 36.966 - 36.966X &= 35.453 \\
-1.997X &= -1.513 \\
X &= 0.7576 \\
\text{and } (1 - X) &= 0.2424
\end{aligned}
$$

Therefore the percentage of ^{35}Cl is 75.76 and the percentage of ^{37}Cl is 24.24.

Problems (Atomic Weights from Isotopic Abundance)

3-21. The naturally occurring form of the element fluorine consists of a single isotope of mass 18.9984. What is the atomic weight of fluorine?

3-22. The element boron is composed of two isotopes. ^{10}B constitutes 19.60% of all boron atoms and has a mass of 10.01294. ^{11}B constitutes 80.40% and has a mass of 11.00931. Calculate the atomic weight to the proper number of significant figures.

3-23. Potassium has a chemical atomic weight of 39.10. It is composed of three isotopes of mass 38.96, 39.96, and 40.96. The isotope of mass 39.96 is present in a negligibly small amount insofar as making a significant contribution to the atomic weight. What are the percentages of the other two isotopes?

3-24. Ordinary oxygen gas consists of three isotopes with the abundances listed below. (a) What is the atomic weight of oxygen? (b) What is the molecular weight of oxygen gas?

O Isotope	Exact Mass	% Abundance
16	15.99491	99.759
17	16.99914	0.037
18	17.99916	0.204

General Problems

3-25. Determine the molecular or formula weight of each of the following substances (see inside front cover for atomic weights):
(a) HNO_2 (c) $(NH_4)_2SO_3$ (e) $Ca_3(PO_4)_2$
(b) $Sr(OH)_2$ (d) KOH (f) $MgNH_4PO_4$

3-26. 43.5 grams of H_2SeO_4 represent how many moles of H_2SeO_4?

3-27. 182 g of KCl represent how many moles of KCl?

3-28. How many moles of LiCl are there in 21.2 g of LiCl?

3-29. (a) How many moles are there in 75.0 g of each of the following?
(i) H_3PO_4 (iii) NH_4NO_3
(ii) $CuSO_4$ (iv) $Fe_2(CO_3)_3$
(b) Determine the number of moles of each element in 75.0 g of the compounds in (a).

3-30. How many grams of strontium are there in 0.125 moles of $Sr(OH)_2$?

3-31. Determine the number of grams in each of the following:
(a) 3.80 moles H_2SO_4 (c) 1.75 moles $K_2Cr_2O_7$
(b) 0.87 mole $KClO_3$ (d) 0.375 mole $C_{12}H_{22}O_{11}$

3-32. Determine the number of moles in 0.425 kg of each of the following:
(a) LiOH (c) H_2TeO_4
(b) $BaCO_3$ (d) H_3AsO_4

3-33. Determine the number of grams in each of the following:
(a) 0.68 mole HI (c) 2.60 moles $Be(OH)_2$
(b) 0.911 mole $CdSO_4$ (d) 8.30 moles $HClO_4$

3-34. How many moles of O are there in 4.9 g of H_2SO_4?

3-35. 87.0 grams of Al represent how many (a) moles, (b) atoms?

3-36. How many moles of each element are present in 36.8 g of NaCl?

3-37. 4.50×10^{24} atoms of sodium would be how many moles?

3-38. 3.01×10^{22} molecules of HCl would be how many moles?

3-39. Calculate the mass in grams of 1.50×10^{25} molecules of H_2SO_4.

3-40. How many moles of CCl_4 will contain 12.8 moles of chlorine atoms?

3-41. What mass in grams of H_2SO_4 will contain 1.50 moles of O atoms?

3-42. How many chloride ions, Cl^-, are contained in 1.00 lb of table salt, NaCl?

3-43. How many atoms of each element are present in 11.7 g of $Mg(OH)_2$?

3-44. How many molecules of H_2O are contained in 2.48 g of $Na_2S_2O_3 \cdot 5H_2O$?

3-45. An eight ounce glass of H_2O contains about 2.5×10^2 g of H_2O. How many molecules of H_2O are contained therein?

3-46. Could you lift 2.50×10^{29} atoms of Fe? (Calculate the mass in grams and compare this to your lifting ability.)

3-47. A typical 5.0 grain tablet of aspirin ($C_9H_8O_4$) contains (a) how many grams of aspirin, (b) how many moles of aspirin, and (c) how many aspirin molecules? (1.00 grain = 0.0648 g).

3-48. Ozone, O_3, is found in the upper atmosphere. Its presence there is important since O_3 absorbs ultraviolet radiation from the sun. This radiation would be dangerous to humans if not screened before reaching the earth's surface. 144 g of O_3 would represent (a) how many moles of ozone molecules, O_3, (b) how many molecules of O_3, (c) how many moles of oxygen atoms, and (d) how many oxygen atoms?

3-49. 35.2 g of vitamin C (ascorbic acid, $C_6H_8O_6$) contain how many (a) moles of $C_6H_8O_6$, (b) moles of carbon atoms, (c) grams of carbon, (d) molecules of $C_6H_8O_6$, and (e) atoms total?

3-50. The average mass of a copper atom in naturally occurring copper metal is 1.055×10^{-22} g. What is the atomic weight of copper?

3-51. Element *X* has been found to consist of a number of isotopes whose atoms have an average mass of 1.594×10^{-22} g. What is the atomic weight of *X*?

3-52. A pure sample of $Fe(NH_4)_2(SO_4)_2$ contains 8.25×10^{24} ammonium ions. Calculate the mass in grams of the entire sample.

3-53. Natural gallium, whose atomic weight is 69.72, is a mixture of the two isotopes ^{69}Ga and ^{71}Ga. The exact masses of these isotopes are 68.93 amu and 70.93 amu, respectively. Determine the percentage of each isotope in the naturally occurring mixture.

3-54. Determine the atomic weight of lithium, which is a mixture of 7.5% ^{6}Li and 92.5% ^{7}Li. The exact masses of these isotopes are 6.01 amu and 7.02 amu.

3-55. Calculate the atomic weight of the element zirconium, which consists of five isotopes whose exact masses and percentage abundances are as follows:

Mass	% Abundance
89.90	51.46
90.91	11.23
91.90	17.11
93.91	17.40
95.91	2.80

3-56. Antimony consists of two isotopes whose exact masses are 120.904 and 122.904. The atomic weight of antimony is 121.754. What is the percentage of each isotope in the natural element?

3-57. A sample of uranium is prepared in the laboratory by mixing ^{235}U (mass 235.04) and ^{238}U (mass 238.05). What would the atomic weight of this synthetic sample appear to be if 36.75% of the atoms were ^{238}U and the balance were ^{235}U?

3-58. Iridium, Ir, with an atomic weight of 192.2, is composed of two isotopes, ^{191}Ir and ^{193}Ir, with respective masses 191.04 and 193.04. Determine the percentage of each isotope present in the natural mixture.

3-59. Thallium metal, whose atomic weight is 204.37, consists of two isotopes, ^{203}Tl and ^{205}Tl. The exact mass of ^{203}Tl is 202.97 and the exact mass of ^{205}Tl is 204.97. What are the atom percentages of ^{205}Tl and ^{203}Tl?

3-60. Naturally occurring copper, atomic weight 63.54, is made up of two isotopes with masses 62.93 and 64.93, respectively. Determine the percentage of each isotope in the naturally occurring mixture.

3-61. Determine the atomic weight of magnesium from the following data:

Mass of isotope	% Abundance
23.99	79.0
24.99	10.0
25.98	11.0

Chapter 4

Formulas and Percentage Composition

4-A. Percentage Composition from Formulas

A chemical formula is a good source of information because it gives the number and types of atoms present in the compound. With the aid of atomic weights, the percentage by weight of each element present may be determined. The percentage by weight of any element is the number of parts of mass of that element in 100 parts of the compound; it is computed by dividing the mass of the element by the total mass of the compound, then multiplying by 100.

Example 4-a. Calculate the percentage by weight of each element in sucrose, $C_{12}H_{22}O_{11}$. The atomic weights of C, H, and O are 12, 1, and 16, respectively.

Solution. First determine the relative mass of each element in one molecular weight of the compound. Then divide the mass of each element by the molecular weight of the compound to obtain the fraction of each element and multiply the fraction by 100 to obtain the percentage.

$$\begin{aligned} &\text{12 atoms of C have relative mass } 12 \times 12 = 144 \\ &\text{22 atoms of H have relative mass } 22 \times 1 = 22 \\ &\text{11 atoms of O have relative mass } 11 \times 16 = \underline{176} \\ &\qquad\qquad\qquad\qquad\qquad\qquad\quad \text{MW} = 342 \end{aligned}$$

$$\text{Percent carbon} = \frac{144}{342} \times 100 = 42.1\%\ \text{C}$$

$$\text{Percent hydrogen} = \frac{22}{342} \times 100 = 6.4\%\ \text{H}$$

$$\text{Percent oxygen} = \frac{176}{342} \times 100 = 51.5\%\ \text{O}$$

Total 100.0%

Example 4-b. Analysis of a compound containing iron and sulfur indicated that 5.19 g of compound contained 2.79 g of iron and 2.40 g of sulfur. Calculate the percentage by weight of each element in this compound.

Solution.

$$\%\text{ by weight of element} = \frac{\text{Grams of element}}{\text{Grams of compound}} \times 100$$

Hence

$$\%\text{ by weight Fe} = \frac{2.79\text{ g Fe}}{5.19\text{ g cmpd}} \times 100 = 53.8\%\text{ Fe}$$

$$\%\text{ by weight S} = \frac{2.40\text{ g S}}{5.19\text{ g cmpd}} \times 100 = 46.2\%\text{ S}$$

Example 4-c. What mass of Fe is contained in 675 g of Fe_2O_3? The atomic weights of Fe and O are 55.8 and 16.0.

Solution. First determine the percentage by weight of Fe in the compound. Then, using units, convert from g Fe_2O_3 to g Fe.

$$\%\text{ by weight Fe in Fe}_2\text{O}_3 = \frac{2 \times 55.8}{159.6} \times 100 = 69.9\%\text{ Fe}$$

Now 69.9% Fe implies 0.699 g Fe per 1.00 g Fe_2O_3. Hence

$$675\ \cancel{\text{g Fe}_2\text{O}_3} \times \frac{0.699\text{ g Fe}}{1.00\ \cancel{\text{g Fe}_2\text{O}_3}} = 472\text{ g Fe}$$

Example 4-d. A silver coin of mass 2.50 g is analyzed by dissolving in dilute nitric acid and then precipitating and weighing the silver as AgCl. If the precipitate AgCl has mass 2.99 g, what is the percentage by weight of silver in the coin?

Solution

$$\% \text{ by weight Ag in AgCl} = \frac{107.9}{143.4} \times 100 = 75.2\%$$

$$\text{Mass of Ag in coin} = 2.99 \text{ g } \cancel{\text{AgCl}} \times \frac{0.752 \text{ g Ag}}{1.00 \text{ g } \cancel{\text{AgCl}}} = 2.25 \text{ g Ag}$$

$$\% \text{ by weight Ag in coin} = \frac{2.25 \text{ g Ag}}{2.50 \text{ g coin}} \times 100 = 90.0\% \text{ Ag}$$

Problems (Percentage Composition)

4-1. Calculate the percentage composition of each of the following compounds:
(a) KBr
(b) Sb_2S_5
(c) C_5H_{10}

4-2. Calculate the percentage composition of each of the following compounds:
(a) K_2SO_4
(b) $Na_3Fe(CN)_6$
(c) $Ce(NO_3)_4$

4-3. 4.86 g of magnesium is allowed to react with oxygen gas. Upon completion, 8.06 g of pure magnesium oxide remain. Calculate the % by weight of each element in the compound.

4-4. What mass of nitrogen is contained in 45.0 g of each of the following:
(a) HNO_3 (c) NH_4NO_3
(b) NH_4Br (d) $(NH_4)_2HPO_4$

4-5. If 3.27 g of the compound ammonium nitrite, NH_4NO_2, decomposes to give only H_2O and N_2 gas, what mass of N_2 is produced?

4-6. 0.483 g of a magnesium alloy is analyzed for Mg by precipitating and weighing $MgNH_4PO_4$. If 0.966 g of the latter is obtained, what is the percentage by weight of Mg in the alloy?

4-7. 3.621 g of an ore containing barium carbonate ($BaCO_3$) yielded 1.754 g of $BaSO_4$ in an analysis for barium. What is the percentage by weight of barium carbonate in the sample?

4-B. Formulas from Percentage Composition Data

An **empirical formula**, which shows the relative number of atoms of each element in a formula unit, may be determined from the percentage composition data of a pure compound or from the direct weight ratios of the elements comprising the compound. To convert the empirical formula to a **molecular formula**, an experimental measurement must be made of the molecular weight. These measurements are discussed in Chapters 6 and 11.

The following steps may be followed to obtain empirical and molecular formulas.

1. If percentage composition data are given, assume you have 100 g of the compound. This gives a number of grams of each element numerically equal to its percentage in the compound. If direct weight data are presented, use the exact masses of the compound and elements.

2. Determine the number of moles of atoms of each element.

3. Divide each value obtained in Step 2 by the smallest of these values. If this procedure yields results that approximate whole numbers, proceed to Step 5. If not, proceed to Step 4.

4. Multiply the numbers obtained in Step 3 by the smallest integer possible to approximate all whole number values. Proceed to Step 5.

5. Write these whole numbers as subscripts of the elements in an empirical formula.

6. If the molecular weight is known, determine how many empirical formula weights are required to obtain the molecular weight. Use this factor to multiply the number of atoms in the empirical formula to obtain the number of atoms of each element in the true molecular formula.

Examples 4-e, 4-f, and 4-g illustrate these steps in detail. Study them carefully.

Example 4-e. Analysis of a compound of boron and hydrogen showed it to contain 78.2% B and 21.8% H. An experimentally determined value of the molecular weight was 27.6. What are the empirical and molecular formulas of the compound?

Solution. Assume that you have 100.0 g of compound. Then 100.0 g of compound will contain 78.2 g B and 21.8 g H (Step 1). Converting these masses to moles of atoms (Step 2):

$$78.2\ \cancel{\text{g B}} \times \frac{1 \text{ mole B atoms}}{10.8\ \cancel{\text{g B}}} = 7.24 \text{ moles B atoms}$$

$$21.8\ \cancel{\text{g H}} \times \frac{1 \text{ mole H atoms}}{1.01\ \cancel{\text{g H}}} = 21.6 \text{ moles H atoms}$$

The ratio of moles of H atoms to moles of B atoms in 100 g of the compound must be the same as the ratio of H to B atoms in one molecule of the compound. That is, the empirical formula could be written

$$B_{7.24}H_{21.6}$$

However, atoms in formulas usually have whole-number subscripts, and, to obtain them, we must divide through by the smallest of the two values (Step 3), that is, 7.24. The formula then is

$$B_{\frac{(7.24)}{(7.24)}}H_{\frac{(21.6)}{(7.24)}} = BH_{2.98}$$

The values obtained at this point should be very close to whole numbers. Any slight difference results from experimental error and we round these values to the nearest whole number, in this case BH_3 (Step 5). BH_3 is therefore the **empirical** formula for this compound, and the formula weight is 10.8 + 3.0 = 13.8. The molecular formula will either be the same as the empirical formula or some multiple of it such as B_2H_6 or B_3H_9. Because the molecular weight of the compound was determined experimentally to be 27.6 (or 2 × the empirical formula weight), the **molecular** formula must be B_2H_6 (Step 6).

Example 4-f. A compound is found to contain 11.2% N, 3.20% H, 41.2% Cr, and 44.4% O. Determine the empirical formula.

Solution. Calculate the number of moles of atoms of each element in 100 g of the compound.

$$11.2\ \cancel{g\ N} \times \frac{1 \text{ mole N atoms}}{14.0\ \cancel{g\ N}} = 0.80 \text{ mole N atoms}$$

$$3.20\ \cancel{g\ H} \times \frac{1 \text{ mole H atoms}}{1.01\ \cancel{g\ H}} = 3.17 \text{ moles H atoms}$$

$$41.2\ \cancel{g\ Cr} \times \frac{1 \text{ mole Cr atoms}}{52.0\ \cancel{g\ Cr}} = 0.79 \text{ mole Cr atoms}$$

$$44.4\ \cancel{g\ O} \times \frac{1 \text{ mole O atoms}}{16.0\ \cancel{g\ O}} = 2.78 \text{ moles O atoms}$$

We now divide through by the smallest of these values to obtain

$$N_{\frac{(0.80)}{(0.79)}}H_{\frac{(3.17)}{(0.79)}}Cr_{\frac{(0.79)}{(0.79)}}O_{\frac{(2.78)}{(0.79)}} \text{ or } NH_4CrO_{3.5}$$

Since the 3.5 value for O is not a whole number we must multiply all values in the formula by **two** to obtain the empirical formula

$$N_2H_4Cr_2O_7$$

Example 4-g. In Example 4-b we saw that 5.19 g of a pure compound contained 2.79 g of Fe and 2.40 g of S. Determine the empirical formula of the compound.

Solution. Determine the number of moles of atoms of Fe and S.

$$2.79\ \text{g Fe} \times \frac{1\ \text{mole Fe atoms}}{55.8\ \text{g Fe}} = 0.0500\ \text{mole Fe atoms}$$

$$2.40\ \text{g S} \times \frac{1\ \text{mole S atoms}}{32.0\ \text{g S}} = 0.0750\ \text{mole S atoms}$$

Find the mole ratio of Fe and S, i.e.,

$$Fe_{\frac{(0.050)}{(0.050)}}S_{\frac{(0.075)}{(0.050)}} = FeS_{1.5}$$

Multiply all values in the formula by **two** to obtain the empirical formula

$$Fe_2S_3$$

Example 4-h. A hydrate of ferric thiocyanate, $Fe(SCN)_3$, was found to contain 19.0% H_2O. What is the formula for the salt hydrate?

Solution. This question is analogous to asking how many moles of H_2O there are per mole of $Fe(SCN)_3$. In principle this problem is worked in a manner similar to the above examples. First calculate the number of moles of H_2O and $Fe(SCN)_3$ in 100 g of the compound. Then determine the number of moles of H_2O per mole of $Fe(SCN)_3$.

$$19.0\ \text{g H}_2\text{O} \times \frac{1\ \text{mole H}_2\text{O}}{18.0\ \text{g H}_2\text{O}} = 1.06\ \text{moles H}_2\text{O}$$

$$81.0\ \text{g Fe(SCN)}_3 \times \frac{1\ \text{mole Fe(SCN)}_3}{229.8\ \text{g Fe(SCN)}_3} = 0.352\ \text{mole Fe(SCN)}_3$$

Number of moles of H_2O/mole $Fe(SCN)_3$

$$= \frac{1.06\ \text{moles H}_2\text{O}}{0.352\ \text{mole Fe(SCN)}_3} = 3.01\ \frac{\text{moles H}_2\text{O}}{\text{mole Fe(SCN)}_3}$$

The formula for the compound is therefore $Fe(SCN)_3 \cdot 3H_2O$.

Problems (Formulas from Percentage Composition Data)

4-8. A compound is analyzed and found to contain 22.55% P and 77.45% Cl. What is its empirical formula?

4-9. A pure compound of mercury and iodine results on reaction of 21.66 g of mercury with 27.42 g of iodine. What is the empirical formula for the compound?

4-10. A compound of bromine and fluorine is formed by direct reaction of the two elements. By a careful experiment, it is found that 3.02 g of bromine react with 2.15 g of fluorine. (a) What is the empirical formula for the compound? (b) If the molecular weight is 137, what is the molecular formula for the compound?

4-11. A compound of aluminum and chlorine is composed of 7.20 g of aluminum for every 28.4 g of chlorine. (a) What is the empirical formula? (b) If the molecular weight is 267, what is the molecular formula?

4-12. Determine the empirical formula for each of the following compounds from the percentage composition data.

Weight Percentage

Compound	C	H	O	N
(*a*)	52.2	13.0	34.8	—
(*b*)	—	2.13	68.0	29.8
(*c*)	48.6	8.10	43.3	—
(*d*)	58.5	4.07	26.0	11.4
(*e*)	49.3	9.60	21.9	19.2

4-13. Nicotine, a compound found in tobacco leaves, was analyzed and found to contain 74.0% C, 8.70% H, and 17.3% N. (a) Determine the empirical formula for nicotine. (b) If the molecular weight is 162, determine the molecular formula.

4-14. Praseodymium perchlorate, $Pr(ClO_4)_3$, forms a hydrate containing six molecules of water per formula unit of $Pr(ClO_4)_3$. What is the percentage of water in the hydrate?

4-15. The formula for a hydrate of hydrazine, N_2H_4, is $N_2H_4 \cdot H_2O$. What is the percentage of water in the hydrated compound?

4-16. Gadolinium chloride, $GdCl_3$, forms a hydrate containing 70.7% $GdCl_3$. What is the formula for the hydrate?

4-17. Perchloric acid, $HClO_4$, forms a crystalline hydrate melting at 50°C. It contains 15.2% H_2O and 84.8% $HClO_4$. What is the formula of the hydrate?

4-18. In 1933 Professor L. Pauling predicted the existence of a certain compound of the rare gas xenon and fluorine. Subsequently, the compound was prepared almost simultaneously by research workers in several different laboratories. The compound was found to consist of 53.5% Xe and 46.5% F. (a) What is its empirical formula? (b) If its molecular weight is 245, what is its molecular formula?

General Problems

4-19. Calculate the percentage by weight of oxygen in each of the following compounds:
(a) CdO
(b) $(NH_4)_2Cr_2O_7$
(c) $Cd(NO_3)_2$
(d) $Co_2(SO_4)_3$
(e) $Ba_3(GaO_3)_2$
(f) $BaZrO_3$

4-20. It was found that 4.06 g of Hg reacted with oxygen gas to form 4.38 g of compound. Determine the percentage by weight of Hg and O in the compound.

4-21. A compound containing only copper and sulfur was analyzed. It was found that 5.44 g of the compound contained 4.34 g of copper. Determine the percentage by weight of Cu and S in the compound.

4-22. What mass of oxygen gas (O_2) may be obtained by electrolysis of H_2O when 0.915 kg of H_2O is completely converted to hydrogen gas and oxygen gas?

4-23. Assuming 100% conversion, what mass of P would be needed to produce 5.00×10^2 g of H_3PO_4?

4-24. How many grams of As are present in 58.5 g of As_2S_3?

4-25. A compound is found to contain 31.55% Ca, 24.38% P, and 44.08% O. What is the empirical formula for the compound?

4-26. A pure compound is analyzed and found to contain 27.0% Si and 73.0% F. (a) What is the empirical formula for the compound? (b) If the experimentally determined molecular weight is 104, what is the molecular formula?

4-27. A substance with an empirical formula of CH_2 and a known molecular weight of 84 must have what molecular formula?

4-28. From the following analytical data, determine the empirical formula for each of the following compounds:
(a) 92.2% C; 7.8% H
(b) 78.6% Ni; 21.4% O
(c) 22.3% Na; 46.6% O; 31.1% S
(d) 65.2% Sc; 34.8% O
(e) 35.5% Ti; 11.9% O; 52.6% Cl

4-29. A compound of aluminum was found experimentally to have a molecular weight of 144. By chemical analysis, it was shown to consist of 37.4% Al, 50.0% C, and 12.6% H. What is its molecular formula?

4-30. Determine the percentage water of hydration for each of the following hydrates:
(a) $Na_2SO_4 \cdot 10H_2O$
(b) $CuSO_4 \cdot 5H_2O$
(c) $Zn(ClO_4)_2 \cdot 6H_2O$
(d) $CrCl_3 \cdot 6H_2O$
(e) $NH_4Fe(SO_4)_2 \cdot 12H_2O$
(f) $Nd(BrO_3)_3 \cdot 9H_2O$

4-31. If 7.58 g of a hydrate of $FeSO_4$ lose 3.44 g H_2O when heated, what is the formula for the hydrate?

4-32. Determine the formula of a hydrated salt that has the composition 5.92% Be, 46.65% Cl, and 47.42% H_2O.

4-33. If 5.60 g of a hydrate of Na_2CO_3, when heated, leave a residue of mass 2.07 g, determine the formula of the hydrated salt.

4-34. $ZnCl_2$ combines with NH_3 to form a complex salt. If 3.250 g of $ZnCl_2$ combine with 1.616 g of NH_3, what is the formula for the complex?

4-35. If, on being heated, 2.500 g of a hydrate of $Li_2B_4O_7$ leave a residue of $Li_2B_4O_7$ of mass 1.508 g, determine the formula for the hydrated salt.

4-36. On heating 4.64 g of a compound, 4.32 g of silver and 0.320 g of oxygen were obtained. What is the simplest formula for the compound?

4-37. Potassium chlorate, when heated, yields oxygen gas (O_2) and potassium chloride. If 2.76 g of potassium chlorate yield 1.08 g of oxygen gas and 1.68 g of potassium chloride, determine the simplest formula for potassium chlorate.

4-38. A compound formed between xenon, oxygen, and fluorine has the following composition: Xe = 70.8%, O = 8.7%. What is the empirical formula?

4-39. If 1.75 g of Fe metal is burned in oxygen and the product has mass 2.50 g, determine the empirical formula for the oxide from these data.

4-40. On heating solid calcium carbonate, $CaCO_3$, gaseous carbon dioxide, CO_2, and solid calcium oxide, CaO, are formed. A 3.75 g sample of $CaCO_3$ is heated until 3.25 g of solid remain. What percentage of the $CaCO_3$ has decomposed?

4-41. $MgCO_3$ when heated yields MgO and CO_2. If 2.25 g of $MgCO_3$ are heated until the residue has mass 1.95 g, what percentage of the $MgCO_3$ remains undecomposed?

4-42. In 1966, chemists prepared a compound containing molecular nitrogen and ammonia coordinated to ruthenium (II) chloride. Analysis showed that for every 1.000 g of ruthenium there were 0.843 g ammonia, 0.702 g of chlorine, and 0.277 g of molecular nitrogen (N_2). Determine the simplest formula.

4-43. Chemists in several locations around the world have isolated compounds of zero-valent platinum. One compound was found to consist of 15.68% Pt and 84.32% triphenylphosphine [$P(C_6H_5)_3$]. Determine the number of triphenylphosphine molecules for each platinum atom.

4-44. The compound described in Problem 4-43 reacts with molecular oxygen in benzene solution to form bright orange needleshaped crystals.

Analysis of this new compound gave the following results: 25.95% Pt; 69.79% $P(C_6H_5)_3$; 4.26% O. Determine the number of triphenylphosphine molecules and the number of oxygen atoms for each platinum atom in the compound.

4-45. A compound contains 20.86% Si and 79.14% Cl. (a) Deterime the empirical formula for the compound. (b) If the experimentally determined molecular weight is 269, write the true formula for the compound.

4-46. In a laboratory experiment to determine the formula for the compound formed between Cu and S, a student heated a weighed amount of copper with an excess of sulfur. After the reaction was completed, the excess sulfur was volatilized off by heating until the crucible and its contents had achieved a constant weight. The following data were obtained:

Mass of crucible	19.732 g
Mass of crucible plus Cu	27.304 g
Mass of crucible plus the Cu–S compound	29.214 g

What is the simplest formula for the compound?

4-47. The first true chemical compound of a rare gas element was prepared in 1962 by Professor Neil Bartlett while he was a member of the faculty of the University of British Columbia. He succeeded in preparing a yellow compound by direct reaction of xenon gas with platinum hexafluoride which, on chemical analysis, was found to be composed of 29.8% Xe, 44.3% Pt, and 25.9% F. What is the empirical formula for the yellow compound?

4-48. A sample of a pure compound containing K, Cl, and O had a mass of 0.855 g. The sample was first heated to drive off all O. The residue had a mass of 0.522 g. The entire residue was then treated with $AgNO_3$ to precipitate all the chlorine as AgCl. The obtained AgCl had a mass of 1.003 g. From these data determine the empirical formula of the initial pure compound.

4-49. A certain hydrated salt has the general formula $Co_xCl_y \cdot zH_2O$. A 0.928 g sample of the hydrate was heated to drive off all H_2O. The anhydrous residue had a mass of 0.507 g. The residue is then treated with silver nitrate to precipitate the chloride ion as AgCl. 1.117 g of AgCl were obtained. Determine the simplest formula of the hydrate—that is, the values of x, y, and z.

4-50. The following nitrogen compounds are frequently added to the soil as fertilizers: (a) $(NH_4)_2HPO_4$, (b) CON_2H_4, (c) $(NH_4)_2SO_4$, (d) NH_4NO_3. The approximate costs per 1.00 kilogram of these in bulk quantities are $0.260, $0.545, $0.218, and $0.394, respectively. For each of these compounds, determine the cost to furnish 2.50×10^3 kg of nitrogen to the soil.

Chapter 5

Methods of Balancing Equations

A balanced equation is one in which the same number of atoms of each element and the same total of charges appear on the left and right sides of the equation. For this to be true, the proper coefficients or numbers appearing immediately before the formulas in the equation must be used. The process of obtaining the proper coefficients is called balancing. Note that only coefficients (and not subscripts) can be adjusted in this process because placing a coefficient in front of a formula simply changes the amount of a substance; changing a subscript **changes the identity of the material.** In balancing an equation, the identity of the material cannot be altered.

Equations involving only a few reactants and products may be balanced by simple inspection; this includes some simple oxidation–reduction reactions that, by definition, involve changes in oxidation numbers of some of the substances. Often, however, oxidation–reduction equations are difficult to balance by inspection; fortunately, systematic methods are available for balancing them. These are considered after a brief discussion of the inspection method of balancing equations.

5-A. Balancing by Inspection

1. Molecular Equations

A good general rule to follow is: Starting with the most complicated formula, balance elements by adjusting coefficients of the formulas as necessary, leaving hydrogen and oxygen until last.

Example 5-a. Balance the equation

$$Al(OH)_3 + H_2SO_4 \longrightarrow Al_2(SO_4)_3 + H_2O$$

Solution. Pick out the most complicated formula which in this case is obviously $Al_2(SO_4)_3$. Since two Al atoms are present in this formula, place the coefficient 2 in front of $Al(OH)_3$ on the left side. Likewise three S atoms (or three SO_4 radicals) appear in the formula, hence a coefficient 3 is needed for H_2SO_4 on the left side. Next balance H atoms; since twelve H atoms now appear on the left side, a coefficient of 6 is required for H_2O on the right. The balanced equation becomes:

$$2\,Al(OH)_3 + 3\,H_2SO_4 \longrightarrow Al_2(SO_4)_3 + 6\,H_2O$$

Example 5-b. Balance the equation

$$BaBr_2 + H_3PO_4 \longrightarrow Ba_3(PO_4)_2 + HBr$$

Solution. Start with $Ba_3(PO_4)_2$ and observe that three Ba appear in the formula; therefore use a coefficient of 3 for $BaBr_2$. Likewise two P atoms (or two PO_4 radicals) call for a coefficient of 2 for H_3PO_4. Now six Br atoms on the left call for a coefficient of 6 for HBr on the right. The balanced equation becomes:

$$3\,BaBr_2 + 2\,H_3PO_4 \longrightarrow Ba_3(PO_4)_2 + 6\,HBr$$

Example 5-c. Balance the equation

$$C_2H_6 + O_2 \longrightarrow CO_2 + H_2O$$

Solution. In this case start with C_2H_6 and note that two C appear in the formula. Therefore place the coefficient 2 in front of CO_2. Similarly, six H require the coefficient 3 in front of H_2O. Now the seven O atoms on the right require the coefficient 3.5 for O_2. Hence:

$$C_2H_6 + 3.5\,O_2 \longrightarrow 2\,CO_2 + 3\,H_2O$$

Although balanced, the equation is not in conventional form. Convention indicates that a balanced equation contains the smallest possible set of **whole** number co-efficients. Hence we multiply all coefficients in the above equation by two to obtain:

$$2\,C_2H_6 + 7\,O_2 \longrightarrow 4\,CO_2 + 6\,H_2O$$

2. Ionic Equations

The same general rules employed for balancing molecular equations apply to ionic equations, but, **in addition to balancing atoms, the charges must also be balanced.**

In writing ionic equations, write as ions those substances that exist as ions in solution; write complete formulas for nonionized substances, solids, and gases. Generally only those substances undergoing a net chemical change are included in ionic equations.

Example 5-d. Balance the equation

$$Al + H^+ \longrightarrow Al^{3+} + H_2$$

Solution. If a coefficient of 2 is used for H^+, the equation is balanced atomically but not electrically, since there would be two positive charges on the left and three positive charges on the right. Inspection reveals that an even number of H^+—and thus positive charges—must be used on the left because every molecule of H_2 on the right has two atoms of hydrogen. Since the Al^{3+} ion has three positive charges, the smallest coefficient for Al must be 2, so that an even number of charges are obtained on the right. The coefficient of H^+ must be 6 for the charges to balance. Hence the balanced equation becomes:

$$2\,Al + 6\,H^+ \longrightarrow 2\,Al^{3+} + 3\,H_2$$

Molecular Equations (Balance by Inspection)

5-1. $Mg + O_2 \rightarrow MgO$
5-2. $NaI + Cl_2 \rightarrow NaCl + I_2$
5-3. $MgO + HBr \rightarrow MgBr_2 + H_2O$
5-4. $AgNO_3 + AlCl_3 \rightarrow AgCl + Al(NO_3)_3$
5-5. $CaBr_2 + H_3PO_4 \rightarrow Ca_3(PO_4)_2 + HBr$
5-6. $AlBr_3 + Cl_2 \rightarrow AlCl_3 + Br_2$
5-7. $BaCl_2 + Na_3AsO_4 \rightarrow Ba_3(AsO_4)_2 + NaCl$
5-8. $(NH_4)_2S + HgCl_2 \rightarrow HgS + NH_4Cl$
5-9. $Al(OH)_3 + HNO_3 \rightarrow Al(NO_3)_3 + H_2O$
5-10. $K + H_2O \rightarrow KOH + H_2$
5-11. $C_4H_{10} + O_2 \rightarrow CO_2 + H_2O$
5-12. $Th(NO_3)_4 + K_3PO_4 \rightarrow Th_3(PO_4)_4 + KNO_3$
5-13. $PBr_3 + H_2O \rightarrow H_3PO_3 + HBr$
5-14. $C_2H_2 + O_2 \rightarrow CO_2 + H_2O$
5-15. $Al + HCl \rightarrow AlCl_3 + H_2$
5-16. $KNO_3 \rightarrow KNO_2 + O_2$
5-17. $C_6H_6 + O_2 \rightarrow CO_2 + H_2O$
5-18. $CaH_2 + H_2O \rightarrow Ca(OH)_2 + H_2$
5-19. $ZnCO_3 + HNO_3 \rightarrow Zn(NO_3)_2 + CO_2 + H_2O$
5-20. $CaNCN + H_2O \rightarrow CaCO_3 + NH_3$
5-21. $C_2H_5OH + O_2 \rightarrow CO_2 + H_2O$
5-22. $NH_4NO_3 + Ca(OH)_2 \rightarrow Ca(NO_3)_2 + NH_3 + H_2O$
5-23. $NaMnO_4 + H_2SO_4 \rightarrow Na_2SO_4 + Mn_2O_7 + H_2O$

5-24. $P_4 + O_2 + H_2O \rightarrow H_3PO_4$
5-25. $PbS + O_2 \rightarrow PbO + SO_2$
5-26. $Fe_2O_3 + CO \rightarrow Fe_3O_4 + CO_2$
5-27. $BaS + H_2O_2 \rightarrow BaSO_4 + H_2O$
5-28. $Cu(NO_3)_2 \rightarrow CuO + NO_2 + O_2$
5-29. $H_2S + H_2SO_3 \rightarrow S + H_2O$
5-30. $Li_2O_2 + H_2O \rightarrow LiOH + O_2$
5-31. $Al + NaOH + H_2O \rightarrow NaAl(OH)_4 + H_2$

Ionic Equations (Balance by Inspection)

5-32. $NH_4^+ + OH^- \rightarrow NH_3 + H_2O$
5-33. $Cr + H^+ \rightarrow Cr^{2+} + H_2$
5-34. $PbCO_3 + H^+ \rightarrow Pb^{2+} + CO_2 + H_2O$
5-35. $As_2S_5 + S^{2-} \rightarrow AsS_4^{3-}$
5-36. $CdS + H^+ \rightarrow H_2S + Cd^{2+}$
5-37. $Ba_3(PO_4)_2 + H^+ \rightarrow H_3PO_4 + Ba^{2+}$
5-38. $Th^{4+} + H_2PO_4^- \rightarrow Th_3(PO_4)_4 + H^+$
5-39. $Fe_2O_3 + H^+ \rightarrow Fe^{3+} + H_2O$
5-40. $CdS + As^{3+} \rightarrow As_2S_3 + Cd^{2+}$
5-41. $Br^- + MnO_2 + H^+ \rightarrow HBrO + Mn^{2+} + H_2O$
5-42. $Zn + OH^- + H_2O \rightarrow Zn(OH)_4^{2-} + H_2$
5-43. $CuS + SO_4^{2-} + H^+ \rightarrow Cu^{2+} + S + SO_2 + H_2O$
5-44. $S_2O_8^{2-} + Ce^{3+} \rightarrow SO_4^{2-} + Ce^{4+}$

5-B. Balancing Oxidation-Reduction Equations

Oxidation involves a loss of electrons and/or an increase in oxidation number by an atom, group of atoms, or an ion. Similarly, reduction involves a gain of electrons and/or a decrease in oxidation number by a substance. The substance whose oxidation number undergoes an increase is said to be oxidized; the substance whose oxidation number undergoes a decrease is said to be reduced. The substance that loses electrons (i.e., is oxidized) is called the reducing agent, and the substance that gains electrons (i.e., is reduced) is called the oxidizing agent. In a system the total number of electrons lost must equal the total number of electrons gained—that is, the number of electrons in the system must be conserved. Because of this we see that oxidation and reduction must occur simultaneously.

Oxidation numbers are the apparent charges that atoms would have if the electrons in the compound were distributed among atoms in a very arbitrary fashion. These numbers are a useful aid in following electron shifts in oxidation–reduction reactions and are assigned as follows.

1. The oxidation numbers of atoms of elements are zero.

2. In ionic substances, the oxidation numbers of monatomic ions are equal to their charges.

3. In a covalent molecule or a complex ion, the oxidation number is the apparent charge left on an atom when the bonding electrons are assigned completely to the more electronegative of the two bonded atoms; electrons shared between atoms of the same element are divided equally between them. In all these cases, the assignment must be such that there is a conservation of charge on the molecule or ion.

From the preceding rules, Li ion and Cl ion in LiCl are assigned the values $+1$ and -1, respectively. Oxygen in H_2O is assigned the value -2 and hydrogen the value $+1$ because oxygen is the more electronegative of the two atoms. These are the usual values for H and O except in a few instances; O is -1 in peroxides and H is -1 in hydrides.

Using customary oxidation numbers of the simple ions, the less common oxidation numbers of other atoms may be assigned. In H_2SO_4, the contribution of two H to the apparent charge is $+2$ and the contribution of four O atoms is -8. For the sum of the oxidation numbers to be zero, as required for a neutral molecule, the total apparent charge for S must be $+6$. In the complex ion $[PtCl_6]^{2-}$, Cl is given its usual value of -1, and then for the sum of the oxidation numbers to equal the -2 charge of the ion, Pt must be assigned the oxidation number of $+4$.

Oxidation-reduction equations are balanced on the principle of **conservation of charge,** that is, the number of electrons lost by the reducing agent must equal the number of electrons gained by the oxidizing agent. One method of balancing these equations is known as the **Ion-Electron Half-Reaction Method** and the other is known as the **Change in Oxidation Number Method.**

1. Ion-Electron Half-Reaction Method

There are a series of steps (or rules) to follow in this method that result in a systematic procedure for balancing redox equations. These steps are outlined below.

1. **The reaction is separated into two half-reactions, one of which shows the oxidation reaction and the other the reduction reaction.**

2. **Balance each half-reaction atomically**. Formulas for the reacting species are written to represent as closely as possible those present in large amounts; formulas for the products are written similarly. It is often necessary to add H_2O, H^+, or OH^- to the half-reactions to balance them. It must be remembered, however, that OH^- is not present in large amount in acid solution and that H^+ is not present in large amount in basic solution. Convenient rules to follow are: (a) To balance oxygen atoms, add water if in acidic solution, and OH^- if in basic solution, to the side of the half-reaction deficient in O atoms; (b) balance H atoms after O balancing has been completed.

3. **Balance each half-reaction electrically by adding electrons.**

4. **Equalize electron gain and loss for the system by multiplying the entirety of each half-reaction by the proper coefficient**. This step assures that the number of electrons given up in the oxidation half-reaction is equal to the number accepted in the reduction half-reaction.

5. **Add the two half-reactions to obtain the net overall reaction.** Remember to cancel species common to both sides of the equation.

The following examples illustrate this method. Examples 5-e and 5-g are worked in detail.

Example 5-e. Balance the equation for the oxidation of Cu by NO_3^- in acid solution. Copper is oxidized to Cu^{2+}, and NO_3^- is reduced to NO.

Solution.

Step (1). Write two skeleton half-reactions, one for the oxidation half-reaction and one for the reduction half-reaction.

$$\mathrm{Cu} \longrightarrow \mathrm{Cu^{2+}} \qquad \text{(oxidation)}$$

$$\mathrm{NO_3^-} \longrightarrow \mathrm{NO} \qquad \text{(reduction)}$$

Step (2). Balance the half reactions atomically. It is observed that in the oxidation half-reaction there is one Cu atom on each side of the equation. In the reduction reaction there are two more O atoms on the left than on the right, and in acid solution, O with a -2 oxidation state must become H_2O. Therefore $4\,H^+$ are required on the left and the equation is now written as

$$4\,\mathrm{H^+} + \mathrm{NO_3^-} \longrightarrow \mathrm{NO} + 2\,\mathrm{H_2O}$$

Both half-reactions are now balanced atomically.

Step (3). Balance the charge for each half-reaction by adding negative electrons. The two half-reactions thus become

$$\mathrm{Cu} \longrightarrow \mathrm{Cu^{2+}} + 2\,e^- \qquad \text{(oxidation)}$$

$$3\,e^- + 4\,\mathrm{H^+} + \mathrm{NO_3^-} \longrightarrow \mathrm{NO} + 2\,\mathrm{H_2O} \qquad \text{(reduction)}$$

Step (4). Multiply the oxidation half-reaction by 3 and the reduction half-reaction by 2 to obtain 6 electrons lost and gained

$$3\,Cu \longrightarrow 3\,Cu^{2+} + 6\,e^-$$

$$6\,e^- + 8\,H^+ + 2\,NO_3^- \longrightarrow 2\,NO + 4\,H_2O$$

Step (5). In adding these two half-reactions, the electrons cancel and the net ionic equation for the reaction becomes

$$3\,Cu + 8\,H^+ + 2\,NO_3^- \longrightarrow 3\,Cu^{2+} + 2\,NO + 4\,H_2O$$

Example 5-f. Balance the equation

$$Fe^{2+} + H^+ + MnO_4^- \longrightarrow Fe^{3+} + Mn^{2+} + H_2O$$

Solution.

$$5\,[Fe^{2+} \longrightarrow Fe^{3+} + e^-] \qquad \text{(oxidation)}$$

$$5\,e^- + 8\,H^+ + MnO_4^- \longrightarrow 4\,H_2O + Mn^{2+} \qquad \text{(reduction)}$$

$$5\,Fe^{2+} + 8\,H^+ + MnO_4^- \longrightarrow 5\,Fe^{3+} + Mn^{2+} + 4\,H_2O$$

Example 5-g. Balance the equation for the reaction between $Bi(OH)_3$ and SnO_2^{2-} that occurs in basic solution.

$$Bi(OH)_3 + SnO_2^{2-} \longrightarrow Bi + SnO_3^{2-}$$

Solution.

Step (1). Separate the equation into an oxidation half-reaction and a reduction half-reaction.

$$Bi(OH)_3 \longrightarrow Bi \qquad \text{(reduction)}$$

$$SnO_2^{2-} \longrightarrow SnO_3^{2-} \qquad \text{(oxidation)}$$

Step (2). Balance atomically, but remember that, in basic solution, no H^+ should appear in either half-reaction. To balance atoms for the reduction half-reaction, add 3 OH^- to the right side of the equation.

$$Bi(OH)_3 \longrightarrow Bi + 3\,OH^-$$

There is a deficiency of O atoms on the left side of the oxidation half-reaction, and so hydroxide ions are added to the left to provide them. Twice as many OH^- ions must be added as oxygen atoms needed because the H^+ separated from OH^- must form water by reaction with a second OH^-. When balanced atomically the half-reaction becomes

$$2\,OH^- + SnO_2^{2-} \longrightarrow SnO_3^{2-} + H_2O \qquad \text{(correct)}$$

$$OH^- + SnO_2^{2-} \longrightarrow SnO_3^{2-} + H^+ \qquad \text{(wrong)}$$

The latter equation is incorrect because H^+ is not present in large amount in basic solution.

Step (3). On balancing charge the half-reactions become

$$3\,e^- + Bi(OH)_3 \longrightarrow Bi + 3\,OH^- \quad \text{(reduction)}$$

and

$$2\,OH^- + SnO_2^{2-} \longrightarrow SnO_3^{2-} + H_2O + 2\,e^- \quad \text{(oxidation)}$$

Step (4). Multiply the reduction half-reaction by 2 and the oxidation half-reaction by 3 to make electrons gained equal to electrons lost.

$$6\,e^- + 2\,Bi(OH)_3 \longrightarrow 2\,Bi + 6\,OH^-$$

$$6\,OH^- + 3\,SnO_2^{2-} \longrightarrow 3\,SnO_3^{2-} + 3\,H_2O + 6\,e^-$$

Step (5). Add the two half-reactions to obtain the final balanced equation. Note that the electrons and OH^- ions cancel; such cancellations should be made whenever possible.

$$2\,Bi(OH)_3 + 3\,SnO_2^{2-} \longrightarrow 3\,SnO_3^{2-} + 2\,Bi + 3\,H_2O$$

Example 5-h. Balance the equation for the oxidation of AsO_3^{3-} by MnO_4^- in basic solution:

$$MnO_4^- + AsO_3^{3-} + H_2O \longrightarrow MnO_2 + AsO_4^{3-} + OH^-$$

Solution. Separate the completed equation into half-reactions and balance each step separately. Since the reaction is in basic solution, hydrogen ion should not appear in either half-reaction. It is seen from the equation that AsO_3^{3-} is oxidized to AsO_4^{3-} and MnO_4^- is reduced to MnO_2. With this information the two half-reactions are

$$3\,[AsO_3^{3-} + 2\,OH^- \longrightarrow AsO_4^{3-} + H_2O + 2\,e^-] \quad \text{(oxidation)}$$

$$2\,[3\,e^- + MnO_4^- + 2\,H_2O \longrightarrow MnO_2 + 4\,OH^-] \quad \text{(reduction)}$$

$$2\,MnO_4^- + 3\,AsO_3^{3-} + H_2O \longrightarrow 2\,MnO_2 + 3\,AsO_4^{3-} + 2\,OH^-$$

Example 5-i. Balance the equation for the reaction of elemental sulfur to form sulfite and sulfide ions in basic solution.

$$S \longrightarrow SO_3^{2-} + S^{2-}$$

Solution. Note in this reaction that sulfur is both oxidized and reduced. When this occurs the substance involved (sulfur) is said to undergo **disproportionation** or **auto-oxidation-reduction.**

$$6\,OH^- + S \longrightarrow SO_3^{2-} + 3\,H_2O + 4\,e^- \quad \text{(oxidation)}$$

$$2\,[2\,e^- + S \longrightarrow S^{2-}] \quad \text{(reduction)}$$

$$6\,OH^- + 3\,S \longrightarrow SO_3^{2-} + 2\,S^{2-} + 3\,H_2O$$

Equations (Ion-Electron Half-Reactions)

5-45. Balance by the ion-electron method. Balance each half-reaction atomically and electrically; then add the two steps to obtain the net ionic equation.

(a) $C_2O_4^{2-} \rightarrow CO_2$
$H^+ + MnO_4^- \rightarrow H_2O + Mn^{2+}$
(b) $Bi_2S_3 \rightarrow Bi^{3+} + S$
$H^+ + NO_3^- \rightarrow H_2O + NO$
(c) $Fe^{2+} \rightarrow Fe^{3+}$
$H^+ + Cr_2O_7^{2-} \rightarrow H_2O + Cr^{3+}$
(d) $Mn + OH^- \rightarrow Mn(OH)_2$
$H_2O \rightarrow H_2 + OH^-$
(e) $S_2O_3^{2-} + H_2O \rightarrow SO_4^{2-} + H^+$
$Br_2 \rightarrow Br^-$
(f) $CrO_2^- + OH^- \rightarrow CrO_4^{2-} + H_2O$
$ClO^- + H_2O \rightarrow Cl^- + OH^-$
(g) $Zn \rightarrow Zn^{2+}$
$HAsO_2 + H^+ \rightarrow AsH_3 + H_2O$
(h) $Pd + Cl^- \rightarrow PdCl_6^{2-}$
$H^+ + Cl^- + NO_3^- \rightarrow NOCl + H_2O$
(i) $Ag + Cl^- \rightarrow AgCl$
$H^+ + ClO_3^- \rightarrow Cl^- + H_2O$
(j) $OH^- + Cl_2 \rightarrow ClO^- + H_2O$
$Cl_2 \rightarrow Cl^-$

5-46. In the following, only the final unbalanced and incomplete equation is given. Separate each reaction into two half-reactions, one oxidation and the other reduction. Balance each step, then add the two steps. Use OH^-, H^+, and H_2O where necessary to balance each half-reaction. Collect all terms and make cancellations where possible.

(a) $I_2 + S_2O_3^{2-} \rightarrow I^- + S_4O_6^{2-}$
(b) $Fe^{2+} + H^+ + NO_3^- \rightarrow Fe^{3+} + NO$
(c) $Zn + CNS^- \rightarrow Zn^{2+} + H_2S + HCN$ (acid solution)
(d) $Fe^{2+} + H_2O_2 + H^+ \rightarrow Fe^{3+} + H_2O$
(e) $CoS + H^+ + NO_3^- + Cl^- \rightarrow Co^{2+} + S + NOCl$
(f) $MnO_4^- + NO_2 + H_2O \rightarrow Mn^{2+} + NO_3^- + H^+$
(g) $Zn + NO_3^- \rightarrow NH_4^+ + Zn^{2+}$ (acid solution)
(h) $Co(OH)_2 + O_2^{2-} \rightarrow Co(OH)_3 + OH^-$
(i) $As + H^+ + NO_3^- \rightarrow AsO_4^{3-} + NO_2$
(j) $H_2O_2 + MnO_4^- + H^+ \rightarrow O_2 + Mn^{2+} + H_2O$
(k) $Ag_2S + NO_3^- \rightarrow Ag^+ + S + NO$ (acid solution)

(l) $MnO_4^- + SnO_2^{2-} \rightarrow SnO_3^{2-} + MnO_2$ (basic solution)
(m) $C_2H_5OH + Cr_2O_7^{2-} \rightarrow CH_3CHO + Cr^{3+}$ (acid solution)
(n) $NO_2 \rightarrow NO + NO_3^-$ (acid solution)
(o) $P_4 \rightarrow PH_3 + H_2PO_2^-$ (basic solution)

2. Balancing from Oxidation Number Changes

The following steps are useful in balancing equations from oxidation number changes.

1. Determine which elements change in oxidation number.
2. Determine the amount of change per formula unit of each substance.
3. Make the gain of electrons equal the loss of electrons by adjusting the coefficients of the formulas for the reducing agent and oxidizing agent in the equation.
4. Complete the balancing of the equations by inspection, keeping in mind that the ratio of oxidizing to reducing agent as determined above must be maintained.

These steps are illustrated in the examples below.

Example 5-j. Balance the equation

$$ZnS + HNO_3 \longrightarrow ZnSO_4 + NO + H_2O$$

Solution. In this reaction, sulfur changes in oxidation number from -2 in ZnS to $+6$ in $ZnSO_4$; a loss of eight electrons per formula unit of ZnS. Meanwhile, nitrogen changes in oxidation number from $+5$ in HNO_3 to $+2$ in NO, a gain of three electrons per molecule of HNO_3. The total gain of electrons must equal the total loss and to balance them, the number 3 is made the coefficient of ZnS and the number 8 becomes the coefficient of HNO_3 in the final equation. Thus 3 ZnS lose twenty-four electrons and 8 HNO_3 gain twenty-four electrons.

(lose $8\,e^-$) $\times$ 3 = 24
(oxidation)

$$\overset{-2}{3\,ZnS} + \overset{+5}{8\,HNO_3} \longrightarrow \overset{+6}{3\,ZnSO_4} + \overset{+2}{8\,NO} + 4\,H_2O$$

(gain $3\,e^-$) $\times$ 8 = 24
(reduction)

Example 5-k. Balance the equation

$$H_2S + K_2Cr_2O_7 + HCl \longrightarrow S + CrCl_3 + KCl + H_2O$$

Solution. Sulfur apparently loses two electrons per molecule of H_2S; each Cr atom in $K_2Cr_2O_7$ apparently gains three electrons, so the electron gain per formula unit of $K_2Cr_2O_7$ is $2 \times 3 = 6$. Hence, to balance electrons, we need a ratio of $6\,H_2S$ to $2\,K_2Cr_2O_7$, or simply 3 to 1. Having fixed the ratio of H_2S and $K_2Cr_2O_7$, the remainder of the balancing follows by inspection.

(lose $2\,e^-$) $\times 3 = 6$

(oxidation)

$$\overset{-2}{3\,H_2S} + \overset{+6}{K_2Cr_2O_7} + 8\,HCl \longrightarrow \overset{0}{3\,S} + \overset{+3}{2\,CrCl_3} + 2\,KCl + 7\,H_2O$$

gain $3\,e^-$/Cr

or gain $6\,e^-/K_2Cr_2O_7$

(reduction)

Example 5-l. Balance the equation

$$AsO_3^{3-} + Cr_2O_7^{2-} + H^+ \longrightarrow AsO_4^{3-} + Cr^{3+} + H_2O$$

Solution.

(lose $2\,e^-$) $\times 3 = 6$

(oxidation)

$$\overset{+3}{3\,AsO_3^{3-}} + \overset{+6}{Cr_2O_7^{2-}} + 8\,H^+ \longrightarrow \overset{+5}{3\,AsO_4^{3-}} + \overset{+3}{2\,Cr^{3+}} + 4\,H_2O$$

(gain $6\,e^-/Cr_2O_7^{2-}$)

(reduction)

Example 5-m. Balance the equation

$$As_4O_6 + Cl_2 + H_2O \longrightarrow H_3AsO_4 + HCl$$

Solution. Lose $2\,e^-$/As or $8\,e^-/As_4O_6$; gain $1\,e^-$/Cl or $2\,e^-/Cl_2$.

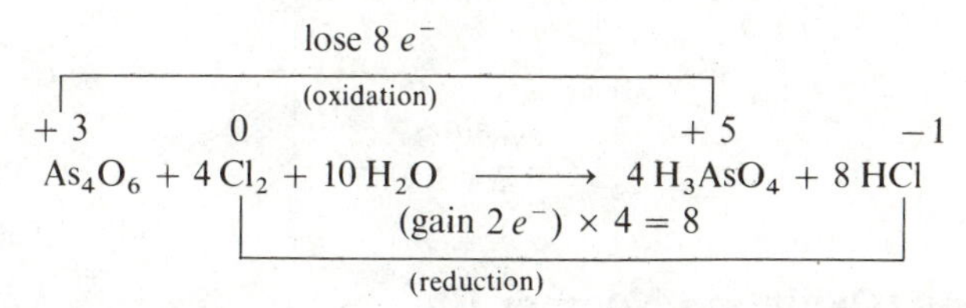

Example 5-n. Occasionally a reaction in which more than two elements change in oxidation number is encountered. We need only obtain the **net change in each substance** to arrive at the proper coefficients. Balance the equation

$$As_2S_3 + Mn(NO_3)_2 + K_2CO_3 \longrightarrow$$
$$K_3AsO_4 + K_2SO_4 + K_2MnO_4 + NO + CO_2$$

Solution.

Apparent electron loss by As per formula unit

$$As_2S_3 = 2 \times 2 = 4 \text{ electrons lost}$$

Apparent electron loss by S per formula unit

$$As_2S_3 = 3 \times 8 = 24 \text{ electrons lost}$$

Net change per formula unit

$$As_2S_3 = 28 \text{ electrons } \textbf{lost}$$

Apparent electron loss by Mn per formula unit

$$Mn(NO_3)_2 = 1 \times 4 = 4 \text{ electrons lost.}$$

Apparent electron gain by N per formula unit

$$Mn(NO_3)_2 = 2 \times 3 = 6 \text{ electrons gained}$$

Net change per formula unit

$$Mn(NO_3)_2 = 2 \text{ electrons } \textbf{gained}$$

To balance the electron gain and loss, we arrive at the ratio 28 to 2 or 14 $Mn(NO_3)_2$ to 1 As_2S_3. The remainder of the equation is balanced by inspection.

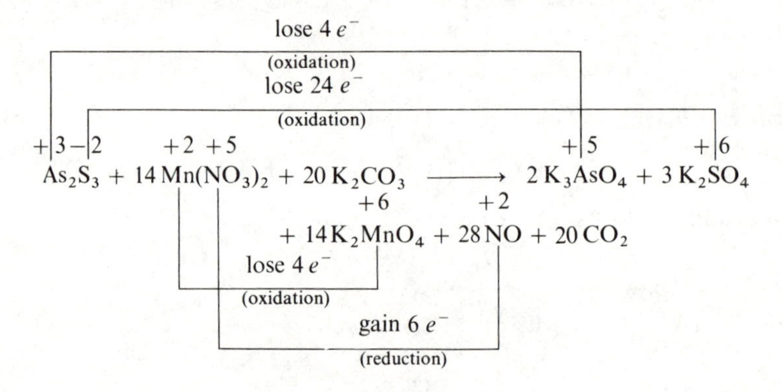

Equations (Oxidation Number Change)

5-47. $Bi(OH)_3 + K_2Sn(OH)_4 \rightarrow Bi + K_2Sn(OH)_6$

5-48. $H_2S + KMnO_4 + HCl \rightarrow S + KCl + MnCl_2 + H_2O$

5-49. $MnO_2 + PbO_2 + HNO_3 \rightarrow Pb(NO_3)_2 + HMnO_4 + H_2O$

5-50. $Fe_2(SO_4)_3 + NaI \rightarrow FeSO_4 + Na_2SO_4 + I_2$

5-51. $Sb + HNO_3 \rightarrow Sb_2O_5 + NO + H_2O$

5-52. $HMnO_4 + AsH_3 + H_2SO_4 \rightarrow H_3AsO_4 + MnSO_4 + H_2O$

5-53. $Bi + HNO_3 \rightarrow Bi(NO_3)_3 + NO + H_2O$

5-54. $FeSO_4 + H_2SO_4 + KMnO_4 \rightarrow$
$Fe_2(SO_4)_3 + K_2SO_4 + MnSO_4 + H_2O$

5-55. $Ag + H^+ + NO_3^- \rightarrow Ag^+ + NO + H_2O$

5-56. $Sn^{2+} + H^+ + IO_3^- \rightarrow Sn^{4+} + I^- + H_2O$

5-57. $As_4 + H_2O + NO_3^- \rightarrow AsO_4^{3-} + NO + H^+$

5-58. $Cr_2O_7^{2-} + C_2H_5OH + H^+ \rightarrow Cr^{3+} + HC_2H_3O_2 + H_2O$

5-59. $Ag_3AsO_4 + Zn + H^+ \rightarrow AsH_3 + Ag + Zn^{2+} + H_2O$

5-60. $Sb_2S_3 + H^+ + NO_3^- \rightarrow Sb_2O_5 + SO_4^{2-} + NO + H_2O$

5-61. $Cr_2S_3 + Mn(NO_3)_2 + Na_2CO_3 \rightarrow$
$Na_2CrO_4 + Na_2MnO_4 + Na_2SO_4 + CO_2 + NO$

General Oxidation-Reduction Equations

Balance by any method.

5-62. $Sn^{2+} + [PtCl_6]^{2-} \rightarrow Sn^{4+} + [PtCl_4]^{2-} + Cl^-$

5-63. $Zn + SbO_4^{3-} + H^+ \rightarrow Zn^{2+} + SbH_3 + H_2O$

5-64. $KMnO_4 + H_2SO_4 + PH_3 \rightarrow$
$K_2SO_4 + MnSO_4 + H_2O + H_3PO_4$

5-65. $Fe^{2+} + PuO_2^+ + H^+ \rightarrow Fe^{3+} + Pu^{4+} + H_2O$

5-66. $Fe(OH)_2 + MnO_4^- + H_2O \rightarrow Fe(OH)_3 + Mn(OH)_2 + OH^-$

5-67. $I^- + Br_2 + H_2O \rightarrow IO_3^- + H^+ + Br^-$

5-68. $KMnO_4 + H_2C_2O_4 + H_2SO_4 \rightarrow$
$K_2SO_4 + MnSO_4 + CO_2 + H_2O$

5-69. $U + NO_3^- + H^+ \rightarrow UO_2^{2+} + NO + H_2O$

5-70. $Cr(OH)_3 + OH^- + IO_3^- \rightarrow CrO_4^{2-} + I^- + H_2O$

5-71. $Am^{3+} + H_2O + S_2O_8^{2-} \rightarrow AmO_2^+ + H^+ + SO_4^{2-}$

5-72. $As_2S_5 + Cl_2 + H_2O \rightarrow H_3AsO_4 + H_2SO_4 + HCl$

5-73. $AsH_3 + H_2O + AuCl_4^- \rightarrow H_3AsO_3 + H^+ + Au + Cl^-$

5-74. $Fe(OH)_2 + Pb(OH)_3^- \rightarrow Fe(OH)_3 + Pb + OH^-$

5-75. $SO_2 + IO_3^- + H_2O \rightarrow SO_4^{2-} + I_2 + H^+$

5-76. $F_2 + H_2O \rightarrow HF + O_3$

5-77. $ZnS + NO_3^- + H^+ \rightarrow Zn^{2+} + S + NH_4^+ + H_2O$

5-78. $Cr(OH)_3 + OH^- + Cl_2 \rightarrow CrO_4^{2-} + Cl^- + H_2O$

5-79. $AsO_3^{3-} + MnO_4^- + H^+ \rightarrow AsO_4^{3-} + Mn^{2+} + H_2O$

5-80. $Fe + HNO_3 \rightarrow Fe(NO_3)_3 + NO + H_2O$

5-81. $VO^{2+} + H_2O + MnO_4^- \rightarrow VO_4^{3-} + H^+ + Mn^{2+}$

5-82. $C_2H_6O_2 + HIO_4 \rightarrow H_2CO_3 + HIO_3 + H_2O$

5-83. $CN^- + MnO_4^- + H_2O \rightarrow CNO^- + MnO_2 + OH^-$

5-84. $Re + NO_3^- + H^+ \rightarrow HReO_4 + NO + H_2O$

5-85. $Na_2S + Ag + H_2O + O_2 \rightarrow NaOH + Ag_2S$

5-86. $H_2O_2 + ClO_2 + OH^- \rightarrow O_2 + ClO_2^- + H_2O$

5-87. $Mn^{2+} + BiO_3^- + H^+ \rightarrow MnO_4^- + Bi^{3+} + H_2O$

5-88. $Br_2 + NH_3 \rightarrow NH_4Br + N_2$

5-89. $Cr_2O_7^{2-} + H_2C_2O_4 + H^+ \rightarrow Cr^{3+} + CO_2 + H_2O$

5-90. $Br_2 + OH^- \rightarrow BrO_3^- + Br^- + H_2O$

5-91. $P_4 + H_2O \rightarrow HPO_3^{2-} + PH_3 + H^+$

5-92. $H^+ + IO_3^- + I^- \rightarrow I_2 + H_2O$

5-93. $H_2O + PbSO_4 \rightarrow Pb + PbO_2 + SO_4^{2-} + H^+$

5-94. $H^+ + I^- + H_5IO_6 \rightarrow I_2 + H_2O$

5-95. $KMnO_4 + MnSO_4 + KOH \rightarrow K_2SO_4 + MnO_2 + H_2O$

5-96. $NiS + H^+ + Cl^- + NO_3^- \rightarrow Ni^{2+} + S + NOCl + H_2O$

5-97. $CH_3OH + MnO_4^- \rightarrow HCOO^- + MnO_2 + OH^- + H_2O$

5-98. $FeS + NO_3^- + H^+ \rightarrow Fe^{3+} + S + NO + H_2O$

5-99. $Fe(OH)_2 + SO_3^{2-} + H_2O \rightarrow FeS + Fe(OH)_3 + OH^-$

5-100. $IPO_4 + H_2O \rightarrow IO_3^- + I_2 + H_2PO_4^- + H^+$

5-101. $As_2S_3 + K_2CO_3 + KNO_3 \rightarrow K_3AsO_4 + K_2SO_4 + NO + CO_2$

5-102. $Sb_2S_3 + HNO_3 \rightarrow Sb_2O_5 + NO_2 + S + H_2O$

5-103. $FeS_2 + Na_2O_2 \rightarrow Fe_2O_3 + Na_2SO_4 + Na_2O$

Chapter 6

Gas Laws

When dealing with gases it is usually more convenient to measure volumes instead of masses for determining the number of moles of gas involved; consequently, it is important to know the relationship between the volume of a gas and moles. We also need to know how the volume of a gas responds to changes in temperature and pressure. The gas laws describe the relationships among all these variables, and an understanding of the laws enables us to make interconversions of volumes to moles, and so on. In this section, we work problems based on the principal laws to which real gases conform quite closely.

Before proceeding to a discussion of the various gas laws and subsequent problems, we should get a firm feeling for the units of the several parameters involved. Kelvin or absolute temperature is always used for the manipulations required in the gas laws. Volume is generally expressed in terms of liters or milliliters. The pressure of a gas (force per unit area) can be expressed in a number of different ways. The most commonly used units of pressure in chemistry are the **torr**, **mm of Hg**, and the **atmosphere** (atm). These three units are related in the following way:

760 torr = 760 mm Hg = 1.00 atm (all to three significant figures)

(Note that 1 torr = 1 mm Hg.)

The accepted SI unit for pressure is the pascal (Pa). The pascal is defined as a force of one newton (N) per square meter (m^2). Since the newton has the units kg m/sec^2, we see that the pascal is a **derived** SI unit—that is, derived from three of the seven basic or defined SI units. The relationship of torr and atmosphere to pascal is presented below:

$$760 \text{ torr} = 1.00 \text{ atm} = 101325 \text{ Pa} = 101.325 \text{ kPa}$$

The International Committee on Weights and Measures allows the usage of the unit atmosphere for pressure but frowns on use of the unit torr. Chemists generally, however, have resisted the change to pascals because of the inconvenience of its size with respect to the atmosphere and the torr. In this problem book we will generally use the units atmosphere and torr and conversion to pascals can then be accomplished, if desired.

The abbreviation STP is often used in connection with gases. This stands for **S**tandard **T**emperature and **P**ressure and the conditions are: pressure = 760 torr, or 1 atmosphere, and temperature = 273 K (0°C).

6-A. Boyle's Law

For a given mass of gas at constant temperature the volume is inversely proportional to the pressure.

Expressed mathematically, $V = \text{constant} \times 1/P$ or $PV = \text{constant}$. It may also be stated in the form $P_1 V_1 = P_2 V_2$, where V_1 is the volume of the gas at a pressure P_1, and V_2 is the volume of the gas at a pressure P_2.

Example 6-a. 500 ml of gas at a pressure of 600 torr will occupy what volume if the pressure is increased to 750 torr at constant temperature?

Solution.

Original conditions	Final conditions
$P_1 = 600$ torr	$P_2 = 750$ torr
$V_1 = 500$ ml	V_2 = new volume

(a) We may reason this way: The new volume will be the volume we start with (in this case 500 ml) multiplied by a correction factor to account for the change in pressure, that is,

$$V_2 = 500 \text{ ml} \times \text{correction factor}$$

The correction factor will be a ratio of the two pressures, that is, 600 torr/750 torr or 750 torr/600 torr. Since the pressure is increasing, the volume must decrease (Boyle's law); therefore, the correction factor must be the ratio less than one, that is, 600/750. Consequently,

$$V_2 = 500 \text{ ml} \times \frac{600 \cancel{\text{torr}}}{750 \cancel{\text{torr}}} = 400 \text{ ml}$$

(b) Or, we may substitute in the equation $P_1 V_1 = P_2 V_2$

$$(600 \text{ torr})(500 \text{ ml}) = (750 \text{ torr})(V_2)$$

$$V_2 = \frac{(600 \cancel{\text{torr}})(500 \text{ ml})}{(750 \cancel{\text{torr}})} = 400 \text{ ml}$$

Example 6-b. What pressure must be applied to 2.0 liters of a gas at 1.0 atmosphere pressure to compress it to 0.80 liter?

Solution.

Original conditions	Final conditions
$V_1 = 2.0$ liters	$V_2 = 0.80$ liter
$P_1 = 1.0$ atm	$P_2 =$ new pressure

We reason that the new pressure will be the original pressure multiplied by a correction for the volume change.

$$P_2 = 1.0 \text{ atm} \times \text{correction for volume}$$

The volume correction must be the ratio of volumes, that is, 2.0 liters/0.80 liter or 0.80 liter/2.0 liters. Since the volume in this problem is decreasing (from 2.0 liters to 0.80 liter), the pressure must increase (Boyle's law); hence we must use the ratio of volumes greater than one, that is, 2.0 liters/0.80 liter.

$$P_2 = 1.0 \text{ atm} \times \frac{2.0 \cancel{\text{liters}}}{0.80 \cancel{\text{liter}}} = 2.5 \text{ atm}$$

Problems (Boyle's Law)

6-1. By what factor must the pressure on a gas be decreased to triple the volume?

6-2. A 288 ml sample of H_2 gas, originally at a pressure of 228 torr, will occupy what volume at standard pressure?

6-3. Oxygen gas in a cylinder is under a pressure of 2250 lb/inch^2. The volume of the cylinder is 6.00 ft^3. If all the gas in the cylinder is allowed to expand to 15.0 lb/inch^2 pressure, what will be the new volume of gas in ft^3?

6-4. A 2.00 liter sample of N_2 gas has a pressure of 76.7 kPa. Assuming no temperature change, what is the volume of this sample at standard pressure?

6-5. 1.00 liter of O_2 gas at 760 torr pressure and 0°C has a mass of 1.42 g. If the pressure on the gas is increased to 8.00 atm at 0°C, what is the density of the O_2 in grams/liter?

6-B. Charles' Law

For a given mass of gas at constant pressure the volume is directly proportional to the absolute temperature.

Expressed mathematically, $V = \text{constant} \times T$, or $V/T = \text{constant}$. Charles' law may also be stated in the form $V_1/T_1 = V_2/T_2$, where V_1 is the volume at absolute temperature T_1, and V_2 is the volume at absolute temperature T_2.

Example 6-c. If 2.00 liters of a gas are heated from 0°C to 91°C at constant pressure, what is the volume of gas at the higher temperature?

Solution.

Original conditions	Final conditions
$V_1 = 2.00$ liters	$V_2 =$ new volume
$T_1 = 273$ K	$T_2 = 91° + 273 = 364$ K

(a) $V_2 = 2.00$ liters × correction factor for temperature.

The correction factor will be a ratio of the absolute temperatures, that is, 364 K/273 K or 273 K/364 K. Since the temperature is increasing, the volume must increase (Charles' law); hence we must use 364 K/273 K, the larger of the two ratios, as the correction factor. Consequently,

$$V_2 = 2.00 \text{ liters} \times \frac{364 \not{K}}{273 \not{K}} = 2.67 \text{ liters}$$

(b) Or, by substituting in the equation $V_1/T_1 = V_2/T_2$, we obtain

$$\frac{2.00 \text{ liters}}{273 \text{ K}} = \frac{V_2}{364 \text{ K}}$$

$$V_2 = \frac{2.00 \text{ liters} \times 364 \not{K}}{273 \not{K}} = 2.67 \text{ liters}$$

Example 6-d. To what temperature (°C) must 1.00 liter of gas at 0°C be heated to expand its volume to 2.75 liters? Assume that there is no change in pressure.

Solution.

Original conditions	Final conditions
$V_1 = 1.00$ liter	$V_2 = 2.75$ liters
$T_1 = 273$ K	$T_2 =$ new temperature

$T_2 =$ initial temperature × correction for volume change.
$T_2 = 273$ K × correction factor

Since the volume increases from 1.00 to 2.75 liters, the temperature must increase; hence the ratio of volumes greater than unity must be used.

$$\text{New temperature} = 273\ \text{K} \times \frac{2.75\ \text{liters}}{1.00\ \text{liter}} = 751\ \text{K}$$

$$\text{Celsius temperature} = 751 - 273 = 478°\text{C}$$

Problems (Charles' Law)

6-6. A sample of gas whose volume at 315 K is 432 ml is heated at constant pressure until the volume becomes 648 ml. What is the final Celsius temperature of the gas?

6-7. 396 ml of gas at 19°C will occupy what volume at −31°C assuming constant pressure?

6-8. If 855 liters of Cl_2 are cooled at standard pressure from 273°C to 0°C, what will the final volume be?

6-9. At −40°C a sample of gas has a volume of 233 ml. On heating, the volume expands to 313 ml. Assuming constant pressure, what is the new Celsius temperature?

6-10. If 75.0 ml of hydrogen gas at 27°C are heated to 527°C, what is the new volume if the pressure does not change?

6-C. Combined Gas Laws

By combining Boyle's and Charles' laws, we may say: **For a given mass of gas, the volume is inversely proportional to pressure and directly proportional to absolute temperature.**

Expressed mathematically, $V = \text{constant} \times T/P$, or $PV/T = \text{constant}$. This relationship may be stated in the form $P_1V_1/T_1 = P_2V_2/T_2$, where V_1 is the volume at pressure P_1 and absolute temperature T_1, and V_2 is the volume at pressure P_2 and absolute temperature T_2.

It follows from the above that **at constant volume, the pressure of a gas is directly proportional to the absolute temperature,** that is,

$$\frac{P}{T} = \text{constant} \quad \text{or} \quad \frac{P_1}{T_1} = \frac{P_2}{T_2}$$

Example 6-e. 50.0 ft^3 of O_2 gas at a pressure of 3.00 atm and a temperature of 20°C will occupy what volume at a pressure of 1.00 atm and a temperature of 50°C?

Solution.

Original conditions	Final conditions
$P_1 = 3.00$ atm	$P_2 = 1.00$ atm
$V_1 = 50.0$ ft^3	V_2 = new volume
$T_1 = 20° + 273 = 293$ K	$T_2 = 50° + 273 = 323$ K

$V_2 = 50.0$ ft^3 × correction for pressure × correction for temperature.

Since pressure is decreasing from 3.00 atm to 1.00 atm, the volume must increase; thus the correction for pressure change must be 3.00 atm/1.00 atm. Since the temperature is increasing from 293 to 323 K, volume must increase; hence the correction factor for temperature change must be 323 K/293 K. Then

$$V_2 = 50.0 \text{ ft}^3 \times \frac{3.00 \text{ atm}}{1.00 \text{ atm}} \times \frac{323 \text{ K}}{293 \text{ K}} = 165 \text{ ft}^3$$

Or, by substituting in the formula

$$\frac{P_1 V_1}{T_1} = \frac{P_2 V_2}{T_2}$$

we obtain

$$\frac{(3.00 \text{ atm})(50.0 \text{ ft}^3)}{293 \text{ K}} = \frac{(1.00 \text{ atm})(V_2)}{323 \text{ K}}$$

$$V_2 = \frac{(50.0 \text{ ft}^3)(3.00 \text{ atm})(323 \text{ K})}{(1.00 \text{ atm})(293 \text{ K})} = 165 \text{ ft}^3$$

Example 6-f. A given mass of a certain gas occupies a volume of 10.0 liters at standard conditions. To what temperature must the gas be heated to change the volume to 50.0 liters at a pressure of 0.300 atm?

Solution.

Original conditions	Final conditions
$P_1 = 1.00$ atm	$P_2 = 0.300$ atm
$V_1 = 10.0$ liters	$V_2 = 50.0$ liters
$T_1 = 273$ K	T_2 = new temperature

$T_2 = 273$ K × correction for volume × correction for pressure.

$$T_2 = 273 \text{ K} \times \frac{50.0 \text{ liters}}{10.0 \text{ liters}} \times \frac{0.300 \text{ atm}}{1.00 \text{ atm}} = 410 \text{ K}$$

Celsius temperature = 410 − 273 = 137°C

Problems (Combined Gas Laws)

6-11. A 648 ml sample of gas is collected at 5.70×10^2 torr and a temperature of 273°C. What volume would this sample of gas occupy at standard conditions?

6-12. 5.87 liters of a gas at a pressure of 62.3 cm of Hg and a temperature of 115°C will occupy what volume at standard conditions?

6-13. A 4.50 ft^3 sample of He, initially at standard conditions, was heated to a final temperature of 273°C. Calculate the final pressure if the final volume was 13.5 ft^3.

6-14. A 15.0 liter sample of O_2 gas was collected at a temperature of −43°C and a pressure of 50.7 kPa. Calculate the volume at STP.

6-15. A 5.88 liter portion of a gas, collected at 27°C and 444 torr, was stored with a volume of 2.61 liters and a pressure of 666 torr. Calculate the storage temperature in K.

6-16. A large steel gas cylinder contained 15.0 liters of O_2 under a pressure of 35.0 atm at a temperature of 68°F. Determine the pressure in the cylinder during a storeroom fire when the temperature of the cylinder reached 530°F.

6-17. A gas occupied 9.0 m^3 at a certain temperature and pressure. Calculate the volume of the gas when its absolute temperature is halved and the pressure over it is increased tenfold.

6-18. A 7.50 liter sample of Kr gas at 0.400 atm and 27°C is compressed until the pressure is 2.60 atm; the temperature is then changed at constant pressure until the volume is 9.50 liters. What is the new temperature?

6-D. Dalton's Law of Partial Pressures

Each gas in a gaseous mixture acts independently of other gases present, and the total pressure of the mixture is the sum of the pressures of the gases present. The pressure exerted by each gas in the mixture is termed its partial pressure.

$$P_t = p_1 + p_2 + p_3 + \cdots$$

where P_t is the total pressure and p_1, p_2, p_3, and so on, represent the respective pressures exerted by each gas in the mixture.

A practical application of this law is in the collection of gases in the laboratory by the displacement of H_2O. When a gas is collected in this way some H_2O is vaporized, and this vapor exerts a part of the total pressure. The vapor pressure due to H_2O is a function of temperature only and not of volume; hence it should be subtracted from the total pressure before other corrections are made.*

$$P_{\text{total}} = p_{\text{gas}} + p_{H_2O}$$

or

$$p_{\text{gas}} = P_{\text{total}} - p_{H_2O}$$

Example 6-g. 362 ml of O_2 gas is collected in the laboratory by the displacement of liquid H_2O at a temperature of 20°C and a pressure of 689.5 torr. Determine the volume of the dry gas at standard conditions. The vapor pressure of H_2O at 20°C is 17.5 torr.

Solution. In collecting a gas over H_2O, the O_2 gas becomes saturated with H_2O vapor. According to Dalton's law

$$\text{Total pressure} = \text{pressure of } O_2 \text{ gas} + \text{pressure of } H_2O \text{ vapor}$$

or

$$p_{O_2} = P_t - p_{H_2O}$$

In this problem $p_{O_2} = 689.5 - 17.5 = 672 \text{ torr} = P_1$

Original conditions	Final conditions
$P_1 = 672$ torr	$P_2 = 760$ torr
$V_1 = 362$ ml	$V_2 =$ new volume
$T_1 = 293$ K	$T_2 = 273$ K

$$V_2 = 362 \text{ ml} \times \frac{672 \text{ torr}}{760 \text{ torr}} \times \frac{273 \text{ K}}{293 \text{ K}} = 298 \text{ ml}$$

Example 6-h. A metal tank was filled with a mixture of four gases. The total pressure in the container was 1480 torr. The pressure exerted by O_2 was 350 torr, by N_2, 520 torr, and by CO_2, 150 torr. What is the partial pressure of the fourth gas in the tank?

Solution.

$$P_t = 1480 \text{ torr} = p_1 + p_2 + p_3 + p_4$$

$$p_1 = 350 \text{ torr}$$

$$p_2 = 520 \text{ torr}$$

$$p_3 = 150 \text{ torr}$$

$$p_4 = ?$$

$$p_4 = 1480 \text{ torr} - (350 \text{ torr} + 520 \text{ torr} + 150 \text{ torr})$$

$$= 460 \text{ torr}$$

* A table of vapor pressures of water at various temperatures is given in the Appendixes. The student will need to make use of this table to solve problems when a correction for the vapor pressure of water is necessary.

Partial Pressure and Mole Fraction

In a mixture of gases, the partial pressure of each gas is proportional to its **mole fraction.** For each gas the mole fraction is simply the ratio of the moles of given gas to the total moles of all gases present; for example, if a mixture of gases contained 2 moles of *A*, 3 moles of *B*, and 5 moles of *C*, the mole fraction of *A* would be 2 $\cancel{\text{mol}}$/(2 $\cancel{\text{mol}}$ + 3 $\cancel{\text{mol}}$ + 5 $\cancel{\text{mol}}$) = 0.2, the mole fraction of *B* would be 3/10 = 0.3, and the mole fraction of *C* would be 0.5.

Example 6-i. The total pressure of a mixture of 4.0 moles of *A* and 1.0 mole of *B* is 7.0 atm. What is the partial pressure of each gas?

Solution.

$$\text{Mole fraction of } A = \frac{4.0}{4.0 + 1.0} = 0.80$$

$$\text{Mole fraction of } B = \frac{1.0}{4.0 + 1.0} = 0.20$$

$$\text{Partial pressure of } A = \text{Mole fraction of } A \times P_t = (0.80)(7.0 \text{ atm}) = 5.6 \text{ atm}$$

$$\text{Partial pressure of } B = \text{Mole fraction of } B \times P_t = (0.20)(7.0 \text{ atm}) = 1.4 \text{ atm}$$

$$\text{Total } P = 7.0 \text{ atm}$$

Example 6-j. A mixture of gases containing 65.0% CO_2 and 35.0% O_2 by weight has a total pressure of 745 torr. What is the partial pressure of each gas?

Solution. Assume 100.0 g of gas that would contain

$$100.0 \cancel{\text{g gas}} \times \frac{65.0 \text{ g } CO_2}{100.0 \cancel{\text{g gas}}} = 65.0 \text{ g } CO_2$$

and

$$100.0 \cancel{\text{g gas}} \times \frac{35.0 \text{ g } O_2}{100.0 \cancel{\text{g gas}}} = 35.0 \text{ g } O_2$$

$$\text{Moles } CO_2 = 65.0 \cancel{\text{g } CO_2} \times \frac{1 \text{ mole } CO_2}{44.0 \cancel{\text{g } CO_2}} = 1.48 \text{ mole } CO_2$$

$$\text{Moles } O_2 = 35.0 \cancel{\text{g } O_2} \times \frac{1 \text{ mole } O_2}{32.0 \cancel{\text{g } O_2}} = 1.09 \text{ mole } O_2$$

$$\text{Mole fraction } CO_2 = 1.48/2.57 = 0.576$$

$$\text{Mole fraction } O_2 = 1.09/2.57 = 0.424$$

$$p_{CO_2} = (0.576)(745 \text{ torr}) = 429 \text{ torr}$$

$$p_{O_2} = (0.424)(745 \text{ torr}) = 316 \text{ torr}$$

$$\text{Total } P = 745 \text{ torr}$$

Problems (Dalton's Law of Partial Pressure)

6-19. 2.0 liters of O_2, 2.0 liters of N_2, and 2.0 liters of H_2 are collected at a pressure of 1.0 atm, each in a different container. The three gases are then forced into a single vessel of 2.0-liter capacity with the temperature kept constant. Determine the resulting pressure in units of atmospheres.

6-20. Hydrogen gas was prepared in the laboratory by the reaction $Zn + H_2SO_4 \rightarrow ZnSO_4 + H_2$. A 355 ml sample of H_2 was collected over liquid H_2O at 27°C at a pressure of 787 torr. What volume would the dry gas occupy at 1.75 atm and 2.00×10^2 K?

6-21. A 2.0-liter container was filled with O_2 and a 1.0-liter container with N_2, at standard conditions. Both gases were then pumped from their containers into a 5.0-liter flask with the final temperature the same as the initial temperature. Calculate the pressure (atm) of each gas in the 5.0-liter flask, and the total gas pressure.

6-22. The volume of a dry gas at 555 torr and −127°C was 17.5 liters. What volume would this gas occupy if stored over water at 25°C and a total pressure of 748 torr?

6-23. If a 0.250 liter sample of Ar gas is collected over H_2O at 22°C at a total pressure of 78.0 cm of Hg, what would be the corresponding volume of dry Ar gas at 48.0 cm Hg and 22°C?

6-24. Determine the partial pressure of each gas in a mixture with a total pressure of 1.50×10^2 atm that contains 2.85 moles A, 3.45 moles B, and 2.15 moles of C.

6-25. A mixture of gases contains 66.0 g CO_2, 2.02 g of H_2, and 16.0 g of O_2. If the total pressure of the mixture is 725 torr, what is the partial pressure of each gas?

6-26. Air, by weight, is essentially a mixture of 79% N_2 and 21% O_2. Calculate the partial pressure of each gas at a total pressure of 99 kPa.

6-E. Graham's Law of Effusion

The rate of effusion of a gas is inversely proportional to the square root of its density.

$$\text{Rate} = \frac{\text{constant}}{\sqrt{d}}$$

Since the density of a gas is proportional to its molecular weight, the latter may be substituted for density. In comparing the relative rates of effusion of two gases, we may use the formulation:

$$\frac{r_1}{r_2} = \sqrt{\frac{M_2}{M_1}}$$

where r_1 is the rate of effusion of the gas with molecular weight M_1, and r_2 is the rate of effusion of the gas with molecular weight M_2.

Example 6-k. What are the rates of effusion of the gas He (MW = 4) relative to the gases, CH_4, HCl, SO_2, and HBr, with the molecular weights 16, 36, 64, and 81, respectively?

Solution. According to Graham's law of effusion, rates are inversely proportional to the square roots of molecular weights. Therefore, He will effuse $\sqrt{16}/\sqrt{4} = 4/2 = 2$ times as fast as CH_4; $\sqrt{36}/\sqrt{4} = 6/2 = 3$ times as fast as HCl; $\sqrt{64}/\sqrt{4} = 4$ times as fast as SO_2; 9/2 as fast as HBr.

Example 6-l. If gas *A* effuses one-third as fast as methane, CH_4, what is the molecular weight of *A*?

Solution.

$$\frac{\text{Rate for } A}{\text{Rate for } CH_4} = \sqrt{\frac{\text{MW of } CH_4}{\text{MW of } A}}$$

$$\frac{1}{3} = \sqrt{\frac{16}{\text{MW of } A}}$$

$$\left(\frac{1}{3}\right)^2 = \frac{16}{\text{MW of } A}$$

$$\text{MW of } A = 9 \times 16 = 144$$

Problems (Effusion and Molecular Weight)

6-27. Two gases, CH_3OF and H_2, are allowed to effuse through the porous walls of a container in two separate experiments. (a) Which one will effuse faster? (b) How many times as fast?

6-28. The two gases HBr and CH_4 have molecular weights of 81 and 16. What is the ***ratio*** of the rate of the effusion of CH_4 to that of HBr?

6-29. Nitrosyl fluoride gas, NOF, effuses through a small opening at a rate of 6.5 micromole/hr. At the same conditions of temperature and pressure, what will be the rate of effusion of tetrafluoroethylene, C_2F_4, expressed in the same units?

6-30. A certain gas effuses through a small opening at a rate exactly one-fifth as great as He. What is the molecular weight of the gas?

6-31. Gas *A* (MW = 81) effuses into one end of a tube 120 cm long, and gas *B* (MW = 36) effuses into the other end. The point at which the gases meet is marked by a white deposit on the walls. Where will the two gases meet?

6-F. Gay-Lussac's Law of Combining Volumes

Whenever gases react or are formed in a reaction, they do so in the ratio of small whole numbers by volume, provided the gases are under the same conditions of temperature and pressure.

In a balanced equation involving gases, the coefficients give the volume relations; for example,

$$2\ CO(g) + O_2(g) \longrightarrow 2\ CO_2(g)$$

Two volumes of CO react with **one** volume of O_2 to yield **two** volumes of CO_2. Any units of volume may be used, as long as the same units are employed for all gases.

Example 6-m. (a) What volume of oxygen is required to burn 1.0×10^2 liters of H_2S according to the equation

$$2\ H_2S(g) + 3\ O_2(g) \longrightarrow 2\ H_2O(g) + 2\ SO_2(g)$$

with all gases under the same conditions of temperature and pressure? (b) What volume of SO_2 will be formed?

Solution. (a) According to Gay-Lussac's law, H_2S and O_2 react in the ratio of two to three by volume; hence the volume of oxygen required is

$$100\ \cancel{\text{liters of H}_2\text{S}} \times \frac{3\ \text{liters of O}_2}{2\ \cancel{\text{liters of H}_2\text{S}}} = 150\ \text{liters of O}_2 = 1.5 \times 10^2\ \text{liters O}_2$$

(b) Since two volumes of H_2S produce two volumes of SO_2, the volume of SO_2 produced is

$$100\ \cancel{\text{liters of H}_2\text{S}} \times \frac{2\ \text{liters of SO}_2}{2\ \cancel{\text{liters of H}_2\text{S}}} = 100\ \text{liters of SO}_2 = 1.0 \times 10^2\ \text{liters SO}_2$$

Problems (Gay-Lussac's Law of Combining Volumes)

6-32. Zirconium metal, Zr, and Cl_2 gas react to form zirconium tetrachloride, $ZrCl_4$. At a temperature of 300°C and a pressure of 0.300 atm, 364 ml of $ZrCl_4$ gas are obtained. What volume of Cl_2 gas was necessary at this same temperature and pressure?

6-33. At 300°C and 1.0 atm, 5.6 liters of H_2 and 2.8 liters of O_2 are mixed in a cylinder fitted with a piston and the reaction is induced with a spark. Assuming the apparatus does not shatter, what is the volume when the resulting gases are cooled to 300°C and 1.0 atm?

6-34. Calculate (a) the volume of O_2 required and (b) the total volume of all products when 25.0 ft^3 of acetylene, C_2H_2, is burned according to the equation $2\,C_2H_2(g) + 5\,O_2(g) \rightarrow 4\,CO_2(g) + 2\,H_2O(g)$. Assume the products are cooled to the same temperature and pressure as the reactants.

6-G. Avogadro's Law and the Molar Volume of a Gas

Equal volumes of gases under the same conditions of temperature and pressure contain the same number of particles (i.e., atom and/or molecules). The volume occupied by 1 mole of any gaseous substance at STP is 22.41 **liters.** This is termed the **molar volume.** Recall that the number of particles in 1 mole of any substance is Avogadro's number, 6.02×10^{23}.

The fact that 22.4 liters of any gas (atoms or molecules) at STP have a mass equal to the atomic or molecular weight expressed in grams allows the conversion of gas volumes to masses and vice versa.

Example 6-n. (a) What is the mass of 1.00×10^2 liters of O_2 gas at STP? (b) What volume at STP would be occupied by 48.0 g of O_2?

Solution. (a) Since the molecular weight of O_2 is 32.0, there are 32.0 g of O_2 per 22.4 liters at STP. The mass of 1.00×10^2 liters of O_2 at STP is obtained by multiplying by 32.0 g O_2/22.4 liters.

$$\text{Mass of } 1.00 \times 10^2 \text{ liters at STP}$$

$$= 1.00 \times 10^2 \cancel{\text{liters}} \times \frac{32.0 \text{ g } O_2}{22.4 \cancel{\text{liters}}} = 143 \text{ g } O_2$$

(b)
$$\text{Number of moles of } O_2 = 48.0 \cancel{\text{g } O_2} \times \frac{1 \text{ mole } O_2}{32.0 \cancel{\text{g } O_2}} = 1.50 \text{ moles } O_2$$

$$\text{Volume of 48.0 g } O_2 = 1.50 \cancel{\text{moles } O_2} \times \frac{22.4 \text{ liters}}{1 \cancel{\text{mole } O_2}}$$

$$= 33.6 \text{ liters}$$

Example 6-o. 3.25 g of a certain gas occupy a volume of exactly 750 ml at a pressure of 2.00 atm and a temperature of 22°C. (a) Calculate the mass of 1.00 liter, or the density, of the gas at STP. (b) What is the molecular weight of the gas?

Solution. (a) First, correct the volume to STP.

$$\text{Volume at STP} = 750 \text{ ml} \times \frac{2.00 \text{ atm}}{1.00 \text{ atm}} \times \frac{273 \text{ K}}{295 \text{ K}}$$

$$= 1390 \text{ ml}$$

Since there is no change in mass, 1390 ml at STP have a mass of 3.25 g. The mass of 1.00 liter is calculated as follows:

$$\text{Density} = \frac{3.25 \text{ g}}{1390 \text{ ml}} \times \frac{1000 \text{ ml}}{1 \text{ liter}} = \frac{2.34 \text{ g}}{\text{liter}}$$

(b) Since 22.4 liters at STP is the volume of 1 mole, the molecular weight is numerically equal to the mass in grams of 22.4 liters of the gas at STP.

Molecular weight (expressed in grams)

$$= \frac{2.34 \text{ g}}{\text{liter}} \times \frac{22.4 \text{ liters}}{1 \text{ mole}} = 52.4 \text{ g/mole}$$

Hence the molecular weight is 52.4.

Example 6-p. How many formaldehyde molecules, CH_2O, are there in a sample of this gas which occupies 11.2 liters at STP?

Solution.

$$\text{Number of moles of } CH_2O = 11.2 \text{ liters} \times \frac{1 \text{ mole}}{22.4 \text{ liters}}$$

$$= 0.500 \text{ mole}$$

Since 1 mole of any substance contains 6.02×10^{23} molecules, the number of CH_2O molecules

$$= 0.500 \text{ mole} \times \frac{6.02 \times 10^{23} \text{ molecules}}{1 \text{ mole}}$$

$$= 3.01 \times 10^{23} \text{ molecules}$$

Problems (Avogadro's Law and Molar Volume)

6-35. If 43.6 g of a gas occupied a volume of 33.6 liters at standard conditions, what was the molecular weight of the gas?

6-36. How many molecules of H_2 are contained in 22.4 liters of the gas at 0°C and a pressure of 1.00×10^{-3} atm?

6-37. The density of a gas measured at standard conditions was 2.31 g/liter. Determine its molecular weight.

6-38. Calculate the mass in grams of 3.75 liters of O_2 at STP.

6-39. How many molecules are present in 0.112 liter of N_2 gas at STP?

6-40. How many atoms are there in 7.56 liters of ozone gas, O_3, at 2.00×10^{-3} torr and $-40°C$?

6-H. The General or Ideal Gas Law (Equation of State)

According to Boyle's and Charles' laws, the volume of a given mass of gas is inversely proportional to the pressure and directly proportional to the absolute temperature, that is,

$$V \alpha \frac{T}{P}, \quad \text{or} \quad PV = \text{constant} \times T$$

Since 1 mole of any gas at 273.1 K and 1.000 atm pressure occupies a volume of 22.41 liters, we may substitute in the equation above and evaluate the constant for 1 mole.

$$\text{Constant} = \frac{PV}{T} = \frac{(1.000 \text{ atm})(22.41 \text{ liters/mole})}{273.1 \text{ K}}$$

$$= 0.08206 \text{ (liter-atm) per (mole- K)}$$

This value (0.08206 liter-atm/mole- K) is termed the **molar gas constant** and is usually designated by the letter R.

$$\text{For 1 mole of gas } PV = RT$$

$$\text{For } n \text{ moles of gas } PV = nRT$$

The latter equation is considered the "equation of state" for an ideal gas. It is sometimes called the "ideal gas law" and is very useful in calculating the number of moles of gas present under any conditions of volume, temperature, and pressure. Many of the problems of the preceding sections are easily solved by means of this equation.

Example 6-q. Exactly 500 ml of a gaseous compound at 70.0 cm of Hg pressure and a temperature of 27°C have a mass of 1.85 g. Determine the molecular weight of the compound.

Solution. (a) Substituting in the formula $PV = nRT$

$$P = 70.0 \text{ cm Hg} \times \frac{1 \text{ atm}}{76.0 \text{ cm Hg}} = 0.921 \text{ atm}$$

$V = 500$ ml $= 0.500$ liter

R = gas constant = 0.0821 (liter-atm) per (mole-K)

$T = 27°\text{C} = 300$ K

n = number of moles of gas = ?

$$(0.921 \text{ atm})(0.500 \text{ liter}) = n\left(\frac{0.0821 \text{ liter-atm}}{\text{mole K}}\right)(300 \text{ K})$$

$$n = 0.0187 \text{ mole}$$

0.0187 mole has a mass of 1.85 g

Therefore, since n = Grams of gas/MW of gas

$$\text{MW of gas} = 1.85 \text{ g}/0.0187 \text{ mole} = 98.9 \text{ g/mole}$$

(b) An alternate solution to the problem would be

$$\text{Volume of gas at STP} = 500 \text{ ml} \times \frac{700 \text{ torr}}{760 \text{ torr}} \times \frac{273 \text{ K}}{300 \text{ K}}$$

$$= 419 \text{ ml}$$

$$\text{MW of gas} = \frac{1.85 \text{ g}}{419 \text{ ml}} \times \frac{22{,}400 \text{ ml}}{1 \text{ mole}} = 98.9 \text{ g/mole}$$

Example 6-r. What volume will be occupied by 33.0 g of CO_2 at 706 torr pressure and 27°C?

Solution. First, tabulating data,

$$P = 706 \text{ torr} \times \frac{1 \text{ atm}}{760 \text{ torr}} = 0.929 \text{ atm}$$

$V = ?$

$$n = 33.0 \text{ g } CO_2 \times \frac{1 \text{ mole } CO_2}{44.0 \text{ g } CO_2} = 0.750 \text{ mole } CO_2$$

$R = 0.0821$ (liter-atm)/(mole- K)

$T = 27°\text{C} = 300$ K

Substituting in the equation $PV = nRT$

$$(0.929 \text{ atm})(V) = (0.750 \text{ mole})\left(\frac{0.0821 \text{ liter-atm}}{\text{mole- K}}\right)(300 \text{ K})$$

$$V = 19.9 \text{ liters}$$

Problems (Ideal Gas Law)

6-41. Calculate the volume of 8.26 g of CO at $-13°C$ and 3.00 atm pressure.

6-42. (a) How many moles of gas are present in 2.80 liters of a gas at STP? (b) If the gas in part (a) has a mass of 9.5 g, what is its molecular weight?

6-43. To what temperature must 0.306 g of water vapor in a 0.250 liter flask be heated to have a pressure of 2.25 atm?

6-44. A 0.450-liter sample of a gaseous compound at 72.5 cm of Hg pressure and at $1.00 \times 10^2 °C$ has a mass of 1.86 g. What is the molecular weight of the compound?

6-45. How many molecules are contained in 4.80 liters of NH_3 gas at 4.00 atm and $-73°C$?

6-46. Substituting the proper value in pascals for 1.000 atm calculate the SI equivalent value and units for *R*, the gas constant in the Ideal Gas Law. Repeat, substituting the unit torr for atmosphere.

6-47. What mass of Cl_2 gas will occupy the same volume at 10°C as that occupied by 3.20 g of SO_2 gas at 40°C if the pressure of the two gases is the same?

6-48. Determine the pressure required to hold 6.50 moles of N_2 gas to a volume of 0.650 liter at a temperature of 27°C.

6-49. A steel cylinder with a volume of 19.5 liters contains N_2 gas at a pressure of 91.0 atm and a temperature of 0°C. Determine the mass of N_2 in the cylinder.

6-50. A 5.00 liter cylinder contains 8.60×10^2 g of Ar gas at 27°C. Calculate the pressure of the gas in the cylinder.

General Problems (Gas Laws)

6-51. 22.5 liters of O_2 gas at 1.00 atm pressure will occupy what volume at 5.00 atm pressure if the temperature remains constant?

6-52. A 0.350 liter sample of a gas at a temperature of 27°C will occupy what volume at standard temperature, assuming there is no change in pressure?

6-53. 19.6 liters of He gas at STP will occupy what volume at 3.50 atm and 546°C?

6-54. Calculate the mass of 2.50 liters of O_2 gas at 685 torr and a temperature of 127°C.

6-55. Compare the relative rates of effusion of CH_4, O_2, HCl, HBr, SO_2, He, and NOF with that of H_2.

6-56. Ten milliliters of gas *A* (MW = 17) effuse through an opening in 4.0 sec. Ten milliliters of gas *B* effuse through the same opening in 12 sec. Calculate the molecular weight of gas *B*.

6-57. If you had a sample of hydrogen gas that occupied 39.2 liters at STP, how many hydrogen molecules would there be in the sample?

6-58. Given a 0.584 liter sample of gas collected over water at 22°C and 636 torr, find the volume occupied by the dry gas at STP.

6-59. What volume will be occupied by 99.0 g of CO_2 at STP?

6-60. Octane, C_8H_{18}, is a component of gasoline, and when it undergoes complete combustion the equation for the reaction is

$$2\,C_8H_{18}(g) + 25\,O_2(g) \longrightarrow 16\,CO_2(g) + 18\,H_2O(g)$$

(a) Determine the volume of O_2 at 123°C and 1.0 atm pressure required to produce 6.0 liters of CO_2 at the same temperature and pressure. (b) What volume of water vapor would be produced when 6.0 liters of CO_2 are produced at the same temperature and pressure?

6-61. What is the volume of 4.51×10^{18} molecules of a gas at STP?

6-62. Determine the pressure necessary to compress 476 ft^3 of a gas at 608 kPa pressure to a volume of 23.8 ft^3 at constant temperature.

6-63. 37.5 g of a gas occupy 10.0 liters at 0°C and 2.00 atm pressure. What is the molecular weight of the gas?

6-64. (a) Find the density of a gas, in grams per liter, at STP when 8.48 g occupy 12.0 liters at 655 torr and 227°C. (b) What is the molecular weight?

6-65. A 8.5 liter sample of a H_2—CO_2 mixture was collected in the laboratory at 0°C and a pressure of 1.7×10^2 kPa wherein the partial pressure of CO_2 was determined to be 5.0×10^1 kPa. The CO_2 was removed and the remaining gas was compressed to 1.0 liter at 273°C. Determine the final pressure of H_2 gas.

6-66. The density of air at STP is 1.293 g/liter. Determine the density of air when the barometer exhibits a pressure of 795 torr and the temperature is 15°C.

6-67. Exactly one liter of O_2 at 350 torr and 1.0 liter of H_2 at 410 torr were mixed in a 1.0-liter evacuated flask. Determine the pressure of the mixture of gases.

6-68. 0.25 mole of O_2 gas and 0.35 mole of H_2 gas are mixed in a 1.0-liter container at a temperature of 0°C. What is the total pressure of the mixture?

6-69. What is the volume of 6.02×10^{18} molecules of H_2S at 1.00×10^{-5} atm and 0°C?

6-70. How many molecules of H_2 would be contained in a volume of 9.66 mm^3 at STP?

6-71. 0.250 mole of a compound containing 10.0% H and 90.0% C is burned in oxygen. If 33.6 liters of CO_2 are obtained at STP, (a) what is the empirical formula for the compound, and (b) what is the molecular formula?

6-72. What volume will be occupied at 22°C and 720 torr by 21 g of N_2 and 32 g of Ar?

6-73. A mixture of 12.0 g of He and 20.0 g of Ar is stored in a 2.00 liter cylinder at 25°C. Determine the pressure of the gas mixture in the cylinder.

6-74. A student wished to determine the molar volume of oxygen gas (MW = 32.0) and to do so needed to measure the volume of a given mass of the gas. The student decided to heat a sample of solid HgO, which decomposes to produce Hg and O_2. The mass of the O_2 liberated was determined by weighing the sample tube before and after heating. By collecting the O_2 by displacement of water from an inverted graduated cylinder the student could determine the volume of O_2 at atmospheric pressure when the cylinder is adjusted so that the water level is the same inside and out. The following data were obtained:

Mass of tube plus HgO before heating	24.475 g
Mass of tube plus Hg plus unreacted HgO	24.047 g
Volume of oxygen collected	351 ml
Pressure of atmosphere	709.8 torr
Temperature	22°C

Determine the molar volume at STP.

6-75. In 1928 Professor G. P. Baxter and co-workers at Harvard University determined the atomic weight of Ne from gas-density measurements. At 0°C, the average densities for a large number of measurements at three different pressures are

(a) Density = 0.89990 g/liter at a pressure of 760.000 torr
(b) Density = 0.60004 g/liter at a pressure of 506.667 torr
(c) Density = 0.30009 g/liter at a pressure of 253.333 torr

Calculate the atomic weight of Ne as determined under each set of conditions.

6-76. A sample of dry air on analysis showed the following composition by weight: 75.6% N_2, 23.1% O_2, and 1.3% Ar. Neglecting the other minor constituents present in air, (a) calculate the density of air in grams/liter at STP, and (b) calculate the density of air in grams/liter at 20°C and 725 torr pressure.

6-77. In one method of determining molecular weight, a sample of liquid is introduced into a weighed evacuated flask of known volume at a temperature sufficiently high to volatilize the liquid. Its pressure is measured at a constant temperature and then the flask is weighed again to determine the mass of the enclosed gas. A student obtained the following data on a compound containing only nitrogen and hydrogen:

Mass of evacuated flask	54.764 g
Mass of flask plus compound	54.872 g
Pressure of the gas	83.9 cm of Hg
Temperature of the gas	125°C
Volume of the flask	100.0 ml

(a) What is the molecular weight of the compound? (b) A chemical analysis of the compound indicated that it consisted of 12.5% H and 87.5% N. What is the true molecular formula of the compound?

6-78. Radium metal emits alpha particles that are in fact positively charged helium nuclei. These particles pick up electrons from surrounding matter to become electrically neutral helium atoms. From a particular sample of radium, alpha particles were produced at a rate of 7.02×10^{10} particles per second. Over a period of 145 days all the helium atoms from the radium decay were collected. At 29°C and 1.00 atm pressure, they occupied a volume of 0.0362 ml. Calculate Avogadro's number from these data, assuming the rate of alpha particle emission is constant over the period of time involved and that there is no other source of alpha particles.

Chapter 7

Stoichiometry

Stoichiometry is a study of the relationship between amounts of reactants and products of a chemical change as shown by the chemical equation. The balanced equation gives at once the relationship between **moles** of reactants and products, because the coefficients in the balanced equation represent the number of moles of substances. For example, in the reaction

$$2\ H_2S(g) + 3\ O_2(g) \longrightarrow 2\ SO_2(g) + 2\ H_2O(g)$$

two moles of H_2S gas react with **three** moles of O_2 gas to yield **two** moles of SO_2 gas and **two** moles of H_2O vapor. These amounts of reactants and products are termed **stoichiometric** amounts, that is, the amounts undergoing reaction as given by the chemical equation. Although amounts of reactants in other than stoichiometric ratios may be present, reaction occurs **only** in the stoichiometric ratios.

Amounts of chemicals are usually measured by their masses, although in the case of gases, the measurement may be by volume. Since we now know the relationship between **moles** and mass or volume, it is a simple matter to convert grams or liters (for a gas) to moles or vice versa. Careful attention to the units used is essential in working problems of the type described below.

The principal types of problems based on equations are: (a) weight–weight, (b) weight–volume, and (c) volume–volume. All of these follow the schematic process outlined below. The mass (or volume if a gas) of *A* is converted to moles of *A*. **From the balanced equation** the number of moles of B produced (or reacted with A) is calculated. Finally, the mass of B (or the volume of gaseous B) is calculated. We see from this discussion that the "link" allowing a solution of these problems is the mole relationship between A and B as obtained from the balanced equation.

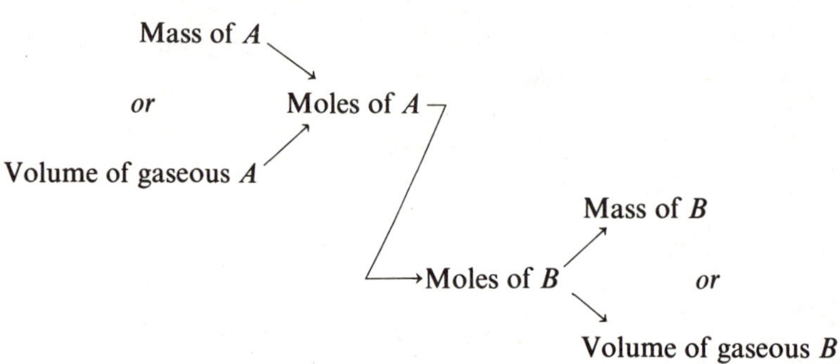

Further, we should note that solution of some problems may require use of only part of the above scheme. For example, if we are given **moles** of A and asked to find **grams** of B, the first step shown above is not required. Similarly, if we are given **moles** of A and asked to find **moles** of B, only the middle step in the scheme is necessary. We will consider these simpler examples initially.

7-A. Mole–Mole, Mole–Weight, and Weight–Weight Calculations

Mole–mole problems are solved simply by using the coefficients in the balanced equation to construct the required conversion factor. Mole-weight problems require the same first step plus a conversion between moles and grams for the desired species. Interconversions between moles and grams were discussed in Chapter 3. Weight–weight problems involve the determination of an unknown mass of a reactant or product from a given mass of some substance in a chemical change.

Example 7-a. Preparation of sodium hydroxide (also called "lye") can be accomplished by the very vigorous reaction of Na metal with H_2O. The balanced equation is:

$$2\ Na(s) + 2\ H_2O(l) \longrightarrow 2\ NaOH(s) + H_2(g)$$

If 2.50 moles of Na are reacted: (a) how many moles of NaOH are produced, (b) how many moles of H_2 are produced, and (c) how many moles of H_2O must have been consumed?

Solution. (a) According to the balanced equation, 2 moles of Na will produce 2 moles of NaOH. Hence

$$\text{Moles NaOH produced} = 2.50\ \cancel{\text{moles Na}} \times \frac{2\ \text{moles NaOH}}{2\ \cancel{\text{moles Na}}} = 2.50\ \text{moles NaOH}$$

(b) Similarly,

$$\text{Moles H}_2\ \text{produced} = 2.50\ \cancel{\text{moles Na}} \times \frac{1\ \text{mole H}_2}{2\ \cancel{\text{moles Na}}} = 1.25\ \text{moles H}_2$$

(c) Again, according to the equation, 2 moles Na require 2 moles H_2O. Hence

$$\text{Moles H}_2\text{O consumed} = 2.50\ \cancel{\text{moles Na}} \times \frac{2\ \text{moles H}_2\text{O}}{2\ \cancel{\text{moles Na}}} = 2.50\ \text{moles H}_2\text{O}$$

Example 7-b. Methyl alcohol (which is sometimes called wood alcohol and is very poisonous when taken internally) has been suggested for use as an alternate fuel in the automobile. In the laboratory, methyl alcohol can be synthesized by the reaction between carbon monoxide and hydrogen gas at a high temperature and pressure. The balanced equation is:

$$\text{CO} + 2\,\text{H}_2 \longrightarrow \text{CH}_3\text{OH}$$

How many grams of CH_3OH can be prepared by the complete reaction of 5.00 moles of H_2?

Solution. Using the balanced equation we first determine how many moles of CH_3OH are formed and then convert this value into grams of CH_3OH.

Stepwise, we have:

$$\text{Moles CH}_3\text{OH produced} = 5.00\ \cancel{\text{moles H}_2} \times \frac{1\ \text{mole CH}_3\text{OH}}{2\ \cancel{\text{moles H}_2}}$$
$$= 2.50\ \text{moles CH}_3\text{OH}$$

Next:

$$\text{Grams CH}_3\text{OH produced} = 2.50\ \cancel{\text{moles CH}_3\text{OH}} \times \frac{32.0\ \text{g CH}_3\text{OH}}{1\ \cancel{\text{mole CH}_3\text{OH}}}$$
$$= 80.0\ \text{g CH}_3\text{OH}$$

To save time we could combine or join the two conversion factors to yield the desired answer:

$$\text{Grams CH}_3\text{OH} = 5.00\ \cancel{\text{moles H}_2} \times \underbrace{\frac{1\ \cancel{\text{mole CH}_3\text{OH}}}{2\ \cancel{\text{moles H}_2}}}_{\substack{\text{Converts to}\\ \text{moles of CH}_3\text{OH}}} \times \underbrace{\frac{32.0\ \text{g CH}_3\text{OH}}{1\ \cancel{\text{mole CH}_3\text{OH}}}}_{\substack{\text{Converts to}\\ \text{grams of CH}_3\text{OH}}}$$
$$= 80.0\ \text{g CH}_3\text{OH}$$

Example 7-c. According to the following equation

$$4\,NH_3(g) + 5\,O_2(g) \longrightarrow 4\,NO(g) + 6\,H_2O(g)$$

what mass of O_2 in grams will be required to react with exactly 100 g of NH_3? The molecular weight of NH_3 is 17.0 and that of O_2 is 32.0.

Solution.

Number of moles of NH_3 used

$$= 100\ \cancel{g\ NH_3} \times \frac{1\text{ mole }NH_3}{17.0\ \cancel{g\ NH_3}} = 5.88\text{ moles }NH_3$$

According to the equation, 4 moles of NH_3 require 5 moles of O_2.

Number of moles of O_2 required

$$= 5.88\ \cancel{\text{moles }NH_3} \times \frac{5\text{ moles }O_2}{4\ \cancel{\text{moles }NH_3}}$$

$$= 7.35\text{ moles }O_2$$

$$\text{Mass of }O_2\text{ required} = 7.35\ \cancel{\text{moles }O_2} \times \frac{32.0\text{ g }O_2}{1\ \cancel{\text{mole }O_2}} = 235\text{ g }O_2$$

Performing the calculation in one step by joining the factors together, we have:

$$\text{Mass }O_2 = 100\ \cancel{g\ NH_3} \times \underbrace{\frac{1\ \cancel{\text{mole }NH_3}}{17.0\ \cancel{g\ NH_3}}}_{\substack{\text{Converts to}\\ \text{moles }NH_3}} \times \underbrace{\frac{5\ \cancel{\text{moles }O_2}}{4\ \cancel{\text{moles }NH_3}}}_{\substack{\text{Converts to}\\ \text{moles }O_2}} \times \underbrace{\frac{32.0\text{ g }O_2}{1\ \cancel{\text{mole }O_2}}}_{\substack{\text{Converts to}\\ \text{grams }O_2}} = 235\text{ g }O_2$$

Problems (Mole–Mole, Mole–Weight, and Weight–Weight)

7-1. SO_2, an air pollutant, can be removed from smokestack emissions according to the following reaction:

$$2\,CaCO_3 + 2\,SO_2 + O_2 \longrightarrow 2\,CaSO_4 + 2\,CO_2$$

How many kilograms of $CaCO_3$ are required to remove 5.00×10^3 moles of SO_2?

7-2. From the balanced equation

$$2\,H_2S(g) + 3\,O_2(g) \longrightarrow 2\,SO_2(g) + 2\,H_2O(g)$$

determine the following: (a) moles of O_2 necessary to react with 0.345 mole of H_2S, (b) moles of SO_2 produced from 0.345 mole of H_2S, (c) grams of O_2 required for 0.345 mole of H_2S, (d) mass of SO_2 produced if 272 g of H_2S are used.

7-3. Benzene, C_6H_6, burns in O_2 as described by the equation

$$2\ C_6H_6(g) + 15\ O_2(g) \longrightarrow 12\ CO_2(g) + 6\ H_2O(g)$$

(a) Calculate the number of moles of O_2 necessary to burn 15.0 moles of C_6H_6. (b) What number of grams of O_2 are necessary in (a)? (c) How many moles of products are produced in (a)? (d) How many grams of C_6H_6 must be burned to produce 6.60 g of CO_2?

7-4. Balance the following equation for the combustion of acetylene, C_2H_2, in oxygen.

$$C_2H_2(g) + O_2(g) \longrightarrow CO_2(g) + H_2O(g)$$

(a) How many moles of O_2 will be required to burn 7.55 moles of C_2H_2? (b) How many moles of CO_2 will be formed? (c) If 104 g of C_2H_2 are burned, how many moles of O_2 are required? (d) How many moles of CO_2 will be formed from 104 g of C_2H_2? (e) How many grams of O_2 will be required to react with 104 g of C_2H_2? (f) How many grams of CO_2 will be formed from 104 g of C_2H_2? (g) How many moles of C_2H_2 could be burned by 96.0 g of O_2? (h) How many grams of C_2H_2 could be burned by 96.0 g of O_2?

7-5. Given the balanced equation

$$4\ NH_3(g) + 5\ O_2(g) \longrightarrow 4\ NO(g) + 6\ H_2O(g)$$

(a) How many moles of NH_3 will be required to form 17.5 moles of NO? (b) How many grams of NH_3 will be required in (a)? (c) How many grams of O_2 will be required in (a)? (d) How many grams of NH_3 will be required to react with 48.0 g of O_2?

7-6. Complete combustion of PH_3 is shown by the equation

$$PH_3(g) + 2\ O_2(g) \longrightarrow H_3PO_4(s)$$

For 85.0 g of PH_3, determine (a) the moles of O_2 required, (b) the grams of O_2 required, (c) the moles of H_3PO_4 formed, and (d) the grams of H_3PO_4 formed.

7-7. Silver nitrate reacts with $CaCl_2$ as shown by the equation

$$2\ AgNO_3 + CaCl_2 \longrightarrow Ca(NO_3)_2 + 2\ AgCl(s)$$

Calculate (a) the mass of $AgNO_3$ required to produce 64.5 g of AgCl, (b) the mass of $CaCl_2$ required to produce 64.5 g of AgCl, and (c) the mass of $Ca(NO_3)_2$ produced when 64.5 g of AgCl are produced.

7-8. In the reaction

$$5\,KI + KIO_3 + 6\,HNO_3 \longrightarrow 6\,KNO_3 + 3\,I_2 + 3\,H_2O$$

25.4 g of I_2 are produced. Determine the number of (a) grams of KI required, (b) grams of KIO_3 required, (c) grams of HNO_3 required, and (d) grams of KNO_3 produced.

7-9. Exactly eighteen grams of CuS are treated with excess dilute HNO_3, and the reaction

$$3\,CuS + 8\,HNO_3 \longrightarrow 3\,Cu(NO_3)_2 + 2\,NO + 4\,H_2O + 3\,S$$

occurs. (a) How many grams of $Cu(NO_3)_2$ are produced? (b) How many grams of S are produced? (c) What is the minimum number of grams of HNO_3 required?

7-10. Two of the several reactions in a blast furnace are

$$2\,C(s) + O_2(g) \longrightarrow 2\,CO(g)$$

$$Fe_2O_3(s) + 3\,CO(g) \longrightarrow 2\,Fe + 3\,CO_2(g)$$

Assume that these are the reactions primarily responsible for the production of Fe. (a) How many kilograms of C are required to reduce 2.50×10^2 kg of Fe_2O_3? (b) How many kilograms of Fe are produced from 2.50×10^2 kg of Fe_2O_3? (c) How many kilograms of CO_2 are produced? (d) How many kilograms of O_2 are required?

7-B. Weight–Volume Calculations

Weight–volume calculations refer to stoichiometric calculations where one of the reactants or products is a gas. As in weight–weight calculations, the balanced chemical equation is used to relate the number of moles of a substance to the number of moles of other reactants or products. The conversion of moles or grams of a gas to volume at some given temperature and pressure is most readily accomplished by use of the ideal gas equation. Alternatively, since 1 mole of gas occupies 22.4 liters at STP, the volume of a given number of moles under these conditions may be calculated and subsequently converted to the desired conditions by use of the P, V, T relations. The conversion of the volume of a gas to moles may be accomplished by the reverse of the process just described. (See Chapter 6.)

Example 7-d. 1.00×10^2 liters of H_2S at STP are bubbled into an aqueous solution of $SbCl_3$. Sb_2S_3 is precipitated according to the equation

$$2\,SbCl_3 + 3\,H_2S(g) \longrightarrow Sb_2S_3 + 6\,HCl$$

Calculate the mass of precipitated Sb_2S_3 in grams if an excess of $SbCl_3$ is present. The formula weight of Sb_2S_3 is 3.40×10^2.

Solution.

Number of moles of H_2S

$$= 1.00 \times 10^2 \cancel{\text{liters } H_2S} \times \frac{1 \text{ mole } H_2S}{22.4 \cancel{\text{liters } H_2S}}$$

$$= 4.46 \text{ moles } H_2S$$

The chemical equation requires that 1 mole of Sb_2S_3 precipitate for each 3 moles of H_2S gas; therefore the number of moles of Sb_2S_3 is

$$4.46 \cancel{\text{moles } H_2S} \times \frac{1 \text{ mole } Sb_2S_3}{3 \cancel{\text{moles } H_2S}} = 1.49 \text{ moles } Sb_2S_3$$

The number of grams of Sb_2S_3 is

$$1.49 \cancel{\text{moles } Sb_2S_3} \times \frac{3.40 \times 10^2 \text{ g } Sb_2S_3}{1 \cancel{\text{mole } Sb_2S_3}} = 507 \text{ g } Sb_2S_3$$

In one step we have:

$$1.00 \times 10^2 \cancel{\text{liters } H_2S} \times \frac{1 \cancel{\text{mole } H_2S}}{22.4 \cancel{\text{liters } H_2S}} \times \frac{1 \cancel{\text{mole } Sb_2S_3}}{3 \cancel{\text{moles } H_2S}} \times \frac{3.40 \times 10^2 \text{ g } Sb_2S_3}{1 \cancel{\text{mole } Sb_2S_3}}$$

$$= 507 \text{ g } Sb_2S_3$$

Example 7-e. Using the equation in Example 7-c, determine the *volume* of O_2 at 27°C and 1.00 atm that would be required to react with 1.00×10^2 g of NH_3.

Solution.

$$\text{Number of moles of } NH_3 = 1.00 \times 10^2 \cancel{\text{g } NH_3} \times \frac{1 \text{ mole } NH_3}{17.0 \cancel{\text{g } NH_3}}$$

$$= 5.88 \text{ moles } NH_3$$

Since 4 moles of NH_3 require 5 moles of O_2,

$$\text{Number of moles of } O_2 = 5.88 \cancel{\text{moles } NH_3} \times \frac{5 \text{ moles } O_2}{4 \cancel{\text{moles } NH_3}}$$

$$= 7.35 \text{ moles } O_2$$

In one step we have:

$$1.00 \times 10^2 \cancel{\text{g } NH_3} \times \frac{1 \cancel{\text{mole } NH_3}}{17.0 \cancel{\text{g } NH_3}} \times \frac{5 \text{ moles } O_2}{4 \cancel{\text{moles } NH_3}} = 7.35 \text{ moles } O_2$$

The volume of O_2 required may be calculated from the ideal gas equation.

$$V = \frac{nRT}{P}$$

$$\text{Volume of } O_2 = 7.35 \text{ moles } O_2 \times 0.0821 \frac{\text{liter-atm}}{\text{mole-K}} \times \frac{300 \text{ K}}{1.00 \text{ atm}}$$

$$= 181 \text{ liters } O_2$$

Problems (Weight-Volume)

7-11. In the dissolution of excess Hg by HNO_3, NO gas is produced.

$$6\,Hg + 8\,HNO_3 \longrightarrow 3\,Hg_2(NO_3)_2 + 2\,NO(g) + 4\,H_2O$$

Determine the volume of NO produced at STP if 47.8 g of HNO_3 are allowed to react with excess Hg.

7-12. Ethyl alcohol can be obtained by fermentation of the sugar glucose according to the following balanced equation:

$$C_6H_{12}O_6 \longrightarrow 2\,C_2H_5OH + 2\,CO_2(g)$$

If 42.4 liters of CO_2 at STP were obtained, how many grams of C_2H_5OH must have been produced?

7-13. In the Solvay process for the production of $NaHCO_3$, a solution saturated with both NaCl and NH_3 is treated with CO_2 gas and the following reaction occurs:

$$NaCl + NH_3(g) + H_2O(l) + CO_2(g) \longrightarrow NaHCO_3(s) + NH_4Cl$$

(a) Calculate the volume of $NH_3(g)$ at STP required to produce 6.55×10^2 g of $NaHCO_3$. (b) What volume of $CO_2(g)$ at 20°C and 1.45 atm is required to produce 2.55×10^2 g of $NaHCO_3$?

7-14. The gas phosphine, PH_3, is prepared from P_4 by the reaction

$$P_4(s) + 3\,NaOH + 3\,H_2O(l) \longrightarrow 3\,NaH_2PO_2 + PH_3(g)$$

(a) What volume of PH_3 at STP can be prepared from 62.0 g of phosphorus? (b) Determine the volume of PH_3 at STP that can be prepared from 30.0 g of NaOH. (c) What volume of PH_3 at 37°C and 0.850 atm is produced if 71.0 g of NaH_2PO_2 are produced?

7-15. In one step of the commercial production of Cu metal, the reaction

$$Cu_2S + 2\,Cu_2O \longrightarrow 6\,Cu + SO_2(g)$$

is very important. How many liters of SO_2 at 327°C and 65.8 cm of Hg are produced from 3.38 kg of Cu_2S?

7-C. Volume–Volume Calculations

It was pointed out in Section 6-F, that Gay–Lussac's law of combining volumes states "Whenever gases react or are formed in a reaction, they do so in the ratio of small whole numbers by volume, provided the gases are under the same conditions of temperature and pressure." The coefficients in the balanced equation give the volume relations for the gaseous substances. This follows from the facts that the coefficients give the number of moles of each gas involved and that 1 mole of each gas occupies the same volume at the same temperature and pressure.

When two or more of the gases are at different temperatures and pressures, one of the following procedures must be employed. All gases can be treated **as if** the reaction was carried out at the particular temperature and pressure of one of the gases; volume corrections are then made for the differences in conditions of the two gases involved by use of the P, V, T relations. **Alternatively**, the volume of one of the gases can be converted to moles by use of the ideal gas equation; then the number of moles of the second gas can be calculated; and, finally, the volume of the second gas is calculated under the desired conditions. Such a problem is illustrated in Example 7-g.

Example 7-f. What volume of oxygen would be required to burn exactly 500 liters of acetylene gas, C_2H_2, according to the equation

$$2\,C_2H_2(g) + 5\,O_2(g) \longrightarrow 4\,CO_2(g) + 2\,H_2O(g),$$

if all gases are at the same temperature and pressure?

Solution. According to the equation, two volumes of C_2H_2 require five volumes of O_2. Therefore, the volume of O_2 required is

$$500\ \cancel{\text{liters } C_2H_2} \times \frac{5 \text{ liters } O_2}{2\ \cancel{\text{liters } C_2H_2}} = 1250 \text{ liters } O_2 \text{ or } 1.25 \times 10^3 \text{ liters } O_2$$

Example 7-g. H_2S burns in O_2 according to the equation

$$2\,H_2S(g) + 3\,O_2(g) \longrightarrow 2\,H_2O(g) + 2\,SO_2(g)$$

Determine (a) the volume of O_2 at STP required to burn 20.0 liters of H_2S at STP, and (b) the volume of SO_2 obtained at a pressure of 70.0 cm of Hg and a temperature of 773 K.

Solution. (a) Two volumes of H_2S require three volumes of O_2; therefore, the volume of oxygen required for 20.0 liters of H_2S is

$$20.0 \text{ } \cancel{\text{liters } H_2S} \times \frac{3 \text{ liters } O_2}{2 \text{ } \cancel{\text{liters } H_2S}} = 30.0 \text{ liters } O_2 \text{ at STP}$$

(b) Two volumes of H_2S yield two volumes of SO_2; therefore 20.0 liters of H_2S at STP will yield 20.0 liters of SO_2 at STP. At 70.0 cm Hg pressure and 773 K the volume of SO_2 will be

$$20.0 \text{ liters} \times \frac{76.0 \text{ } \cancel{\text{cm}}}{70.0 \text{ } \cancel{\text{cm}}} \times \frac{773 \text{ } \cancel{\text{K}}}{273 \text{ } \cancel{\text{K}}} = 61.5 \text{ liters } SO_2$$

Alternate Solution. From the ideal gas equation find the number of moles of H_2S to be burned; determine the moles of SO_2 produced; calculate the volume of SO_2 produced at 70.0 cm of Hg and 773 K.

Problems (Volume-Volume)

7-16. The compound C_2H_6 reacts in a limited amount of O_2 to form CO and H_2O. (a) Balance the equation for the reaction

$$C_2H_6(g) + O_2(g) \longrightarrow CO(g) + H_2O(l)$$

(b) What volume of O_2 reacts with 11.2 liters of C_2H_6 at STP? (c) Determine the volume of CO measured at STP produced from 11.2 liters of C_2H_6 at STP.

7-17. $P_4(g)$ reacts with $Cl_2(g)$ to form $PCl_3(g)$. (a) Balance the equation for the reaction

$$P_4(g) + Cl_2(g) \longrightarrow PCl_3(g)$$

(b) Calculate the volume of Cl_2 required to react with 17.5 liters of gaseous P_4 at a given temperature. (c) What volume of PCl_3 is produced from 17.5 liters of P_4 if the absolute temperature of the PCl_3 is twice that of the reactants and the pressure is constant?

7-18. $SiCl_4(g)$ reacts with $H_2O(g)$ at elevated temperatures as shown in the equation

$$SiCl_4(g) + 2\,H_2O(g) \longrightarrow SiO_2(s) + 4\,HCl(g)$$

If 35.5 liters of $SiCl_4$ at 300.0°C and 0.750 atm pressure react with H_2O, (a) what volume of H_2O at this temperature and pressure will be consumed, (b) what volume of HCl will be produced, (c) what is the volume of that amount of HCl gas at STP?

7-D. Limiting Reactant Calculations

In systems in which chemical reactants are mixed, it often happens that the reactants are not added in stoichiometric amounts, that is, one or more of the reactants is in excess, and hence cannot be completely consumed in the reaction. When performing stoichiometric calculations on these systems, one must first determine which substance is completely consumed in the reaction—that is, which species is the **limiting reactant**. All subsequent calculations are then based on the amount of limiting reactant available. Hence these problems are essentially identical to those previously discussed except that we are careful here to base our calculations on substances completely consumed in a reaction and not those in excess.

Example 7-h. Acetylene, C_2H_2, is completely burned in a welder's torch according to the following reaction:

$$2\,C_2H_2(g) + 5\,O_2(g) \longrightarrow 4\,CO_2(g) + 2\,H_2O(g)$$

If there are available for reaction 78.0 g of C_2H_2 and 288 g of O_2, is there sufficient O_2 present to burn all the C_2H_2?

Solution. Since the coefficients in the equation relate moles of C_2H_2 and O_2, our first step is to convert to moles for each species. Hence:

$$\text{Moles } C_2H_2 = 78.0\ \cancel{\text{g } C_2H_2} \times \frac{1 \text{ mole } C_2H_2}{26.0\ \cancel{\text{g } C_2H_2}} = 3.00 \text{ moles } C_2H_2$$

$$\text{Moles } O_2 = 288\ \cancel{\text{g } O_2} \times \frac{1 \text{ mole } O_2}{32.0\ \cancel{\text{g } O_2}} = 9.00 \text{ moles } O_2$$

Next we choose one of the reactants (it doesn't matter which) and calculate how many moles of the second reactant are required to completely consume the first. We can then determine which of the reactants is in excess. We choose here to determine how many moles of O_2 are required to completely consume 3.00 moles of C_2H_2. Hence:

$$\text{Moles } O_2 \text{ required} = 3.00\ \cancel{\text{moles } C_2H_2} \times \frac{5 \text{ moles } O_2}{2\ \cancel{\text{moles } C_2H_2}} = 7.50 \text{ moles } O_2$$

Since there are 9.00 moles of O_2 available we conclude there is sufficient O_2 present to burn all the C_2H_2.

Suppose we had chosen O_2 instead of C_2H_2. We would calculate how many moles of C_2H_2 would be required to completely consume the 9.00 moles of O_2 present:

$$\text{Moles } C_2H_2 \text{ required} = 9.00\ \cancel{\text{moles } O_2} \times \frac{2 \text{ moles } C_2H_2}{5\ \cancel{\text{moles } O_2}} = 3.60 \text{ moles } C_2H_2$$

Since we have only 3.00 moles of C_2H_2 available, it is obvious that not all the O_2 can be consumed and hence there is sufficient O_2 present to burn all the C_2H_2. If all the C_2H_2 is consumed and the reaction is completed, the system would then contain 6.00 moles of CO_2, 3.00 moles of H_2O, and 1.50 moles of O_2.

Example 7-i. Silicon carbide, SiC, is an extremely hard substance often used as blade material in special cutting instruments. It is prepared by a reaction for which the equation is:

$$SiO_2(s) + 3\ C(s) \longrightarrow SiC(s) + 2\ CO(g)$$

If 30.05 g of SiO_2 and 24.0 g of C are available for reaction, (a) what is the maximum mass (in grams) of SiC that can be produced, and (b) how many moles of the excess reactant remain at the conclusion of the reaction?

Solution. (a) We first determine the moles of each reactant available:

$$\text{Moles } SiO_2 = 30.05\ \cancel{\text{g } SiO_2} \times \frac{1 \text{ mole } SiO_2}{60.1\ \cancel{\text{g } SiO_2}} = 0.500 \text{ mole } SiO_2$$

$$\text{Moles C} = 24.0\ \cancel{\text{g C}} \times \frac{1 \text{ mole C}}{12.0\ \cancel{\text{g C}}} = 2.00 \text{ moles C}$$

Next we choose to determine how many moles of SiO_2 are required to completely consume the 2.00 moles of C:

$$2.00\ \cancel{\text{moles C}} \times \frac{1 \text{ mole } SiO_2}{3\ \cancel{\text{moles C}}} = 0.667 \text{ moles } SiO_2 \text{ required}$$

Since we have available only 0.500 mole SiO_2, the reactant carbon is in excess and hence SiO_2 is the limiting reactant.

The stoichiometric calculation of the maximum number of grams of SiC obtainable easily follows, using the limiting reactant as the basis for the calculation.

$$\text{Grams SiC} = 0.500 \text{ mole } SiO_2 \times \frac{1\ \cancel{\text{mole SiC}}}{1\ \cancel{\text{mole } SiO_2}} \times \frac{40.1 \text{ g SiC}}{1\ \cancel{\text{mole SiC}}} = 20.0 \text{ g SiC}$$

(b) Here we calculate how many moles of C actually underwent reaction and subtract this from the moles of C originally available.

$$\text{Moles C used} = 0.500\ \cancel{\text{mole } SiO_2} \times \frac{3 \text{ moles C}}{1\ \cancel{\text{mole } SiO_2}} = 1.50 \text{ moles C}$$

2.00 moles C available − 1.50 moles C used = 0.50 mole C in excess.

Problems (Limiting Reactant)

7-19. Ammonium nitrate, a fertilizer, can be prepared by the careful neutralization of nitric acid with ammonia:

$$HNO_3(aq) + NH_3(aq) \longrightarrow NH_4NO_3(aq)$$

If 189 g of HNO_3 and 104 g of NH_3 are available for reaction, (a) which reactant is in excess, (b) what is the maximum mass of NH_4NO_3 obtainable, and (c) how many grams of excess reactant remain unreacted?

7-20. Ammonia gas can be recovered from NH_4Cl by the reaction:

$$CaO(s) + 2\ NH_4Cl(s) \longrightarrow CaCl_2(s) + H_2O(g) + 2\ NH_3(g)$$

If we have 75.0 g NH_4Cl and 50.0 g CaO, calculate the maximum mass of NH_3 obtainable.

7-21. Iron can be obtained in a blast furnace by reaction of the ore mineral hematite, Fe_2O_3, with carbon in a series of reactions whose overall net equation may be represented as:

$$2\ Fe_2O_3(s) + 3\ C(s) \longrightarrow 3\ CO_2(g) + 4\ Fe(s)$$

Suppose a company has a stockpile of 6.55×10^6 kg of hematite and 5.85×10^5 kg of carbon. (a) Is there enough carbon to process all the hematite available? (b) Calculate the maximum mass of Fe that can be produced based on the limiting reactant.

7-E. Percent Yield Calculations

Not all chemical reactions are as simple or straightforward as might be indicated by looking at a balanced chemical equation. In some cases one of the reactants might undergo side reactions to yield products in addition to those predicted by the balanced equation. In other cases the reaction just does not seem to go to completion. Even though reactants are still present, product amounts do not increase with time. For such reactions the amount of product obtained is less than that predicted by stoichiometry.

These complicating factors are taken into account by introducing the concept of a **percent yield** for chemical reactions. Percent yield is calculated in the following way:

$$\%\,\text{Yield} = \frac{\text{Actual yield}}{\text{Theoretical yield}} \times 100$$

Theoretical yield is the yield predicted by stoichiometry.

Example 7-j. Aspirin can be formed by allowing salicylic acid and acetic anhydride to react in an appropriate solvent according to the balanced equation:

$$\underset{\text{salicylic acid}}{HOC_6H_4COOH} + \underset{\text{acetic anhydride}}{(CH_3CO)_2O} \longrightarrow \underset{\text{aspirin}}{CH_3OCOC_6H_4COOH} + \underset{\text{acetic acid}}{CH_3COOH}$$

If 0.450 moles, each, of salicylic acid and acetic anhydride are allowed to react and 59.0 g of aspirin are obtained, calculate the % yield for aspirin in the reaction.

Solution. The theoretical yield of aspirin in grams is:

$$0.450 \text{ mole salicylic acid} \times \frac{1 \text{ mole aspirin}}{1 \text{ mole salicylic acid}} \times \frac{1.80 \times 10^2 \text{ g aspirin}}{1 \text{ mole aspirin}}$$

$$= 81.0 \text{ g aspirin}$$

The % yield is calculated as:

$$\frac{59.0 \text{ aspirin (actual)}}{81.0 \text{ aspirin(theoretical)}} \times 100 = 72.8\% \text{ yield}$$

Problems (Percent Yield)

7-22. The pesticide DDT is prepared by a reaction described by the following equation:

$$CCl_3CHO + 2\,C_6H_5Cl \longrightarrow \underset{\text{(DDT)}}{(ClC_6H_4)_2CHCCl_3} + H_2O$$

When 22.5 g of C_6H_5Cl and excess CCl_3CHO are reacted together, 19.8 g of DDT are obtained. Calculate the % yield for DDT.

7-23. Consider the burning of octane, C_8H_{18} (a component in gasoline), in O_2 to yield CO_2 and H_2O. (a) Balance the equation. (b) On burning 17.5 g of C_8H_{18} in excess oxygen, 53.5 g of CO_2 were obtained. Calculate the % yield for CO_2.

7-24. Consider the balanced equation:

$$2\,H_2S(g) + 3\,O_2(g) \longrightarrow 2\,SO_2(g) + 2\,H_2O(g)$$

On reaction of 51.0 g of H_2S with excess O_2, 28.3 liters of SO_2 were obtained at STP. Calculate the % yield of SO_2.

General Problems (Weight-Weight)

7-25. Nitric acid, HNO_3, may be prepared by distillation from a mixture of solid $NaNO_3$ and concentrated H_2SO_4.

$$NaNO_3(s) + H_2SO_4(l) \longrightarrow NaHSO_4(s) + HNO_3(g)$$

(a) How many grams of H_2SO_4 are required to produce 2.75 moles of HNO_3? (b) How many grams of $NaHSO_4$ are produced when 2.75 moles of HNO_3 are produced? (c) How many moles of HNO_3 are produced from 255 g of $NaNO_3$? (d) How many grams of HNO_3 are produced from 255 g of $NaNO_3$?

7-26. Sodium carbonate, washing soda, is produced commercially by heating baking soda, $NaHCO_3$. This salt decomposes on heating as indicated by the equation

$$2\ NaHCO_3(s) \xrightarrow{\Delta} Na_2CO_3(s) + CO_2(g) + H_2O(g)$$

(a) How many moles of $NaHCO_3$ are required for each mole of CO_2 produced? (b) How many grams of CO_2 are produced for each mole of $NaHCO_3$ decomposed? (c) How many moles of each product are obtained for each mole of $NaHCO_3$ decomposed? (d) What is the total mass of products for each kilogram of $NaHCO_3$ used? (e) How many grams of $NaHCO_3$ will be required to produce 265 g of Na_2CO_3? (f) How many moles of CO_2 would be produced simultaneously in (e)?

7-27. Calcium carbide, CaC_2, is produced commercially by heating lime, CaO, and carbon, C, together in an electric furnace. The reaction taking place is

$$CaO(s) + 3\ C(s) \longrightarrow CaC_2(s) + CO(g)$$

If 165 kg of CaC_2 are produced, (a) how many kilograms of CaO, and (b) how many kilograms of C are required?

7-28. Calculate the quantity (in kilograms) of C (coke) necessary to produce 542 kilograms of CaC_2 according to the equation

$$CaO(s) + 3\ C(s) \longrightarrow CaC_2(s) + CO(g)$$

7-29. (a) Calculate the number of grams of O_2 obtainable from the decomposition of 3.88 moles of $KClO_3$ according to the equation

$$2\ KClO_3(s) \longrightarrow 2\ KCl(s) + 3\ O_2(g)$$

(b) How many grams of KCl will be obtained in (a)?

7-30. All heavy metal carbonates are unstable toward heat and yield CO_2 at elevated temperatures

$$CuCO_3(s) \xrightarrow{\Delta} CuO(s) + CO_2(g)$$

$$BaCO_3(s) \xrightarrow{\Delta} BaO(s) + CO_2(g)$$

Calculate the mass of $BaCO_3$ that yields the same mass of CO_2 as 65.0 g of $CuCO_3$.

7-31. If 1.06 moles of $FeCl_3$ are mixed with 1.22 moles of Na_2CO_3, what is the maximum number of moles of $Fe_2(CO_3)_3$ that can be formed?

7-32. A 312.5 g sample of $BaCl_2$ in aqueous solution is mixed with 224 g $(NH_4)_3PO_4$. Determine (a) which reactant is present in excess and (b) the maximum number of moles of precipitate produced.

$$3\ BaCl_2(aq) + 2\ (NH_4)_3PO_4 \longrightarrow Ba_3(PO_4)_2(s) + 6\ NH_4Cl(aq)$$

7-33. Iron metal can be prepared by passing H_2 gas over Fe_3O_4 at elevated temperatures.

$$Fe_3O_4(s) + 4\ H_2(g) \longrightarrow 3\ Fe(s) + 4\ H_2O(g)$$

(a) If 116 g of Fe_3O_4 and 18.1×10^{23} molecules of H_2 are available for reaction, calculate the mass of Fe produced. (b) How many grams of excess reactant remain?

7-34. Consider the following equations:

$$2\ KClO_3(s) \longrightarrow 2\ KCl(s) + 3\ O_2(g)$$

$$C_3H_8(g) + 5\ O_2(g) \longrightarrow 3\ CO_2(g) + 4\ H_2O(g)$$

How many grams of $KClO_3$ are required to supply enough O_2 to burn 154 g of C_3H_8?

7-35. Consider the following equations for the production of H_2SO_4 from SO_2.

$$2\ SO_2(g) + O_2(g) \longrightarrow 2\ SO_3(g)$$

$$SO_3(g) + H_2O(g) \longrightarrow H_2SO_4(l)$$

These reactions do occur to some extent in air and hence are partly responsible for "acidic" rain in industrial areas. If 96.2 g of SO_2 react in the presence of excess O_2 and H_2O, and 8.20 g of H_2SO_4 are produced, calculate the % yield of H_2SO_4.

7-36. Cl_2 can undergo various reactions in H_2O. A reaction responsible for the bleaching properties of aqueous chlorine solutions is:

$$Cl_2(g) + H_2O(l) \longrightarrow HCl(aq) + HOCl(aq)$$

where the bleaching properties result from the presence of HOCl. 17.8 g of Cl_2 are added to excess H_2O and analysis of the products shows the presence of 4.35 g HOCl. What is the % yield for HOCl?

7-37. The reaction between $N_2(g)$ and $H_2(g)$ at 500°C results in the formation of NH_3 gas. The balanced equation is:

$$N_2(g) + 3\ H_2(g) \longrightarrow 2\ NH_3(g)$$

If, at 500°C, 21.0 g N_2 and 6.00 g H_2 yield 8.80 g NH_3, calculate the % yield of NH_3 in this reaction.

7-38. Phosphorus may be produced commercially by heating together in an electric furnace the following: phosphate rock, $Ca_3(PO_4)_2$; sand, SiO_2; and coke, C. Phosphorus, P_4, is distilled away from the reaction mixture. The overall equation for the reaction is

$$2\ Ca_3(PO_4)_2(s) + 6\ SiO_2(s) + 5\ C(s) \longrightarrow P_4(g) + 6\ CaSiO_3(s) + 5\ CO_2(g)$$

Determine the mass of each reactant required to yield 1.00 kg of phosphorus.

7-39. In a certain metallurgical operation, the mineral pyrite, FeS_2, is roasted in air according to the equation

$$4\ FeS_2(s) + 11\ O_2(g) \longrightarrow 2\ Fe_2O_3(s) + 8\ SO_2(g)$$

SO_2 is then converted into H_2SO_4 by the reactions

$$2\ SO_2(g) + O_2(g) \longrightarrow 2\ SO_3(g)$$

$$SO_3(g) + H_2SO_4(l) \longrightarrow H_2S_2O_7(l)$$

$$H_2S_2O_7(l) + H_2O(l) \longrightarrow 2\ H_2SO_4(l)$$

Assuming that an ore is 27.5% FeS_2 and the remainder is inert, what mass of H_2SO_4 is produced per 1.00×10^3 kg of ore used if the processes are 100% efficient?

7-40. Red lead is an oxide of Pb that is useful as a paint pigment. When treated with H_2 gas, it is reduced to metallic Pb and water vapor. Starting with 1.714 g of the oxide, it was found that 1.554 g of Pb remained after complete reduction. (a) What is the formula for red lead? (b) Write an equation for the reaction with H_2.

7-41. To determine the composition of an oxide of Fe, a student passed H_2 gas over a weighed quantity of the heated oxide in a glass tube. Under these conditions the oxide was reduced to elemental Fe. The following data were collected:

Mass of oxide taken	16.00 g
Mass of Fe produced	11.20 g

(a) Determine the empirical formula of the oxide. (b) What is the percentage of Fe in the compound? (c) What mass of H_2O is produced?

7-42. When Mg is burned in air, a mixture of MgO and Mg_3N_2 (magnesium nitride) is formed. If 2.697 g of Mg produce 4.266 g of the mixture, calculate the weight percentage of each compound in the mixture.

7-43. In 1905, the atomic weight of chlorine was determined by Professor T. W. Richards and his co-workers at Harvard University. A known mass of Ag was dissolved in dilute HNO_3, and HCl was then added to precipitate AgCl. The ratio by weight of AgCl to Ag was determined to be 1.328667. Previously, the atomic weight of Ag had been shown to be 107.870. Calculate the atomic weight of chlorine.

7-44. Professors Richards and Willard computed the atomic weight of lithium in 1910 by determining the mass of AgCl that could be prepared from a known mass of LiCl. They found that the ratio of the LiCl used to AgCl produced was 0.295786. Assuming that the atomic weights of Ag and Cl are known to be 107.870 and 35.453, respectively, calculate the atomic weight of Li.

General Problems (Weight-Volume and Volume-Volume)

7-45. Acetylene, C_2H_2, may be produced from the reaction

$$CaC_2(s) + 2\,H_2O(l) \longrightarrow Ca(OH)_2(s) + C_2H_2(g)$$

What volume of C_2H_2 (at STP) will be produced if 7.50 g of CaC_2 are used?

7-46. Chlorine gas is produced commercially by the electrolysis of an aqueous NaCl solution as shown by the equation

$$2\,NaCl + 2\,H_2O \longrightarrow 2\,NaOH + Cl_2(g) + H_2(g)$$

(a) How many kilograms of Cl_2 are obtained if 3.75 kg of NaCl are used? (b) What volume of Cl_2 gas is obtained at STP in (a)? (c) Find the volume of H_2 gas obtained at STP in (a). (d) What is the mass of H_2 gas obtained in (a)?

(e) How many kilograms of NaOH (lye) are obtained for each kilogram of Cl_2 produced?

7-47. Gasoline, with an average composition of C_8H_{18}, burns in oxygen or air according to the equation

$$2\ C_8H_{18}(l) + 25\ O_2(g) \longrightarrow 16\ CO_2(g) + 18\ H_2O(g)$$

(a) Starting with 1.000 liter of liquid C_8H_{18} (density 0.800 g/ml), determine the volume of O_2 at STP required for complete combustion. (b) What volume of CO_2 at STP is produced?

7-48. Consider the reaction

$$2\ H_2S(g) + 3\ O_2(g) \longrightarrow 2\ H_2O(g) + 2\ SO_2(g)$$

(a) Calculate the number of grams of O_2 required to react with 127 g of H_2S. (b) Find the volume of O_2 at STP required in (a). (c) Assuming all gases are under the same conditions of temperature and pressure, what volume of O_2 is required to burn 434 m^3 of H_2S?

7-49. In the commercial production of nitric acid from ammonia, the first step is the oxidation of ammonia according to the equation

$$4\ NH_3(g) + 5\ O_2(g) \longrightarrow 4\ NO(g) + 6\ H_2O(g)$$

(a) Determine the volume of O_2 at STP required to oxidize 2.50 moles of NH_3. (b) If all gases are under the same conditions of temperature and pressure, what volume of NO will be formed for each liter of NH_3 oxidized? (c) How many grams of O_2 are required in (a)? (d) Calculate the volume of NO at 137°C and 1.00 atm pressure produced from 15.0 liters of NH_3 at 27°C and 1.00 atm.

7-50. $N_2(g)$ and $H_2(g)$ react to form $NH_3(g)$. If 2.25 kg N_2 are consumed in the reaction, calculate the volume of NH_3 produced at STP.

7-51. HBr(g) may be prepared in the laboratory by the following reaction:

$$2\ NaBr(s) + H_3PO_4 \longrightarrow Na_2HPO_4 + 2\ HBr(g)$$

How many grams of NaBr and H_3PO_4 are required to produce 925 ml of HBr(g) at 27°C and 65.0 cm of Hg pressure?

7-52. Propane, C_3H_8, is burned as a fuel in the stoves of many trailers and recreational vehicles. Complete combustion is represented by the equation:

$$C_3H_8(g) + 5\ O_2(g) \longrightarrow 3\ CO_2(g) + 4\ H_2O(g)$$

(a) If 38.5 g of C_3H_8 and 89.6 liters of O_2 at STP are available for reaction, how many grams of H_2O can be prepared? (b) How many grams of excess reactant remain?

7-53. Oxygen is prepared commercially by electrolysis of water. A 4.00 liter steel cylinder is to be filled with oxygen gas. To attain a final pressure of 1.50×10^2 atm at a temperature of 22°C, how many kilograms of H_2O will have to be electrolyzed to furnish the necessary O_2 for filling the cylinder?

7-54. Assume you are going to prepare $KClO_3$ in the laboratory using the following sequence of reactions:

$$2\,KMnO_4 + 16\,HCl \longrightarrow 2\,KCl + 2\,MnCl_2 + 5\,Cl_2(g) + 8\,H_2O$$

$$3\,Cl_2 + 6\,KOH \longrightarrow KClO_3 + 5\,KCl + 3\,H_2O$$

(a) How many moles of $KClO_3$ are produced per mole of $KMnO_4$ used? (b) Calculate the masses of $KMnO_4$ and KOH necessary to produce 35.0 g of $KClO_3$. (c) What volume of Cl_2, in the first reaction, measured at STP, is produced per mole of HCl used?

7-55. SO_2 is a by-product of the roasting of sulfide ores, for example,

$$2\,PbS(s) + 3\,O_2(g) \longrightarrow 2\,PbO(s) + 2\,SO_2(g)$$

What volume of SO_2 at 546°C and 1.00 atm pressure is obtained on roasting 2.50×10^3 kg of ore that contains 47.8% PbS?

7-56. Black gunpowder consists of a mixture of S, C, and KNO_3. An analytical method used for the determination of nitrates in such mixtures depends on the reduction of the nitrate radical to NO(*g*) and the subsequent measurement of the volume of NO gas under known conditions in a **nitrometer**. Metallic Hg is used as the reducing agent in H_2SO_4 solution. The equation for the reaction is

$$6\,Hg + 2\,HNO_3 + 3\,H_2SO_4 \longrightarrow 3\,Hg_2SO_4 + 2\,NO(g) + 4\,H_2O$$

In a typical analysis, the following data were obtained:

Mass of sample	0.1046 g
Volume of NO collected over H_2O	20.0 ml
Temperature	20.0°C
Pressure of gas in collecting bulb	69.8 cm of Hg

Calculate the weight percentage of KNO_3 in the mixture, assuming this compound is the only source of nitrate in the mixture.

7-57. An alloy of Zn and Cd was analyzed by dissolving in dilute HCl and collecting the liberated H_2 gas by displacement of H_2O. Equations for the reactions are

$$Zn + 2\,HCl \longrightarrow ZnCl_2 + H_2(g)$$

$$Cd + 2\,HCl \longrightarrow CdCl_2 + H_2(g)$$

The following data were obtained:

Mass of alloy taken for analysis	2.138 g
Volume of gas (wet)	702 ml
Temperature	20.0°C
Pressure in collecting vessel	697.5 torr

Calculate the weight percentage of each element in the mixture

7-58. A compound of xenon and fluorine was prepared when a mixture of Xe and F_2 was heated to 400°C for 1 hr. A reaction mixture was prepared at 0°C by introduction of F_2 gas at 0.60 atm and Xe gas at 0.12 atm into a 0.25 liter nickel container. After the reaction was completed, the mixture was cooled to 0°C, where it was found that all the Xe had been removed from the gas phase as a solid Xe–F compound. The pressure remaining in the nickel container was 0.36 atm. What is the formula for the compound prepared?

7-59. 0.30 mole of H_2 and 0.30 mole of O_2 are placed in a 5.0 liter container at 327°C. (a) Find the pressure of the mixture. (b) A reaction between H_2 and O_2 to produce H_2O is induced by an electric spark. What is the composition of the final mixture in terms of moles of the components? (c) What is the total pressure of the final mixture at 327°C? Assume all substances are gases under the conditions of this experiment.

Chapter 8

Solids

Solids are characterized by rigidity and definite crystalline or geometric structures corresponding to relatively simple polyhedra. X-ray studies of metals and other crystalline substances have shown that these structures result because the atoms (or ions or molecules) of the substance occupy certain fixed geometrical positions with respect to each other in a three-dimensional network or space lattice. The smallest portion of a crystal that has the structure characteristic of the space lattice is called the **unit cell**. The space lattice may, in principle, be generated by extensive repetition of the unit cell in three dimensions. The lengths of the edges of the unit cell, or the unit cell dimensions, are the distances between the nearest identical positions along the axes of the space lattice.

The unit cells for the recognized crystal types may be characterized in terms of a parallelepiped (a solid with six faces) as shown in Figure 8-1. The lengths of the three axes are a, b, and c, and the angles between axes are α, β, and γ.

Crystals belonging to the cubic system may be further classified as belonging to one of the unit cell arrangements shown in Figure 8-2. The simplest arrangement is depicted in Figure 8-2*a* where there are atoms only at the corners of each cube. The structure shown in *b* is the body-centered cubic structure in which a cube contains an atom at its center, as implied by its name, as well as an atom at each corner. The third type shown in *c* is the face-centered arrangement that has an atom located at the center of each face as well as at the corners of the cube.

Unit Cell	Axis Length	Angles between Axes
cubic	$a = b = c$	$\alpha = \beta = \gamma = 90°$
tetragonal	$a = b \neq c$	$\alpha = \beta = \gamma = 90°$
orthorhombic	$a \neq b \neq c$	$\alpha = \beta = \gamma = 90°$
monoclinic	$a \neq b \neq c$	$\alpha = \gamma = 90°$ $\beta \neq 90°$
triclinic	$a \neq b \neq c$	$\alpha \neq \beta \neq \gamma \neq 90°$
hexagonal	$a = b \neq c$	$\alpha = \beta = 90°$ $\gamma = 120°$
rhombohedral	$a = b = c$	$\alpha = \beta = \gamma \neq 90°$

Although Figure 8-2 does show the relative positions of the atoms in the various cubic arrangements, it does not give the correct perspective regarding distance between atom (or ion or molecule) centers. This perspective is shown more correctly in Figure 8-3.

To determine the **number of atoms per unit cell,** we must consider how atoms are shared between adjacent unit cells. The following rules apply.

1. Atoms on a face are shared equally between two adjacent unit cells and hence are counted one-half for each.
2. Atoms on an edge are shared equally among four adjacent unit cells and hence are counted one-fourth for each.
3. Atoms at a corner of a unit cell are shared among eight adjacent unit cells and hence are counted one-eighth for each.
4. Atoms completely within the unit cell are not shared with others and hence are counted only for the cell in which located.

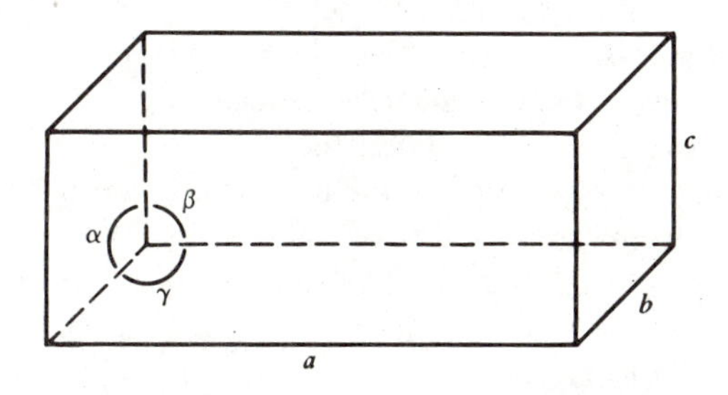

Figure 8-1 Generalized unit cell.

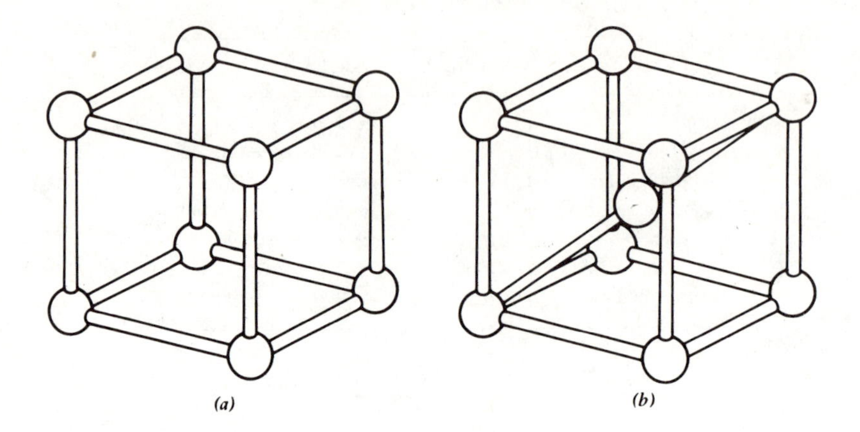

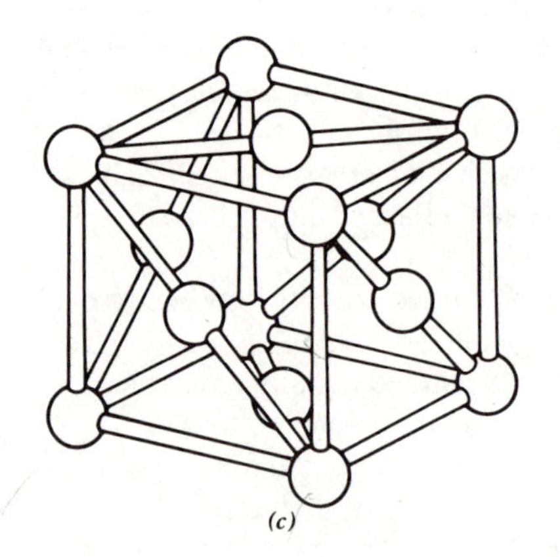

Figure 8-2 Geometric arrangements of the cubic system: (a) simple cubic; (b) body-centered cubic; and (c) face-centered cubic.

Atoms that surround a particular atom at the shortest distance between atoms are called **nearest neighbors**, and the number of nearest neighbors is termed the **crystallographic coordination number**. For example, in a body-centered cubic system, a metal atom at the center of the cube is surrounded by eight other metal atoms at the same distance; hence it has eight nearest neighbors or a coordination number of eight.

Crystals are often classified on the basis of the units that occupy the various sites in the space lattice and the bonding between these units. On this basis, there are four types of crystals: metallic or atomic, ionic, molecular, and covalent. Only the first two types are discussed here.

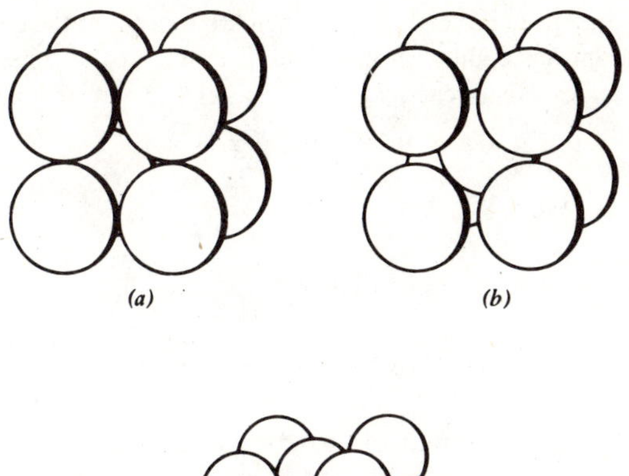

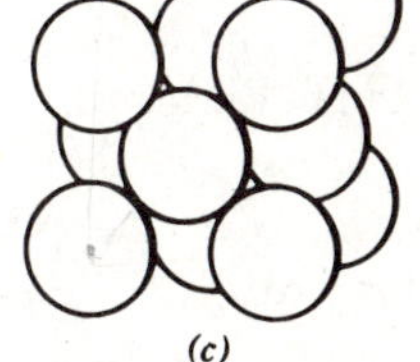

Figure 8-3 Center-to-center distance perspective for the cubic system: (a) simple cubic; (b) body-centered cubic; and (c) face-centered cubic.

8-A. Metallic or Atomic Crystals

The preceding discussion is directly applicable to this type of crystal. Some representative problems follow.

Example 8-a. At 20°C, iron metal crystallizes in a form known as α-Fe that has a body-centered cubic structure with a unit cell dimension of 0.286 nm. (a) How many atoms are there per unit cell? (b) How many nearest neighbors does a given iron atom have? (c) What is the volume, in cubic centimeters, of the unit cell? (d) What is the density of α-Fe in grams per cubic centimeter?

Solution. (a) In a body-centered cubic unit cell there are eight corner sites and one center site. The contribution of each of these occupied sites to the unit cube must be calculated.

$$\begin{aligned} 8 \text{ corner sites} \times \tfrac{1}{8} \text{ (atom/corner)} &= 1 \text{ atom} \\ 1 \text{ center site} \times 1 \text{ atom/site} &= \underline{1 \text{ atom}} \\ \text{Total} &= 2 \text{ atoms per unit cell} \end{aligned}$$

(b) Each center atom has eight nearest neighbor atoms at the corners of the cube. Each corner atom has eight nearest neighbor atoms at the centers of each of the eight adjacent cubes. Hence each iron atom has eight nearest neighbors.

(c) Volume $= a \times b \times c = (0.286 \text{ nm})^3 = 2.34 \times 10^{-2} \text{ nm}^3$/unit cell

Since $1.00 \text{ nm} = 1.00 \times 10^{-7} \text{ cm}$

$$\text{Volume} = 2.34 \times 10^{-2} \cancel{\text{nm}^3} \times \left(\frac{1.00 \times 10^{-7} \text{ cm}}{1.00 \cancel{\text{nm}}}\right)^3$$

$$= 2.34 \times 10^{-23} \text{ cm}^3/\text{unit cell}$$

(d) Because density is the ratio of mass to volume and since the volume of the unit cell represents the volume occupied by two atoms, the **mass** of two atoms must be determined before the density can be calculated.

$$\text{Mass} = \frac{55.8 \text{ g Fe}}{1 \cancel{\text{mole Fe}}} \times \frac{1 \cancel{\text{mole Fe}}}{6.02 \times 10^{23} \cancel{\text{Fe atoms}}} \times \frac{2 \cancel{\text{Fe atoms}}}{\text{Unit cell}}$$

$$= 1.85 \times 10^{-22} \text{ g Fe/unit cell}$$

$$\text{Density} = \frac{1.85 \times 10^{-22} \text{ g Fe}}{1 \cancel{\text{unit cell}}} \times \frac{1 \cancel{\text{unit cell}}}{2.34 \times 10^{-23} \text{ cm}^3}$$

$$= 7.91 \text{ g Fe/cm}^3$$

Example 8-b. A unit cell of Pd metal is a face-centered cube with an edge of 0.389 nm. (a) What is the distance between centers of nearest neighbors? (b) What is the radius of a Pd atom? (c) What is the volume of one mole of Pd (called the molar volume)? Note that corner atoms here do not "touch" one another.

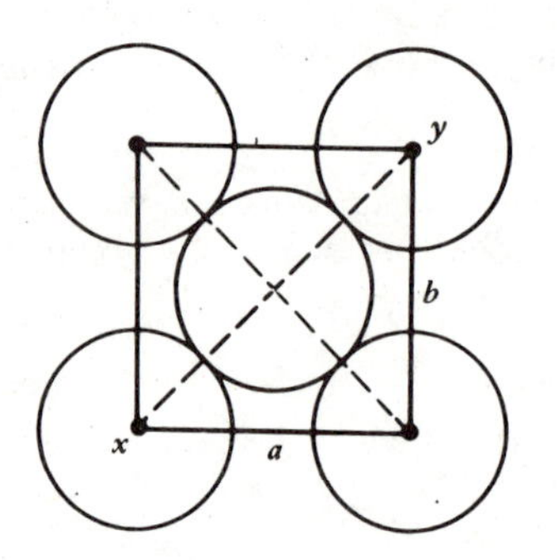

Solution. (a) First visualize the face of the unit cell as it is shown. Each black dot represents the center of a Pd atom. The distance between nearest neighbors is one-half the distance d from atom x to atom y. We know from the theorem of Pythagoras that the square of the hypotenuse of a right triangle is equal to the sum of the squares of the sides. Therefore

$$d^2 = a^2 + b^2 = (0.389\text{ nm})^2 + (0.389\text{ nm})^2 = 2(0.389\text{ nm})^2$$

$$d = \sqrt{2}(0.389\text{ nm}) = 0.550\text{ nm}$$

The distance between centers of nearest neighbors is thus

$$\tfrac{1}{2}(0.550\text{ nm}) = 0.275\text{ nm}$$

(b) If atoms are assumed to be hard spheres that touch along the axis of closest approach, the radius is one-half the distance between centers of atoms or one-half of 0.275 nm = 0.138 nm.

(c) From the unit cell dimensions and the number of atoms per unit cell, the volume per atom may be calculated. By multiplying by Avogadro's number, the molar volume may be obtained.

Following the process used in Example 8-a,

$$\text{Volume} = (0.389\text{ nm})^3/\text{unit cell} = 5.89 \times 10^{-23}\text{ cm}^3/\text{unit cell}$$

In a face-centered cubic unit cell there are eight corner sites and six face sites.

$$8\text{ corner sites} \times (\tfrac{1}{8})\text{ atom/corner site} = 1\text{ atom}$$
$$6\text{ face sites} \times (\tfrac{1}{2})\text{ atom/face site} = 3\text{ atoms}$$
$$\text{Total} = 4\text{ atoms/unit cell}$$

$$\text{Molar volume} = \frac{5.89 \times 10^{-23}\text{ cm}^3}{1\text{ \sout{unit cell}}} \times \frac{1\text{ \sout{unit cell}}}{4\text{ \sout{atoms}}} \times \frac{6.02 \times 10^{23}\text{ \sout{atoms}}}{1\text{ mole}}$$

$$= 8.86\text{ cm}^3/\text{mole}$$

Problems (Metallic or Atomic Crystals)

8-1. How many atoms per unit cell are there in elements crystallizing in each of the following systems: (a) simple cubic, (b) face-centered cubic, (c) simple tetragonal, (d) face-centered tetragonal, (e) body-centered tetragonal, (f) simple orthorhombic.

8-2. For each of the cubic systems whose structures are shown in Figure 8-2, how many nearest neighbors would an atom of a metal have?

8-3. Na metal crystallizes in a body-centered cubic system with a distance of 0.371 nm between nearest neighbors. (a) What is the radius of the Na atom? (b) What is the length of the edge of the unit cell? (c) What is the molar volume of Na? The Na atoms "touch" only along the inner diagonal of the cube. Corner atoms do not "touch" one another. (*Hint*: See Example 8-c.)

8-4. Ag metal crystallizes in the cubic system with a face-centered unit cell. (a) If the distance of closest approach of two Ag atoms is 0.289 nm, what is the length of the edge of the unit cell? (b) What is the density of Ag metal?

8-5. Rb metal crystallizes in a body-centered cubic system with a unit cell dimension of 0.572 nm. (a) What is the distance between the centers of nearest neighbors? (b) What is the volume of the unit cell? (c) What is the molar volume? (d) What is the density of Rb metal? (*Hint*: See Example 8-c.)

8-6. Tin metal exists in two allotropic forms, α and β. β-Sn crystallizes with a simple tetragonal unit cell having dimensions $a = b = 0.302$ nm and $c = 0.318$ nm. (a) How many atoms are there per unit cell in β-Sn? (b) What is the density of β-Sn? What is the molar volume?

8-7. Indium metal crystallizes with a face-centered tetragonal unit cell with the dimensions $a = b = 0.459$ nm and $c = 0.494$ nm. (a) How many atoms are there per unit cell? (b) What is the volume of the unit cell? (c) What is the molar volume? (d) What is the density of In metal?

8-B. Ionic Crystals

Ionic crystals are those in which positive and negative ions occupy the various sites in the space lattice. Depending on the types of ions in the compound, the unit cells are more or less complicated. Two relatively simple types, exhibited by numerous compounds belonging to the cubic system, are those shown by sodium chloride and cesium chloride. The unit cell structures of each of these are illustrated in Figure 8-4. In the sodium chloride type, each

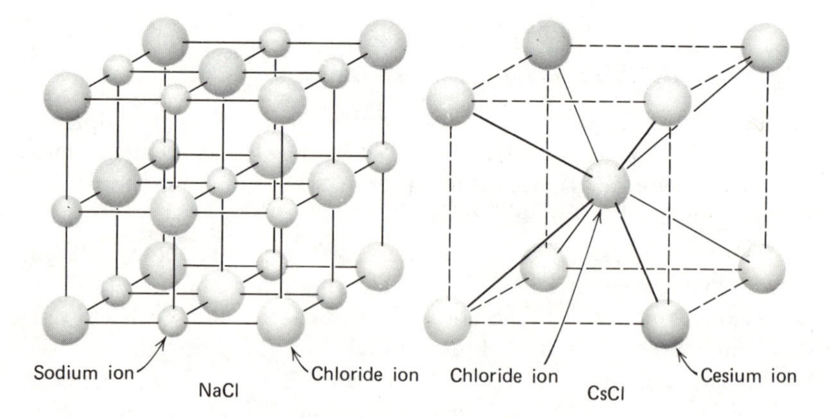

Figure 8-4 Unit cell arrangements.

positive ion is surrounded by six nearest neighbor chloride ions and each chloride ion is surrounded by six nearest neighbor sodium ions. In the cesium chloride type, each positive and negative ion has eight ions of opposite charge for nearest neighbors.

Example 8-c. The edge of the unit cell of CsCl is 0.411 nm. (a) What is the distance between the center of a Cs^+ ion and the center of the nearest Cl^- ion? (b) If the radius of Cl^- is 0.181 nm, what is the radius of Cs^+? (c) How many ions are there per unit cell?

Solution. (a) The shortest distance between centers of a Cs^+ ion and a Cl^- ion is along the diagonal of the unit cube. Let the distance between opposite corners of the CsCl unit cell be d as shown below. From geometry

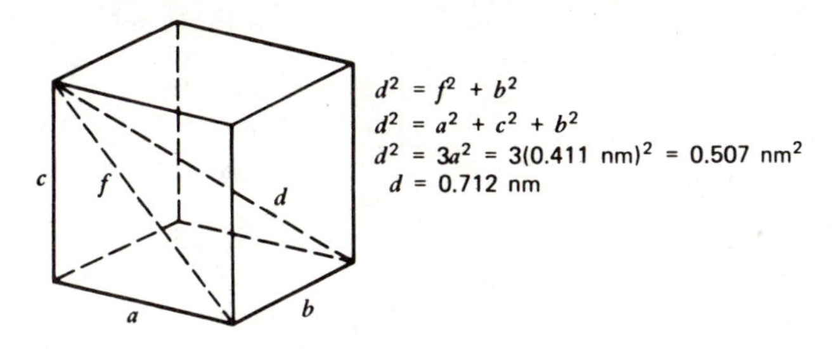

$$d^2 = f^2 + b^2$$
$$d^2 = a^2 + c^2 + b^2$$
$$d^2 = 3a^2 = 3(0.411\ \text{nm})^2 = 0.507\ \text{nm}^2$$
$$d = 0.712\ \text{nm}$$

The distance between centers of Cs^+ and Cl^- is 0.712 nm/2 = 0.356 nm.

(b) Assume that the distance between the centers of Cs^+ and Cl^- is the sum of their radii. Then,

$$(r_{Cs^+}) + (r_{Cl^-}) = 0.356\ \text{nm}$$

$$(r_{Cs^+}) = 0.356 - 0.181 = 0.175\ \text{nm}$$

(c) The number of ions per unit cell is calculated as follows:

$$1\ \text{center}\ Cl^- = 1\ Cl^-$$
$$8\ \text{corner sites} \times \tfrac{1}{8}(Cs^+/\text{corner site}) = 1\ Cs^+$$

There is, therefore, one CsCl per unit cell.

Example 8-d. Crystals may have vacant or unoccupied sites in the crystal lattice. These are said to be **lattice defects**. If a crystal does have unoccupied sites, its density should be less than that calculated from unit cell dimensions with all sites filled. By comparing observed and theoretical densities, the fraction of sites unoccupied may be determined; for example, the observed density of NaCl is 2.165 g/cm^3 and the unit cell of NaCl is 0.5628 nm along each edge. Determine the theoretical density and then calculate the percentage of unoccupied sites.

Solution.

$$\text{Volume of unit cell} = (0.5628\ \text{nm})^3 \times \left(\frac{1.00 \times 10^{-7}\ \text{cm}}{1.00\ \text{nm}}\right)^3$$

$$= 1.783 \times 10^{-22}\ \text{cm}^3$$

Since there are 4 Na^+ and 4 Cl^- per unit cell, the mass per unit cell

$$= \frac{58.44\ \text{g NaCl}}{1\ \cancel{\text{mole NaCl}}} \times \frac{1\ \cancel{\text{mole NaCl}}}{6.022 \times 10^{23}\ \cancel{\text{units NaCl}}} \times \frac{4\ \cancel{\text{units NaCl}}}{1\ \text{unit cell}}$$

$$= 3.882 \times 10^{-22}\ \text{g NaCl/unit cell}$$

$$\text{Density (theoretical)} = \frac{3.882 \times 10^{-22}\text{g}}{1.783 \times 10^{-22}\ \text{cm}^3}$$

$$= 2.177\ \text{g/cm}^3$$

$$\text{Percentage of unoccupied sites} = \frac{2.177 - 2.165}{2.177} \times 100$$

$$= 0.55\%$$

Problems (Ionic Crystals)

8-8. The space lattice of NaCl (Figure 8-4) was shown as two interpenetrating face-centered cubes of Na^+ and Cl^-. The edge of the unit cell is 0.5628 nm. (a) How many ions of each type are present in the unit cell? (b) What is the distance between the center of a Na^+ and the center of the nearest Cl^-? (c) What is the distance between centers of the nearest two Na^+ ions? (d) What is the volume of the unit cell?

8-9. KF crystallizes in the NaCl type structure. (a) If the radius of K^+ is 0.132 nm and F^- is 0.135 nm, what is the shortest K-F distance? (b) What is the length of the edge of the unit cell? (c) What is the shortest K-K distance? (d) What is the volume of the unit cell?

8-10. RbI crystallizes in the CsCl type structure. (a) If the radius of Rb^+ is 0.149 nm and I^- is 0.217 nm, what is the shortest Rb-I distance? (b) What is the shortest Rb-Rb distance? (c) How many Rb^+ ions are there at this distance around a given Rb^+ ion? (d) What is the volume of the unit cell?

8-11. RbCl crystallizes in the NaCl structure. (a) What are the unit cell dimensions if the Rb-Cl distance is 0.327 nm? (b) What is the density of RbCl?

8-12. CsI crystallizes in the CsCl-type structure. (a) What is the length of the unit cell if the Cs-I distance is 0.395 nm? (b) What is the molar volume of CsI?

8-13. CsBr crystallizes in the CsCl structure. (a) If the density of CsBr is 4.438 g/cm^3, what is the length of the edge of the unit cube? (b) What is the shortest Cs-Br distance?

General Problems

8-14. In tetragonal β-Sn, having a simple tetragonal unit cell, the dimensions are $a = b = 0.3022$ nm and $c = 0.3181$ nm. There are two different distances between the centers of Sn atoms. Sketch the unit cell. (a) What are these distances? (b) How many nearest neighbors does a particular atom have at each distance?

8-15. Cs metal crystallizes in a body-centered cubic system with a unit cell length $a = 0.608$ nm. What is the radius of the Cs atom?

8-16. Ni metal crystallizes in a face-centered cubic system with a unit cell length $a = 0.352$ nm. (a) What is the radius of the Ni atom? (b) How many close neighbors does a Ni atom have?

8-17. Indium crystals have a face-centered tetragonal unit cell of dimensions $a = b = 0.459$ nm and $c = 0.494$ nm, and there are two different distances between indium atoms. Sketch the unit cell. What are these distances?

8-18. The ionic radii of Ba^{2+} and O^{2-} ions are 0.136 and 0.140 nm. Barium oxide crystallizes in the NaCl-type structure. (a) What is the volume of the unit cell? (b) What is the shortest distance between centers of adjacent Ba^{2+} ions? (c) How many Ba^{2+} ions are there around a particular Ba^{2+} at this distance? (d) What is the density of BaO?

8-19. PbS crystallizes in the NaCl-type structure. If the shortest distance between centers of Pb^{2+} and S^{2-} ions is 0.2962 nm, (a) what is the length of the unit cell edge? (b) What is the volume of the unit cell? (c) What is the distance between nearest Pb^{2+} ions? (d) What is the molar volume of PbS?

8-20. Calculate Avogadro's number from the fact that Au belongs to the face-centered cubic system and has a unit cell dimension $a = 0.4078$ nm. Its density is 19.30 g/cm^3.

8-21. SrO has a density of 5.05 g/cm^3 and there are 4 Sr^{2+} and 4 O^{2-} ions per unit cell. What is the length of the unit cell if it is cubic?

8-22. LiH crystallizes in a cubic structure and has a density of 0.77 g/cm^3. If the edge of the unit cube is 0.4086 nm, how many Li^+ and H^- ions are there in a unit cell?

8-23. Calculate Avogadro's number from the fact that CsBr crystallizes in the CsCl-type structure and has a unit cell edge length of 0.4289 nm. The density of CsBr is 4.477 g/cm^3.

8-24. (a) For each of the three unit cubes in Figure 8-3, calculate the fraction of the cube actually occupied by atoms. Assume that the atoms are hard spheres and touch along the shortest interatomic distance. (b) Which of the three exhibits the most efficient packing of atoms?

8-25. Data are recorded in the table below for a number of compounds that crystallize in the cubic system, either as NaCl or CsCl types. From the information given for each compound, carry out calculations to obtain data to fill in the blanks.

Compound	Edge of Unit Cell in nm	Distance Between Centers of Like Ions (Atoms) in nm	Distance Between Centers of Unlike Ions (Atoms) in nm	Molar Volume (ml)	Density Theoretical	Density g/ml Observed
NaCl Type						
(a) NaF	0.4620	______	0.231	______	2.84	2.79
(b) LiI	______	0.424	______	32.5	______	4.06
(c) RbBr	0.6854	______	______	______	3.41	3.35
(d) CaO	______	0.340	0.2405	16.75	______	3.35
(e) KBr	0.66000	______	______	______	______	2.75
(f) LiF	______	______	0.201	9.76	2.66	2.60
(g) AgCl	0.5547	______	______	______	5.57	5.56
CsCl Type						
(h) CsBr	0.4286	______	______	47.4	______	4.438
(i) CsI	______	0.4567	______	57.3	4.53	4.51
(j) TlCl	______	0.3834	______	______	7.07	7.00
(k) AgCd	0.333	______	______	______	______	______
(l) LiAg	0.3168	0.3168	______	16.2	______	______
(m) MgSr	______	0.390	0.337	______	3.14	______

8-26. Use the table above in answering this question. By comparing the observed density with that calculated from unit cell dimensions, determine the extent of lattice defect—that is, the percentage of unoccupied sites in the crystal lattice for the compounds NaF, AgCl, and CsI.

Chapter 9

Concentrations and Stoichiometry of Solutions

The concentration of a solution may be expressed in a number of ways, the more important of which are listed below.

9-A. Percentage (By Weight)

% Solute = Parts by weight of solute per 100 parts by weight of solution

or

$$\% \text{ Solute} = \frac{\text{Parts of solute}}{\text{Parts solute} + \text{Parts solvent}} \times 100$$

For example, a solution containing 10 g of solute in 100 g of solution (10 g solute, 90 g solvent) is a 10% solution; one containing 50 g solute in 50 g of solvent is a 50% solution, and so on. Note that the denominator in this expression is the mass of **solution**, which includes both solute and solvent.

Example 9-a. Calculate the percentage of K_2CO_3 in a solution that is made by dissolving 15 g of K_2CO_3 in 60 g of H_2O.

Solution.

$$\text{Percentage } K_2CO_3 = \frac{\text{Grams } K_2CO_3}{\text{Grams } K_2CO_3 + \text{Grams } H_2O} \times 100$$

$$= \frac{15 \text{ g}}{15 \text{ g} + 60 \text{ g}} \times 100 = 20\%$$

Example 9-b. How many grams of NaCl would have to be dissolved in 54.0 g H_2O to give a 10.0% solution?

Solution.

$$\text{Percentage NaCl} = \frac{\text{Grams NaCl}}{\text{Grams NaCl} + \text{Grams } H_2O} \times 100$$

Let

$$X = \text{Grams of NaCl}$$

Then

$$10.0 = \left(\frac{X}{X + 54.0 \text{ g}}\right)100$$

or

$$0.100 = \frac{X}{X + 54.0 \text{ g}}$$

$$0.100X + 5.40 \text{ g} = X$$

$$X = \frac{5.40 \text{ g}}{0.900} = 6.00 \text{ g}$$

Problems (Percentage by Weight)

9-1. How many grams of NaCl are contained in 0.600 kg of a 30.0% solution?

9-2. In making up a 24.8% aqueous solution of sucrose, $C_{12}H_{22}O_{11}$, how many grams of sugar and water would be used to prepare 1.00 kg of solution?

9-3. A certain solution is 10.0% $C_{12}H_{22}O_{11}$. How many grams of sugar must be added to 0.550 kg of this solution to bring the percentage of sugar to 20.0%.

9-B. Mole Fraction

The **mole fraction** is defined as the ratio of moles of a given component to the total moles of all components in a solution; that is,

$$\text{Mole fraction of component } A = X_A$$

$$= \frac{\text{Moles of } A}{\text{Total moles of all components}}$$

For a solution containing a single solute in a solvent

$$\text{Mole fraction of solute} = \frac{\text{Moles of solute}}{\text{Moles of solute} + \text{Moles of solvent}}$$

$$\text{Mole fraction of solvent} = \frac{\text{Moles of solvent}}{\text{Moles of solute} + \text{Moles of solvent}}$$

Obviously, the sum of the two fractions must be unity.

Example 9-c. What is the mole fraction of sugar, $C_{12}H_{22}O_{11}$, in a solution made by dissolving 17.1 g of $C_{12}H_{22}O_{11}$ in 89.1 g of H_2O?

Solution. The molecular weight of $C_{12}H_{22}O_{11}$ is 342 and that of H_2O is 18.0.

$$\text{Moles } C_{12}H_{22}O_{11} = 17.1 \text{ g } \cancel{C_{12}H_{22}O_{11}} \times \frac{1 \text{ mole } C_{12}H_{22}O_{11}}{342 \text{ g } \cancel{C_{12}H_{22}O_{11}}}$$

$$= 0.0500 \text{ mole } C_{12}H_{22}O_{11}$$

$$\text{Moles } H_2O = 89.1 \text{ g } \cancel{H_2O} \times \frac{1 \text{ mole } H_2O}{18.0 \text{ g } \cancel{H_2O}} = 4.95 \text{ moles } H_2O$$

$$\text{Mole fraction } C_{12}H_{22}O_{11} = \frac{0.0500}{0.050 + 4.95} = 0.0100$$

Example 9-d. What is the mole fraction of NaCl in a solution containing 1.00 mole of solute in 1.00 kg of H_2O?

Solution.

$$\text{Moles } H_2O = 1.00 \times 10^3 \text{ g } \cancel{H_2O} \times \frac{1 \text{ mole } H_2O}{18.0 \text{ g } \cancel{H_2O}} = 55.5 \text{ moles } H_2O$$

$$\text{Mole fraction NaCl} = \frac{1.00 \text{ mole}}{1.00 \text{ mole} + 55.5 \text{ moles}}$$

$$= \frac{1.00 \text{ mole}}{56.5 \text{ moles}} = 0.0177$$

Problems (Mole Fraction)

9-4. Calculate the mole fractions of solute and solvent in a solution prepared by dissolving 142 g of NaCl in 3.00 kg of H_2O.

9-5. Calculate the mole fraction of H_2SO_4 in a solution containing 7.50 moles in 1.00 kg of H_2O.

9-6. If 0.54 g of aspirin (MW = 180) is dissolved in 2.5×10^2 g H_2O, calculate the mole fraction of aspirin in the solution.

9-C. Molarity (Formality)

Molarity (M) is defined as the number of moles of solute per liter of solution.* A solution that contains 1 mole per liter is termed 1 molar.

$$\text{Molarity} = \frac{\text{Moles solute}}{\text{Liters of solution}}$$

Of the concentration expressions introduced in this chapter, molarity is probably the most common unit used by the chemist.

Example 9-e. What is the molarity of a solution that contains 49 g of H_3PO_4 in 2.0 liters of solution? The molecular weight of H_3PO_4 is 98.

Solution.

$$\text{Moles } H_3PO_4 = 49 \,\cancel{\text{g } H_3PO_4} \times \frac{1 \text{ mole } H_3PO_4}{98 \,\cancel{\text{g } H_3PO_4}} = 0.50 \text{ mole}$$

$$\text{Molarity} = \frac{0.50 \text{ mole}}{2.0 \text{ liters}} = 0.25 \text{ moles/liter or } 0.25\, M \text{ (molar)}$$

Example 9-f. How many grams of H_2SO_4 are contained in 0.0500 liter of 0.500 M solution?

Solution.

$$\text{Moles } H_2SO_4 \text{ present} = 0.0500 \,\cancel{\text{liter}} \times \frac{0.500 \text{ mole}}{1 \,\cancel{\text{liter}}} = 0.0250 \text{ mole } H_2SO_4$$

$$\text{Grams } H_2SO_4 \text{ present} = 0.0250 \,\cancel{\text{mole } H_2SO_4} \times \frac{98.0 \text{ g } H_2SO_4}{1 \,\cancel{\text{mole } H_2SO_4}} = 2.45 \text{ g } H_2SO_4$$

In one step we have:

$$\text{Grams } H_2SO_4 = 0.0500 \,\cancel{\text{liter}} \times \frac{0.500 \,\cancel{\text{mole } H_2SO_4}}{1 \,\cancel{\text{liter}}} \times \frac{98.0 \text{ g } H_2SO_4}{\cancel{1 \text{ mole } H_2SO_4}}$$

$$= 2.45 \text{ g } H_2SO_4$$

* A similar term, **formality**, is defined as the number of formula weights of solute per liter of solution. Formality may be used with substances where only the empirical formula is known, such as for ionized substances. In practice, however, there is often no distinction made between molecular and formula weights insofar as these types of calculations are concerned, and hence the two terms molarity and formality may be used interchangeably. In the following discussion, only the term molarity is employed.

Example 9-g. Determine the molarity of a solution prepared by diluting 1.00 liter of 0.500 *M* solution to 2.00 liters.

Solution.

$$\text{Moles of solute in 1.00 liter of 0.500 } M \text{ solution} = 1.00\ \cancel{\text{liter}} \times \frac{0.500 \text{ mole}}{1\ \cancel{\text{liter}}}$$

$$= 0.500 \text{ mole}$$

This will also be the number of moles in the diluted solution. Hence:

$$\text{Molarity of diluted solution} = \frac{0.500 \text{ mole}}{2.00 \text{ liters}} = 0.250\ M$$

Example 9-h. To what volume must 0.250 liter of 0.150 *M* H_2SO_4 be diluted to yield a 0.0250 M solution?

Solution. The original solution contains

$$0.250\ \cancel{\text{liter}} \times \frac{0.150 \text{ mole } H_2SO_4}{1\ \cancel{\text{liter}}} = 0.0375 \text{ mole } H_2SO_4$$

The diluted solution will still contain 0.0375 mole of H_2SO_4, and its volume is:

$$\text{Volume} = 0.0375\ \cancel{\text{mole } H_2SO_4} \times \frac{1 \text{ liter}}{0.0250\ \cancel{\text{mole } H_2SO_4}} = 1.50 \text{ liters}$$

In one step we have:

$$\text{Volume} = 0.250\ \cancel{\text{liter}} \times \frac{0.150\ \cancel{\text{mole } H_2SO_4}}{\cancel{1 \text{ liter}}} \times \frac{1 \text{ liter}}{0.0250\ \cancel{\text{mole } H_2SO_4}} = 1.50 \text{ liters}$$

Problems (Molarity)

9-7. What is the molarity of a solution in which 3.15 g of HNO_3 are dissolved in 0.250 liter of solution?

9-8. Determine the molarity of solutions of each of the following solutes in which 1.00×10^2 g of solute are present in 3.50 liters of solution.

(a) NaOH (c) KNO_3

(b) H_2SO_4 (d) $C_{12}H_{22}O_{11}$

9-9. How many grams of H_3PO_4 must be weighed to prepare 4.250 liters of 0.3650 *M* solution?

9-10. Calculate the molarity of a solution of $Ca(OH)_2$ that contains 0.185 g of solute in 4.75×10^2 ml of solution.

9-11. How many grams of HNO_3 will be required to prepare 2.85 liters of 0.450 *M* solution?

9-12. Calculate the molarity of a solution in which 3.88 g of HCl are dissolved in toluene (an organic solvent) and then diluted to 0.750 liter.

9-13. How many grams of each of the following solutes would be needed to prepare 0.250 liter quantities of 0.0650 *M* solution?

(a) H_3PO_4 (c) KOH

(b) $Al_2(SO_4)_3$ (d) $(NH_4)_2CO_3$

9-14. How many liters of 0.444 *M* KOH solution can be prepared from 127 g of KOH?

9-15. If 38 ml of 16 *M* HNO_3 are diluted to 0.75 liter, what is the molarity of the diluted solution?

9-16. To what volume must 0.150 liter of 0.640 *M* H_3PO_4 be diluted to yield a 0.0800 *M* solution?

9-17. If 225 ml of a solution containing 31.0 g of ethylene glycol (antifreeze), $C_2H_6O_2$, is diluted to a total volume of 1.50 liters, calculate the molarity of the diluted solution.

9-18. A 11.7 g sample of NaCl is dissolved in enough water to yield 0.250 liter of solution. This solution is then diluted to 3.00 liters. (a) How many moles of NaCl are present in the undiluted solution? (b) How many moles of NaCl are present in the diluted solution? (c) Determine the molarity of each solution.

9-D. Molality

Molality (*m*) is defined as moles of solute per kilogram of solvent. A 1 molal solution contains 1 mole of solute in 1 kg of solvent.

$$\text{Molality} = \frac{\text{Moles of solute}}{\text{Kg of solvent}}$$

Molarity varies with temperature since the volume of a solution changes with temperature; molality, which is based on mass instead of volume, is independent of temperature. This method of expressing the concentration of a solution is ordinarily used when dealing with **colligative** properties; such properties of solutions are the boiling and freezing points that are dependent on the **number** of dissolved particles and not on their nature or kind. In dilute solutions, molarity and molality have virtually the same values, but they have wide differences in concentrated solutions.

Example 9-i. What is the molality of a solution in which 1.00×10^2 g of NaOH are dissolved in 0.250 kg of H_2O?

Solution.

$$\text{Moles NaOH} = 1.00 \times 10^2 \,\cancel{\text{g NaOH}} \times \frac{1 \text{ mole NaOH}}{40.0 \,\cancel{\text{g NaOH}}} = 2.50 \text{ moles NaOH}$$

Hence

$$\text{Molality} = \frac{2.50 \text{ moles NaOH}}{0.250 \text{ kg } H_2O} = \frac{10.0 \text{ moles NaOH}}{1 \text{ kg } H_2O} = 10.0\, m$$

Example 9-j. How many grams of KCl must be added to 75.0 grams of water to produce a solution that is 2.25 molal (m)?

Solution. First we determine how many moles of KCl are required and then convert to grams of KCl. In one step we have:

$$\text{Grams KCl} = 0.0750 \,\cancel{\text{kg } H_2O} \times \frac{2.25 \,\cancel{\text{moles KCl}}}{1 \,\cancel{\text{kg } H_2O}} \times \frac{74.6 \text{ g KCl}}{1 \,\cancel{\text{mole KCl}}} = 12.6 \text{ g KCl}$$

Problems (Molality)

9-19. What is the molality of a solution that contains 24.5 g of H_3PO_4 in 1.00×10^2 g of H_2O?

9-20. In how many kilograms of H_2O should 11.7 g of NaCl be dissolved to obtain a 0.125 molal solution?

9-21. How many grams of Na_2SO_4 must be added to 845 grams of water to produce a solution that is 4.22 molal?

9-E. Interconversion Between Concentration Units

Interconversion between % (weight), mole fraction, and molality requires knowledge of the molecular weights of both solute and solvent and use of the proper conversion factors. Interconversion between these three concentration units and molarity requires the density of the solution in addition to the molecular weights. Study the following examples.

Example 9-k. Determine the mole fraction and molality of sucrose, $C_{12}H_{22}O_{11}$, in a 10.0% sucrose solution. The solvent is H_2O. (Molecular weights: $H_2O = 18.0$; $C_{12}H_{22}O_{11} = 342$).

Solution. A 10.0% sucrose solution implies 10.0 g sucrose plus 90.0 g H_2O to yield 100.0 g of solution.

(a) To determine mole fraction we must convert the above masses to moles, that is:

$$\text{Moles } C_{12}H_{22}O_{11} = 10.0 \text{ g } C_{12}H_{22}O_{11} \times \frac{1 \text{ mole } C_{12}H_{22}O_{11}}{342 \text{ g } C_{12}H_{22}O_{11}}$$

$$= 0.0292 \text{ mole } C_{12}H_{22}O_{11}$$

$$\text{Moles } H_2O = 90.0 \text{ g } H_2O \times \frac{1 \text{ mole } H_2O}{18.0 \text{ g } H_2O} = 5.00 \text{ mole } H_2O$$

$$\text{Mole fraction } C_{12}H_{22}O_{11} = \frac{\text{Moles } C_{12}H_{22}O_{11}}{\text{Total moles}} = \frac{0.0292}{5.03} = 0.00581$$

(b) To determine molality we must determine moles $C_{12}H_{22}O_{11}$ per kilogram H_2O. Hence:

$$\text{Molality} = \frac{\text{Moles } C_{12}H_{22}O_{11}}{\text{Kg } H_2O} = \frac{0.0292 \text{ mole}}{0.0900 \text{ kg } H_2O} = 0.325\ m$$

Example 9-I. Determine the molarity of a concentrated HCl solution that is 37.0% by weight HCl. The solvent is H_2O and the density of the solution is 1.19 g/ml.

Solution. 37.0% HCl implies 37.0 g HCl per 100.0 g of solution. Since molarity contains the units moles HCl per liter of solution we must convert mass of solution to volume of solution and grams HCl to moles HCl. In one step we have:

$$\text{Molarity} = \frac{37.0 \text{ g HCl}}{100.0 \text{ g solution}} \times \underbrace{\frac{1.19 \text{ g solution}}{1 \text{ ml solution}} \times \frac{1000 \text{ ml}}{1 \text{ liter}}}_{\text{Converts to volume of solution}} \times \underbrace{\frac{1 \text{ mole HCl}}{36.5 \text{ g HCl}}}_{\text{Converts to moles HCl}}$$

$$= 12.1 \frac{\text{mole HCl}}{1 \text{ liter}} = 12.1\ M$$

Problems (Interconversions)

9-22. What is the mole fraction of $CaCl_2$ in a 8.45 weight percent aqueous solution?

9-23. Calculate the molality of H_2SO_4 in a 1.50 M aqueous solution that has a density of 1.090 g/ml.

9-24. Calculate the weight percent of H_2SO_4 in 9.61 *M* H_2SO_4. The density of this solution is 1.520 g/ml.

9-25. 855 g of a 0.625 molal solution of K_2SO_4 are evaporated to dryness. What is the mass in grams of the K_2SO_4 remaining?

9-F. Solution Stoichiometry

These problems are approached in a manner similar to that described in Chapter 7. The only different or new factor here is that we can use the definition of the concentration unit to convert from volume of solution to moles of solute or vice versa. Relating moles of two substances by using the balanced equation is still the key link in the solution of these problems.

Example 9-m. How many grams of NaOH would be required to react with 25.0 ml of 1.50 *M* HCl solution? The balanced equation is:

$$NaOH(aq) + HCl(aq) \longrightarrow NaCl(aq) + H_2O$$

Solution. We first determine how many moles of HCl are present in 25.0 ml of 1.50 *M* HCl solution.

$$\text{Moles HCl} = 25.0\ \cancel{\text{ml}} \times \frac{1\ \cancel{\text{liter}}}{1000\ \cancel{\text{ml}}} \times \frac{1.50\ \text{mole HCl}}{1\ \cancel{\text{liter}}} = 0.0375\ \text{mole HCl}$$

Using the balanced equation, we convert from moles HCl to moles NaOH and finally grams of NaOH:

$$\text{Grams NaOH} = 0.0375\ \cancel{\text{mole HCl}} \times \frac{1\ \cancel{\text{mole NaOH}}}{1\ \cancel{\text{mole HCl}}} \times \frac{40.0\ \text{g NaOH}}{1\ \cancel{\text{mole NaOH}}}$$

$$= 1.50\ \text{g NaOH}$$

Doing the problem in one step we have:

Grams NaOH

$$= 25.0\ \cancel{\text{ml}} \times \frac{1\ \cancel{\text{liter}}}{1000\ \cancel{\text{ml}}} \times \frac{1.50\ \cancel{\text{mole HCl}}}{1\ \cancel{\text{liter}}} \times \frac{1\ \cancel{\text{mole NaOH}}}{1\ \cancel{\text{mole HCl}}} \times \frac{40.0\ \text{g NaOH}}{1\ \cancel{\text{mole NaOH}}}$$

$$= 1.50\ \text{g NaOH}$$

Example 9-n. How many liters of 0.650 *M* $Pb(NO_3)_2$ solution are required to react completely with 1.55 liters of 2.25 *M* NaCl solution? The balanced equation is:

$$Pb(NO_3)_2(aq) + 2\ NaCl(aq) \longrightarrow PbCl_2(s) + 2\ NaNO_3(aq)$$

Solution. First we determine moles of NaCl present. Using the balanced equation, we then determine moles of $Pb(NO_3)_2$ required, and finally the volume of $Pb(NO_3)_2$ solution required. In one step we have:

Liters of $Pb(NO_3)_2$ solution

$$= 1.55\ \cancel{\text{liters}} \times \frac{2.25\ \cancel{\text{mole NaCl}}}{1\ \cancel{\text{liter}}} \times \frac{1\ \cancel{\text{mole Pb(NO}_3)_2}}{2\ \cancel{\text{mole NaCl}}} \times \frac{1\ \text{liter Pb(NO}_3)_2\ \text{soln}}{0.650\ \cancel{\text{mole Pb(NO}_3)_2}}$$

$$= 2.68\ \text{liters Pb(NO}_3)_2\ \text{solution}$$

Problems (Solution Stoichiometry)

9-26. How many ml of 0.388 *M* H_2SO_4 solution are required to react completely with 0.561 g of KOH? The balanced equation is:

$$H_2SO_4(aq) + 2\,KOH(aq) \longrightarrow K_2SO_4(aq) + 2\,H_2O$$

9-27. How many grams of $BaSO_4$ can be precipitated from solution by adding excess Na_2SO_4 to 75.0 ml of 0.355 *M* $BaCl_2$ solution? The balanced equation is:

$$BaCl_2(aq) + Na_2SO_4(aq) \longrightarrow BaSO_4(s) + 2\,NaCl(aq)$$

9-28. How many ml of 0.565 *M* $CaCl_2$ solution are required to precipitate as AgCl all the silver ion in 0.850 liter of 0.268 *M* $AgNO_3$ solution? The balanced equation is:

$$CaCl_2(aq) + 2\,AgNO_3(aq) \longrightarrow 2\,AgCl(s) + Ca(NO_3)_2(aq)$$

9-29. If 154 ml of 0.200 *M* H_2SO_4 are added to an excess of solid Na_2CO_3, what volume of CO_2 gas is evolved at STP?

9-30. If 25.3 ml of 0.265 *M* $FeCl_3$ solution are mixed with 42.5 ml of 0.515 *M* NaOH solution, determine the maximum mass, in grams, of $Fe(OH)_3$ that will be precipitated.

9-G. Normality

Normality (*N*) is defined as the number of equivalent weights of solute (or equivalents of solute) per liter of solution.

$$\text{Normality} = \frac{\text{Equivalents of solute}}{\text{Liters of solution}}$$

The equivalent weight of a compound depends on the reaction it undergoes, and, further, an equivalent weight of one substance reacts with exactly one equivalent weight of a second substance. We will consider normality in terms of two different kinds of reactions: (1) acid-base, and (2) oxidation-reduction.

(1) Acid-Base Reactions

For acid-base reactions the **equivalent weight** of an acid is that mass in grams that will furnish one mole of H^+ ions. If a sample of an acid contains a mass in grams equal to its equivalent weight, it is said that the sample contains one **equivalent** of the acid. Hence one equivalent of an acid will furnish one Avogadro's number of H^+ ions. The equivalent weight of a base is that mass in grams of the base that will react with one mole of H^+ or one equivalent of acid. An analogous statement would be: one equivalent of a base is that amount that will yield one mole of OH^- ions for reaction.

To illustrate, the equivalent weight of H_2SO_4, assuming reaction of both H^+ ions, is equal to the formula weight/2 since one mole of the acid contains two moles of H^+. For $Al(OH)_3$, assuming reaction of all three OH^- ions, a formula weight, expressed in grams, reacts with three moles of H^+ (or yields three moles of OH^-), and therefore the equivalent weight is the formula weight divided by three.

The equivalent weight is always some simple fraction of the molecular or formula weight (if not equal to it), and the normality of a given solution will always be some simple multiple of the molarity (if not equal to it).

Example 9-o. Determine the normality of a solution of H_2SO_4 that contains 49 g of H_2SO_4 in 2.0 liters of solution. Assume both H^+ ions are available for reaction. The molecular weight of H_2SO_4 is 98.

Solution. First we convert from grams H_2SO_4 to equivalents of H_2SO_4:

$$49\ \cancel{\text{g H}_2\text{SO}_4} \times \frac{1\ \cancel{\text{mole H}_2\text{SO}_4}}{98\ \cancel{\text{g H}_2\text{SO}_4}} \times \frac{2\ \text{equiv H}_2\text{SO}_4}{1\ \cancel{\text{mole H}_2\text{SO}_4}} = 1.0\ \text{equiv H}_2\text{SO}_4$$

Now

$$\text{Normality} = \frac{\text{Equiv H}_2\text{SO}_4}{\text{Liters}} = \frac{1.0\ \text{equiv H}_2\text{SO}_4}{2.0\ \text{liters}}$$

$$= 0.50\ \frac{\text{equiv H}_2\text{SO}_4}{\text{liter}} = 0.50\ N$$

Example 9-p. Calculate the normality of a H_3PO_4 solution that contains 73.5 grams of H_3PO_4 in 0.500 liter of solution. In the reaction to be studied H_3PO_4 is converted to HPO_4^{2-}.

Solution. For this problem, although H_3PO_4 contains three H^+ ions, only two H^+ ions are to be utilized in the reaction. Hence **for this reaction** H_3PO_4 contains two equivalents per mole. The normality of the solution is calculated as follows:

$$N = \frac{73.5 \text{ g } H_3PO_4}{0.500 \text{ liter}} \times \frac{1 \text{ mole } H_3PO_4}{98.0 \text{ g } H_3PO_4} \times \frac{2 \text{ equiv } H_3PO_4}{1 \text{ mole } H_3PO_4}$$

$$= 3.00 \frac{\text{equiv } H_3PO_4}{\text{liter}} = 3.00\, N$$

Considering the above reaction, it might be pertinent to emphasize again that **the equivalent weight of a compound depends on the reaction it undergoes.**

Example 9-q. How many grams of $Ca(OH)_2$ are contained in 0.250 liter of 0.01000 N solution? Assume both OH^- ions undergo reaction. The formula weight of $Ca(OH)_2$ is 74.1.

Solution. In series we convert from volume of solution to equivalents, then moles, and finally grams of $Ca(OH)_2$. Hence

Grams $Ca(OH)_2$

$$= 0.250 \text{ liter} \times \frac{0.01000 \text{ equiv } Ca(OH)_2}{1 \text{ liter}} \times \frac{1 \text{ mole } Ca(OH)_2}{2 \text{ equiv } Ca(OH)_2} \times \frac{74.1 \text{ g } Ca(OH)_2}{1 \text{ mole } Ca(OH)_2}$$

$$= 0.0926 \text{ g } Ca(OH)_2$$

(2) Oxidation-Reduction Reactions

In oxidation-reduction reactions the equivalent weight of an element or compound is defined as the mass in grams of the element or compound required for involvement in the transfer of **one mole of electrons.**

To determine the equivalent weight of a substance in a reaction we must know the particular half-reaction for that substance. Consider the two possible half-reactions listed below for Fe.

$$\text{(a)} \quad Fe \rightarrow Fe^{2+} + 2e^-$$
$$\text{(b)} \quad Fe \rightarrow Fe^{3+} + 3e^-$$

In case (a) one mole of Fe yields two moles of electrons and hence the equivalent weight equals the atomic weight divided by two, or there are two equivalents of Fe per mole for this reaction. However, in (b) one mole of Fe yields three moles of electrons; consequently, the equivalent weight of Fe is equal to the atomic weight divided by three, or in this case there are three equivalents per mole.

For each mole of electrons lost, one mole of electrons must be gained and hence we see that one equivalent of a reducing agent must react with exactly one equivalent of an oxidizing agent and vice versa.

Example 9-r. Consider the following balanced equation:

$$2\,Fe^{3+} + Sn^{2+} \longrightarrow 2\,Fe^{2+} + Sn^{4+}$$

Calculate the normality of a $SnCl_2$ solution (for use in the above reaction) prepared by dissolving 47.4 g $SnCl_2$ in dilute acid solution and diluting with H_2O to a total volume of 2.25 liters.

Solution. We see that the half-reaction for tin is:

$$Sn^{2+} \longrightarrow Sn^{4+} + 2e^-$$

or there are two equivalents per mole for $SnCl_2$. With this knowledge the normality is calculated as follows: we convert from grams to moles, then equivalents, and finally normality for $SnCl_2$:

$$\text{Normality} = \frac{47.4\ \cancel{\text{g SnCl}_2}}{2.25\ \text{liters}} \times \frac{1\ \cancel{\text{mole SnCl}_2}}{189.7\ \cancel{\text{g SnCl}_2}} \times \frac{2\ \text{equiv SnCl}_2}{1\ \cancel{\text{mole SnCl}_2}}$$

$$= 0.222\,\frac{\text{equiv SnCl}_2}{\text{liter}} = 0.222\ N$$

Example 9-s. Consider the following half-reaction for $KMnO_4$:

$$5\,e^- + KMnO_4(aq) + 8\,H^+ \longrightarrow K^+ + Mn^{2+} + 4\,H_2O$$

If a solution of $KMnO_4$, prepared for use as above, is 0.850 N, how many grams of $KMnO_4$ are present in 3.75 liters of this solution?

Solution. According to the balanced half-reaction there are five equivalents of $KMnO_4$ per mole. Converting from volume of solution to equivalents, then moles, and finally grams of $KMnO_4$ we have:

Grams $KMnO_4$

$$= 3.75\ \cancel{\text{liters}} \times \frac{0.850\ \cancel{\text{equiv KMnO}_4}}{1\ \cancel{\text{liter}}} \times \frac{1\ \cancel{\text{mole KMnO}_4}}{5\ \cancel{\text{equiv KMnO}_4}} \times \frac{158\ \text{g KMnO}_4}{1\ \cancel{\text{mole KMnO}_4}}$$

$$= 101\ \text{g KMnO}_4$$

Problems (Normality)

9-31. (a) Determine the normality of a solution that contains 6.86 g of $Ba(OH)_2$ in 0.500 liter of solution. (b) What is the molarity of the solution? Assume reaction of both OH^- ions.

9-32. Assuming reaction of both H^+ ions, 55.0 ml of a 0.275 *N* H_2SO_4 solution would contain how many grams of H_2SO_4?

9-33. How many grams of H_3PO_4 would be needed to prepare 0.750 liter of 0.345 *N* solution assuming the reaction:

$$H_3PO_4 \longrightarrow H_2PO_4^- + H^+?$$

9-34. Consider the reaction in aqueous solution

$$H_2S + Br_2 \longrightarrow 2\,Br^- + 2\,H^+ + S$$

Calculate the normality of a Br_2 solution if 20.0 g of Br_2 are contained in 2.00 liters of solution.

9-35. Consider the oxidation-reduction reaction

$$3\,Fe^{2+} + 4\,H^+ + NO_3^- \longrightarrow 3\,Fe^{3+} + NO + 2\,H_2O$$

(a) How many grams of HNO_3 are contained in 0.750 liter of 1.66 *N* solution? (b) How many moles of HNO_3 are present in (a)?

General Problems

9-36. Determine the molarity of a solution that contains 18.0 g of $H_2C_2O_4$ in 2.00 liters of solution.

9-37. Calculate the molarity of a solution which contains 15.0 g of NaOH in 0.750 liter of solution.

9-38. What mass of $KMnO_4$ will be required to prepare 2.25 liters of 0.148 *M* $KMnO_4$?

9-39. Determine the mass of $MgCl_2 \cdot 6\,H_2O$ required to prepare 0.300 liter of 0.250 *M* solution.

9-40. Calculate the percentage by weight ascorbic acid (vitamin C), if 5.00×10^2 mg of vitamin C is dissolved in 5.00 g of H_2O.

9-41. If 24.0 g of acetic acid ($HC_2H_3O_2$, a component in vinegar) are dissolved in 125 g of H_2O, determine the molality of $HC_2H_3O_2$.

9-42. Determine the mole fractions of solute and solvent in a solution made by dissolving 48.0 g of urea, $CO(NH_2)_2$, in 0.225 kg of H_2O.

9-43. How many grams of selenic acid, H_2SeO_4, are contained in 2.5 liters of 0.20 *N* solution? Assume both protons react.

9-44. How many equivalents of H_3PO_4 are contained in 525 ml of 0.150 *N* solution? Assume only one proton reacts.

9-45. Consider the half-reaction

$$6e^- + Cr_2O_7^{2-} + 14\,H^+ \longrightarrow 2\,Cr^{3+} + 7\,H_2O$$

Calculate the normality of a solution that contains 2.94 g of $K_2Cr_2O_7$ per 0.500 liter of solution.

9-46. How many grams of oxalic acid, $H_2C_2O_4$, are contained in 4.50×10^2 ml of 0.250 *N* $H_2C_2O_4$ solution. Assume the half-reaction

$$H_2C_2O_4 \longrightarrow 2\,CO_2 + 2\,H^+ + 2\,e^-$$

9-47. Find the normality of a solution containing 6.50 g of H_2SO_4 in 1.000 liter of solution. Assume both protons react.

9-48. 39.2 g of H_3PO_4 are dissolved in water and the solution is diluted to 2.00 liters. Calculate (a) the molarity of the solution, and (b) the normality. Assume all three protons react.

9-49. Calculate the mass of NaOH required to prepare 7.50×10^2 g of 1.250 molal solution of NaOH.

9-50. Determine the molarity of a solution containing 12.3 g of $BaCl_2$ in 0.115 liter of solution.

9-51. Calculate the molarity of a solution of NaOH, 0.350 liter of which contains 37.5 g of solute. What is the normality of the same solution?

9-52. 2.50 liters of 0.425 *M* solution of H_2SO_4 contain (a) how many grams of H_2SO_4, and (b) how many moles of H_2SO_4?

9-53. Determine the number of grams of solute in 0.125 liter of each of the following solutions:

(a) 0.516 *M* HCl
(b) 1.62 *M* $Ca(OH)_2$
(c) 0.333 *M* H_2SO_4
(d) 0.950 *M* $MgSO_4 \cdot 7H_2O$
(e) 3.18 *M* H_3PO_4

9-54. To what total volume must each of the following solutions be diluted to give 0.225 *M* solution? (a) 10.0 ml of 10.0 *M* HCl, (b) 25.0 ml of 0.900 *M* HNO_3, (c) 5.00 ml of 6.00 *M* H_2SO_4, and (d) 0.75 ml of 12 *M* NH_3.

9-55. Consider the ionic equation for the reaction between $K_2Cr_2O_7$ and $FeSO_4$:

$$Cr_2O_7^{2-} + 6\,Fe^{2+} + 14\,H^+ \longrightarrow 2\,Cr^{3+} + 6\,Fe^{3+} + 7\,H_2O$$

(a) How many grams of $FeSO_4$ would be required to react with 50.0 ml of 0.100 *M* $K_2Cr_2O_7$ solution? (b) What is the normality of the $K_2Cr_2O_7$ solution? (c) How many grams of $K_2Cr_2O_7$ are available for reaction in (a)?

9-56. How many milliliters of 0.210 M $Pb(NO_3)_2$ solution would be required to precipitate all the I^- in 634 ml of 0.300 M KI solution? The balanced equation is:

$$Pb(NO_3)_2(aq) + 2\,KI(aq) \longrightarrow PbI_2(s) + 2\,KNO_3(aq)$$

9-57. How many grams of $Na_2S_2O_3$ are contained in 0.125 liter of 0.100 N $Na_2S_2O_3$ solution? Assume the half reaction:

$$2\,S_2O_3^{2-} \longrightarrow S_4O_6^{2-} + 2\,e^-$$

9-58. How many grams of Zn metal can be "dissolved" by 25.0 ml of 6.00 M HCl? The balanced equation is:

$$Zn(s) + 2\,HCl(aq) \longrightarrow ZnCl_2(aq) + H_2(g)$$

9-59. The density of a 1.170 M aqueous solution of $ZnSO_4$ is 1.181 g/ml at 15°C. (a) What is the weight percentage of $ZnSO_4$ in the solution? (b) What is the mole fraction of $ZnSO_4$?

9-60. (a) What is the mole fraction of H_2SO_4 in a 10.00 weight percent aqueous solution? (b) If the density of the solution in part (a) is 1.1068 g/ml, what is the molarity of the solution?

9-61. A solution, made by dissolving 20.0 g of pure H_2SO_4 in 80.0 g of water, has a density of 1.143 g/ml. Express the concentration of H_2SO_4 in this solution as (a) weight percentage, (b) molarity, (c) molality, and (d) mole fraction of H_2SO_4.

9-62. Determine the density of an aqueous solution that is 6.00 M and 6.85 molal in HCl.

9-63. What volume of 40.0% HCl (d = 1.20 g/ml) would be required to prepare each of the following solutions: (a) 0.150 liter of 1.25 M solution, (b) 0.100 liter of 4.50 molar solution, and (c) 0.100 kg of 1.30% solution?

9-64. Determine the molality and mole fraction of solute in (a) 11.5 weight percent NaCl solution, (b) a solution of H_2SO_4 (32.0 weight percent) of specific gravity 1.24, and (c) a solution made by dissolving 216 g of CH_3OH in 584 g H_2O.

9-65. What mass of solute must be dissolved in 1.00 kg of 10.0 weight percent solution to yield a 25.0 weight percent solution?

9-66. 54.0 grams of glucose, $C_6H_{12}O_6$, are dissolved in 0.500 kg of a 0.150 molal glucose solution. What are (a) the molality of the resultant solution, and (b) the weight percentage of solute?

9-67. The solubility of $CaCrO_4$ at 100°C is 3.00 weight percent and at 0°C is 12.0 weight percent. If 0.500 kg of solution is saturated at 0°C, what mass of salt will precipitate on heating this solution to 100°C?

9-68. 186 g of $ZnSO_4$ are dissolved in 395 g of H_2O at 15°C to give 0.410 liter of solution. Determine the following: (a) molality, (b) molarity, (c) mole fraction of $ZnSO_4$, (d) density of solution, and (e) weight percentage of $ZnSO_4$.

9-69. A saturated solution of KCl at 20°C contains 296 g KCl per liter of solution. The density is 1.17 g/ml. Calculate the following: (a) weight percentage of KCl, (b) molarity, (c) molality, and (d) mole fraction of KCl. Assume the solvent is H_2O.

9-70. 0.500 liter of 2.50 *M* $HClO_4$ is added to 0.800 liter of 3.75 *M* $HClO_4$. Assuming the total volume of solution after mixing is 1.310 liter, what is the molarity of the resulting solution?

9-71. How many molecules of CO_2 will be generated by the complete reaction of 0.500 liter of 0.750 *M* Na_2CO_3 with excess HCl? The balanced equation is:

$$Na_2CO_3(aq) + 2\,HCl(aq) \longrightarrow 2\,NaCl(aq) + CO_2(g) + H_2O(l)$$

9-72. (a) How many grams of Cu metal can be "dissolved" by 50.0 ml of 12.0 *N* HNO_3? The balanced equation is:

$$3\,Cu + 2\,NO_3^- + 8\,H^+ \longrightarrow 3\,Cu^{2+} + 2\,NO + 4\,H_2O$$

(b) What is the molarity of the HNO_3 solution in (a)?

Chapter 10
Titrations

Frequently it is convenient to determine the concentration of a solution by allowing it to react with a second solution of known concentration. A given volume of the first solution is measured out and the second solution is added from a buret until equivalent amounts of the two solutions are present. The **equivalence point** or **end point** of the reaction may be determined by using an indicator. The process is termed **titration**.

Suppose solution A is being titrated with solution B. The end point of the titration occurs when stoichiometric or equivalent quantities of A and B are present; that is, once the reaction is defined,

$$\text{Equiv of } A = \text{Equiv of } B \tag{1}$$

Since Normality = Equiv/Liters (see Chapter 9), we may substitute in Equation (1) and obtain

$$\text{Normality of } A \times \text{Liters of } A = \text{Normality of } B \times \text{Liters of } B$$

or

$$\text{Normality of } A \times \text{Volume of } A = \text{Normality of } B \times \text{Volume of } B \tag{2}$$

Any units of volume may be used as long as they are the same for both reagents. Equation (2) is very useful in solving titration problems for either unknown volumes or concentrations.

In situations where a mass (instead of a volume) of reagent A is titrated with a solution of reagent B, we relate equivalents of B delivered to equivalents of A by the 1 : 1 relationship between equivalents. Since the reaction has been defined, we can then easily convert from equivalents of A to mass of A.

Calculations involving titrations are conveniently classified into two categories: (a) those that involve oxidation and reduction, and (b) those that do not (mainly acid-base). The need for this division, to repeat, arises because the equivalent weight of a reactant or product depends on the reaction. For acid-base reactions, equivalent weights are defined as discussed in Section 9-G-1. Equivalent weights for oxidation-reduction substances are defined in Section 9-G-2. Once the equivalent weights are known, the calculations for both types of reactions parallel each other.

Finally, it is important to understand that any of the following examples or problems can be worked on the **mole** instead of the **equivalent** basis. If the mole basis is used, the relationship between moles of substance A and moles of substance B is dependent on the coefficients for A and B in the balanced equation—that is, the mole relationship between A and B is not necessarily 1 : 1. (Examples of problems worked on the mole basis were discussed in Section 9-F, Solution Stoichiometry.) In this chapter we have chosen to emphasize in our examples the **equivalent** relationship between substances A and B. This was done because, **once the reaction is defined, one equivalent** of substance A will always react with or produce **one equivalent** of substance B—that is, the reaction is always 1 : 1 on an equivalent basis. Hence five equivalents of A react with or produce five equivalents of B; 0.235 equivalents of A react with or produce 0.235 equivalents of B. This relationship generally makes the "numbers" easier in the required arithmetic of the chemical calculations.

10-A. Titrations (Acid-Base Reactions)

As stated in Section 9-G-1, the equivalent weight of an acid (or base) is that number of grams that provides (or reacts with) one mole of H^+ ions.

Most volumes used in titrations are small and best expressed in milliliters instead of liters. Let's look at an expression for normality using milliliters as volume and a correspondingly smaller unit for the amount of substance present, that is, milliequivalents (meq) of solute as compared to equivalents. Recalling the definition of the prefix milli, we see that 1000 meq = 1 equiv. If a solution is 0.10 N, this imples

$$\frac{0.10\ \cancel{\text{equiv}}}{1\ \cancel{\text{liter}}} \times \frac{1\ \cancel{\text{liter}}}{1000\ \text{ml}} \times \frac{1000\ \text{meq}}{1\ \cancel{\text{equiv}}} = \frac{0.10\ \text{meq}}{1\ \text{ml}}$$

Thus **N is numerically the same whether expressed as equiv/liter or meq/ml.** We make use of this identity in the examples that follow. Remember, one meq of *A* will react with or produce 1 meq of *B*.

Example 10-a. What volume of 0.100 *N* H_2SO_4 would be required to neutralize 50.0 ml of 0.0500 *N* NaOH? Assume both of the H^+ ions in H_2SO_4 react.

Solution. Using dimensional analysis, we have:

$$\text{Volume } H_2SO_4 = 50.0\ \cancel{\text{ml NaOH}} \times \frac{0.0500\ \cancel{\text{meq NaOH}}}{1\ \cancel{\text{ml NaOH}}} \times \frac{1\ \cancel{\text{meq } H_2SO_4}}{1\ \cancel{\text{meq NaOH}}}$$

$$\times \frac{1\ \text{ml } H_2SO_4}{0.100\ \cancel{\text{meq } H_2SO_4}}$$

$$= 25.0\ \text{ml } H_2SO_4 \text{ solution}$$

or using Equation (2)

Volume of H_2SO_4 × Normality of H_2SO_4 = Volume of NaOH × Normality of NaOH

Volume of $H_2SO_4 \times 0.100$ meq/ml = 50.0 ml × 0.0500 meq/ml

$$\text{Volume of } H_2SO_4 = \frac{50.0\ \text{ml} \times 0.0500\ \text{meq/ml}}{0.100\ \text{meq/ml}} = 25.0\ \text{ml}$$

Example 10-b. 35 ml of standard base (0.25 *N*) is used to neutralize 15 ml of a solution of an unknown acid. What is the concentration of the unknown acid?

Solution. Let A = acid and B = base.

$$N_A = 35\ \cancel{\text{ml } B} \times \frac{0.25\ \cancel{\text{meq } B}}{1\ \cancel{\text{ml } B}} \times \frac{1\ \text{meq } A}{1\ \cancel{\text{meq } B}} \times \frac{1}{15\ \text{ml } A}$$

$$= \frac{8.75\ \text{meq } A}{15\ \text{ml}}$$

$$= 0.58\ \frac{\text{meq } A}{\text{ml}} = 0.58\ N$$

Example 10-c. In the Kjeldahl method for determination of nitrogen, the sample is digested with hot concentrated H_2SO_4 to oxidize organic matter and convert nitrogen present to NH_4^+. NaOH is then added and (1) NH_3 is distilled from the solution, (2) NH_3 is absorbed in a measured excess of a standard solution of an acid, and (3) the acid in excess of that neutralized by NH_3 is then titrated with a standard basic solution. These reactions are illustrated as follows:

$$(1) \quad NH_4^+ + OH^- \xrightarrow{\Delta} NH_3\uparrow + H_2O$$

$$(2) \quad NH_3 + H^+ \longrightarrow NH_4^+$$

$$(3) \quad H^+(\text{excess}) + OH^- \longrightarrow H_2O$$

From these data the mass of NH_3 can be calculated, and from this information the percentage of nitrogen in the original sample can be ascertained. Study the following problem and its solution.

The NH_3 obtained from a 1.40 g sample of a protein is absorbed in 45.0 ml of 0.400 *N* HNO_3 solution. Exactly 20.0 ml of 0.150 *N* NaOH solution is required to neutralize the excess acid. What is the percentage of nitrogen (by weight) in the protein sample?

Solution.

$$\text{Meq NaOH delivered} = 20.0\,\cancel{\text{ml}} \times \frac{0.150 \text{ meq NaOH}}{1\,\cancel{\text{ml}}} = 3.00 \text{ meq NaOH}$$

$$\text{Meq HNO}_3 \text{ (total)} = 45.0\,\cancel{\text{ml}} \times \frac{0.400 \text{ meq HNO}_3}{1\,\cancel{\text{ml}}} = 18.0 \text{ meq HNO}_3$$

$$\text{Meq NH}_3 \text{ distilled} = 18.0 - 3.00 = 15.0 \text{ meq NH}_3$$

$$\text{Grams } N = 15.0\,\cancel{\text{meq NH}_3} \times \frac{1\,\cancel{\text{equiv NH}_3}}{1000\,\cancel{\text{meq NH}_3}} \times \frac{17.0\,\cancel{\text{g NH}_3}}{1\,\cancel{\text{equiv NH}_3}} \times \frac{14.0 \text{ g N}}{17.0\,\cancel{\text{g NH}_3}}$$

$$= 0.210 \text{ gram Nitrogen}$$

$$\% \text{N} = \frac{\text{Grams N}}{\text{Grams sample}} \times 100 = \frac{0.210 \text{ g N}}{1.40 \text{ g sample}} \times 100 = 15.0\% \text{N}$$

Problems (Acid-Base Reactions)

10-1. 25.0 ml of an unknown acid solution are titrated with 40.0 ml of NaOH solution that has a normality of 0.750 equiv/liter. Determine the normality of the unknown acid solution.

10-2. How many milliliters of 0.18 *N* NaOH are required to neutralize 35 ml of 0.24 *N* H_2SO_4? (Assume reaction of both protons in H_2SO_4.)

10-3. How many milliliters of 0.300 *M* NaOH are required to neutralize 50.0 ml of 0.150 *M* H_2SO_4? (Assume reaction of both protons in H_2SO_4.)

10-4. A 25.0 ml sample of a solution of H_2SO_4 requires 38.4 ml of 0.195 *N* NaOH for neutralization. (a) Calculate the normality of the H_2SO_4 solution. (b) What is its molarity? (c) How many grams of H_2SO_4 are contained in the 25.0 ml? (Assume reaction of both protons in H_2SO_4.)

10-5. Assuming the reaction $H_3PO_4 + 2\,OH^- \rightarrow HPO_4^{2-} + 2\,H_2O$, a sample containing 1.47 g of H_3PO_4 required 40.0 ml of base solution for neutralization. Calculate the normality of the base solution.

10-6. A series of Kjeldahl determinations gave the following data. In each case determine the percentage of nitrogen in the sample (see Example 10-c.)

Sample	Mass of Sample	Volume of 0.1000 *N* HNO_3 Used	Volume of 0.1000 *N* NaOH to Neutralize Excess Acid
A	1.25 g	100.0 ml	50.0 ml
B	0.245	50.0	40.0
C	0.995	67.00	22.0
D	0.218	25.00	13.00

10-7. The equivalent weights of organic acids are often determined by titration with a standard solution of base. Determine the equivalent weight of benzoic acid if 0.915 g of the latter required 25.0 ml of 0.300 *N* NaOH for neutralization.

10-8. Determine the equivalent weight of succinic acid if 0.553 g requires 44.4 ml of 0.211 *N* base for neutralization.

10-9. An impure sample of aspirin, $CH_3OCOC_6H_4COOH$ (one acidic hydrogen), is titrated with 0.250 *M* NaOH solution to determine the % purity (by weight). If a 2.35 g sample required 36.4 ml of NaOH for neutralization, calculate the % purity (by weight) of the aspirin sample.

10-B. Titrations (Oxidation-Reduction)

These problems are worked similarly to those in Section 10-A. Here, however, the equivalent weight of a substance is related to electron exchange capabilities and not acid-base properties. Recall (Section 9-G-2) that the equivalent weight of a substance involved in oxidation/reduction is defined as that mass that reacts with or supplies one Avogadro's number of electrons (also called one equivalent or 1 mole of electrons). The equivalent weight is therefore determined by the loss or gain of electrons per formula unit; for example, when permanganate ion, MnO_4^-, is used as an oxidizing agent in acid solution, it is reduced to Mn^{2+}, which requires 5 electrons per MnO_4^- ion. If $KMnO_4$ is used as the source of MnO_4^-, then 1 mole of $KMnO_4$ would accept 5 moles of electrons, and an equivalent weight would be equal to the formula weight divided by 5, or in this case, there would be five equivalents of $KMnO_4$ per mole. When MnO_4^-, however, is used for oxidation in basic solution, it is reduced to MnO_2, and only 3 electrons are required per formula unit. The equivalent weight in this case is the formula weight divided by 3, or three equivalents of $KMnO_4$ per mole.

Example 10-d. Arsenites react with permanganates in acid solution as shown by

$$5\,AsO_3^{3-} + 6\,H^+ + 2\,MnO_4^- \longrightarrow 5\,AsO_4^{3-} + 2\,Mn^{2+} + 3\,H_2O$$

If 56 ml of 0.10 *N* $KMnO_4$ react with 28 ml of Na_3AsO_3 solution, (a) what is the normality of the sodium arsenite solution, (b) what mass of Na_3AsO_3 in grams is contained in the 28 ml of solution?

Solution.

(a) $$N_{AsO_3^{3-}} = 56\,\cancel{\text{ml } MnO_4^-} \times \frac{0.10\,\cancel{\text{meq } MnO_4^-}}{1\,\cancel{\text{ml } MnO_4^-}} \times \frac{1\text{ meq } AsO_3^{3-}}{1\,\cancel{\text{meq } MnO_4^-}} \times \frac{1}{28\text{ ml } AsO_3^{3-}}$$

$$= 0.20\text{ meq } AsO_3^{3-}/\text{ml} = 0.20\,N$$

(b) In oxidation to AsO_4^{3-} each mole of AsO_3^{3-} gives up **two** moles of electrons—that is, there are two equivalents of Na_3AsO_3 per mole. This same information can be obtained by noting that in going from AsO_3^{3-} to AsO_4^{3-} the oxidation number of As increases from +3 to +5, again indicating **two** moles of electrons per mole of Na_3AsO_3. Hence grams of Na_3AsO_3 present are:

$$0.028\,\cancel{\text{liter}} \times \frac{0.20\,\cancel{\text{equiv } Na_3AsO_3}}{1\,\cancel{\text{liter}}} \times \frac{1\,\cancel{\text{mole } Na_3AsO_3}}{2\,\cancel{\text{equiv } Na_3AsO_3}} \times \frac{192\text{ g } Na_3AsO_3}{1\,\cancel{\text{mole } Na_3AsO_3}}$$

$$= 0.54\text{ g } Na_3AsO_3$$

Example 10-e. Exactly 2 g of a sample of iron ore, after conversion of all Fe^{3+} to Fe^{2+}, are titrated with potassium dichromate solution. If 53.5 ml of 0.400 *N* potassium dichromate solution are required, what is the percentage of Fe in the original sample? The equation for the reaction during titration is

$$6\,Fe^{2+} + Cr_2O_7^{2-} + 14\,H^+ \longrightarrow 6\,Fe^{3+} + 2\,Cr^{3+} + 7\,H_2O$$

Solution. There is one equivalent per mole of iron since one mole of electrons is given up for each mole of Fe^{2+} oxidized to Fe^{3+}. Hence grams of iron titrated are:

$$\text{Mass Fe} = 0.0535 \cancel{\text{liter}} \times \frac{0.400 \cancel{\text{equiv } Cr_2O_7^{2-}}}{1 \cancel{\text{liter}}} \times \frac{1 \cancel{\text{equiv iron}}}{1 \cancel{\text{equiv } Cr_2O_7^{2-}}} \times \frac{55.8 \text{ g iron}}{1 \cancel{\text{equiv iron}}}$$

$$= 1.19 \text{ g iron}$$

$$\text{Percentage Fe in original sample} = \frac{\text{Mass of Fe}}{\text{Mass of sample}} \times 100$$

$$= \frac{1.19 \text{ g}}{2.00 \text{ g}} \times 100 = 59.5\%$$

Problems (Oxidation-Reduction)

10-10. Consider the reaction in aqueous solution

$$H_2S + I_2 \longrightarrow 2\,I^- + 2\,H^+ + S$$

(a) How many grams of I_2 are contained in 0.750 liter of a solution that is 0.330 N? (b) What volume of 0.330 N I_2 solution would be required to react with 0.225 g of H_2S?

10-11. Consider the reaction

$$Sn^{2+} + 2\,Ce^{4+} \longrightarrow Sn^{4+} + 2\,Ce^{3+}$$

What volume of 0.333 N $Ce(SO_4)_2$ would be required to titrate a sample containing 0.950 g of $SnCl_2$?

10-12. Ceric ion, Ce^{4+}, may be used to titrate reducing solutions such as Fe^{2+}.

$$Fe^{2+} + Ce^{4+} \longrightarrow Fe^{3+} + Ce^{3+}$$

A 0.186 g sample of an iron ore is dissolved in acid and all the Fe^{3+} reduced to the Fe^{2+} state. 48.7 ml of 0.0200 N Ce^{4+} solution are required to reoxidize all the Fe^{2+}. Calculate the percentage of Fe in the sample.

10-13. The copper in a 2.00 g sample of a copper mineral was determined quantitatively by the following reactions:

$$2\,Cu^{2+} + 4\,I^- \longrightarrow 2\,CuI + I_2$$

$$I_2 + 2\,S_2O_3^{2-} \longrightarrow S_4O_6^{2-} + 2\,I^-$$

(a) How many grams of Cu^{2+} are there in 1.00 equiv for the first reaction? (b) What is the equivalent weight of $Na_2S_2O_3$? (c) If 47.5 ml of 0.250 N $Na_2S_2O_3$ were used in the titration, calculate the percentage of Cu in the sample.

General Problems

10-14. A solution of KOH is standardized by titration of an aqueous solution containing a known mass of oxalic acid, $H_2C_2O_4 \cdot 2H_2O$. If 20.0 ml of KOH solution are required to neutralize 0.252 g of $H_2C_2O_4 \cdot 2H_2O$, find the normality of the KOH solution. (Assume reaction of both protons in oxalic acid.)

10-15. 37.45 ml of a solution of 0.1504 *M* HCl are required to neutralize 25.00 ml of a solution of NaOH. What is the normality of NaOH?

10-16. Find the volume of 0.3075 *N* H_2SO_4 required to neutralize 75.00 ml of 0.1217 *N* KOH solution. Assume the reaction

$$KOH + H_2SO_4 \longrightarrow KHSO_4 + H_2O$$

10-17. To what volume must 25.0 ml of 0.650 *N* NaOH be diluted to yield a 0.400 *M* solution?

10-18. If 50.0 ml of 0.150 *N* acid require 37.5 ml of basic solution for neutralization, determine the normality of the base.

10-19. A solution of NaOH was standardized by titration of solid oxalic acid, $H_2C_2O_4 \cdot 2H_2O$. If 45.6 ml of NaOH solution were required to neutralize 1.45 g of the solid acid, calculate the normality of the NaOH. (Assume reaction of both protons in oxalic acid.)

10-20. If 40.4 ml of HCl solution are required to neutralize 0.168 g of pure KOH, what is the molarity of the HCl solution?

10-21. 6.30 g of $Ba(OH)_2 \cdot 8\,H_2O$ are neutralized by 80.0 ml of a solution of H_2SO_4. What is the normality of the H_2SO_4 solution? Assume complete neutralization of both species.

10-22. What volume of 0.450 *N* HCl is required to react completely with 2.12 g of solid Na_2CO_3 according to the equation

$$Na_2CO_3 + 2\,HCl \longrightarrow 2\,NaCl + H_2O + CO_2$$

10-23. Calculate the volume of CO_2 produced at STP when 75.0 ml of 0.225 *N* H_2SO_4 act on K_2CO_3 according to the equation

$$K_2CO_3 + H_2SO_4 \longrightarrow K_2SO_4 + CO_2 + H_2O$$

10-24. If a 7.50 ml sample of vinegar required 37.5 ml of 0.127 *N* base for neutralization, (a) what is the normality of the acid in the vinegar? (b) Assuming that the acidity is due to acetic acid, $HC_2H_3O_2$, determine the percentage by weight of acetic acid if the density of the vinegar is 1.00 g/ml.

10-25. A 5.10 g sample of household ammonia, NH_3 in H_2O, requires 175 ml of 0.241 N H_2SO_4 for neutralization. (a) If the density of the NH_3 solution is 0.960 g/ml, what is the normality of the solution? (b) Calculate the weight percentage of NH_3 in the solution. (Assume reaction of both protons in H_2SO_4.)

10-26. 0.300 g of tartaric acid is dissolved in 60.0 ml of 0.100 N NaOH. The resulting solution requires 0.100 liter of 2.00×10^{-2} N HCl for neutralization. Determine the equivalent weight of tartaric acid without looking up the formula.

10-27. 25.0 ml of an $H_2C_2O_4$ solution require 47.6 ml of 0.105 N $KMnO_4$ in a titration.

$$5\,H_2C_2O_4 + 6\,H^+ + 2\,MnO_4^- \longrightarrow 10\,CO_2 + 2\,Mn^{2+} + 8\,H_2O$$

(a) How many equivalents of $KMnO_4$ are used? (b) How many equivalents of $H_2C_2O_4$ undergo reaction? (c) What is the normality of the $H_2C_2O_4$ solution?

10-28. In a certain redox reaction 31.33 ml of 0.1752 N oxidizing agent are used. (a) How many equivalents of oxidizing agent were used? (b) How many equivalents of reducing agent must be used?

10-29. I_2 may be titrated with sodium thiosulfate, $Na_2S_2O_3$, according to the equation

$$I_2 + 2\,Na_2S_2O_3 \longrightarrow 2\,NaI + Na_2S_4O_6 \text{ (sodium tetrathionate)}$$

If 40.00 ml of $Na_2S_2O_3$ solution are required to react with 1.904 g of I_2, find the normality of the $Na_2S_2O_3$ solution.

10-30. A solution of $KMnO_4$ was standardized by weighing out 0.268 g of $Na_2C_2O_4$ and titrating. 18.5 milliliters of $KMnO_4$ were required. (a) How many meq of $KMnO_4$ were used in the titration where the chemical reaction is illustrated by the equation

$$2\,KMnO_4 + 5\,Na_2C_2O_4 + 8\,H_2SO_4 \longrightarrow 2\,MnSO_4 + 5\,Na_2SO_4 + K_2SO_4 + 8\,H_2O + 10\,CO_2$$

(b) What is the normality of the $KMnO_4$ solution?

10-31. Hypochlorite ion, OCl^-, is the active agent in household chlorine bleaches. OCl^- reacts with iodide ion in the following way

$$OCl^- + 2\,I^- + 2\,H^+ \longrightarrow Cl^- + I_2 + H_2O$$

If the I_2 generated is titrated with standard thiosulfate ion,

$$2\,S_2O_3^{2-} + I_2 \longrightarrow S_4O_6^{2-} + 2\,I^-$$

the concentration of OCl^- in the bleach solution can easily be calculated.

(a) If 46.0 ml of 0.500 N $S_2O_3^{2-}$ solution is required to titrate the I_2 generated from a 25.0 ml sample of household bleach solution, what is the normality of OCl^- in the solution? (b) What is the molarity of OCl^-?

10-32. Vitamin C, ascorbic acid, has the molecular formula $C_6H_8O_6$. It can easily be oxidized according to the half-reaction:

$$C_6H_8O_6 \longrightarrow C_6H_6O_6 + 2\,H^+ + 2\,e^-$$

The ascorbic acid content in a vitamin C tablet can be determined by adding a measured excess of standard I_2 solution to the prepared tablet and then titrating the excess I_2 with standard thiosulfate solution. The equations are:

$$C_6H_8O_6 + \underset{\text{(excess)}}{I_2} \longrightarrow C_6H_6O_6 + 2\,H^+ + 2\,I^-$$

$$I_2 + 2\,S_2O_3^{2-} \longrightarrow 2\,I^- + S_4O_6^{2-}$$

A 50.0 ml sample of 0.100 N I_2 was added to a sample of vitamin C. After reaction, 19.3 ml of 0.200 N $S_2O_3^{2-}$ were required to titrate the excess I_2. Calculate the grams of vitamin C present in the sample.

10-33. Arsenic in a sample may be determined through the following sequence of reactions:

$$As + 5\,HNO_3 \longrightarrow H_3AsO_4 + 5\,NO_2 + H_2O$$

$$H_3AsO_4 + 2\,HI \longrightarrow H_3AsO_3 + I_2 + H_2O$$

$$I_2 + 2\,Na_2S_2O_3 \longrightarrow 2\,NaI + Na_2S_4O_6$$

If 1.50 g of a sample of the As-containing substance ultimately required 31.2 ml of 0.325 N $Na_2S_2O_3$ to titrate to the equivalence point, determine the percentage of As in the sample.

10.34. In a procedure for its quantitative determination, Ca^{2+} may be precipitated as CaC_2O_4, as shown by the equation $Ca^{2+} + C_2O_4^{2-} \rightarrow CaC_2O_4$. The precipitate is filtered, suspended in H_2O, dissolved in acid, and titrated with $KMnO_4$ solution of known concentration. In a particular analysis, 5.00 g of a sample of limestone-bearing rock are dissolved in acid and then Ca^{2+} is precipitated as CaC_2O_4. 50.0 milliliters of 0.400 N $KMnO_4$ are used to titrate the CaC_2O_4. The overall reaction may be represented as

$$5\,CaC_2O_4 + 2\,KMnO_4 + 8\,H_2SO_4 \longrightarrow 5\,CaSO_4 + K_2SO_4 + 10\,CO_2 + 8\,H_2O + 2\,MnSO_4$$

Calculate the percentage of calcium in the rock.

Chapter 11

Colligative Properties of Solutions

Properties of solutions that depend on the **number** of dissolved particles and not on their kind are termed **colligative** properties. Such properties include vapor pressures, freezing and boiling points, and osmotic pressures. These properties are useful for the experimental determination of molecular weights of dissolved substances. Freezing points are particularly useful since they may be determined with considerable precision. Solutions of electrolytes exhibit abnormal colligative properties and are discussed near the end of the chapter. The discussion in Sections A, B, and C applies to solutions of nonelectrolytes.

11-A. Vapor Pressures

The vapor pressure of a solution of a **nonvolatile** solute in a liquid is always less than the vapor pressure of pure solvent. In accordance with Raoult's law, the lowering of the vapor pressure, ΔP, is proportional to the mole fraction of solute:

$$\Delta P = P_0 - P = P_0 X_2 \tag{1}$$

where P_0 is the vapor pressure of pure solvent, P is the vapor pressure of the solution, and X_2 is the mole fraction of solute. **Alternatively**, the vapor pressure of the solution is proportional to the mole fraction of solvent:

$$P = P_0 X_1 \tag{2}$$

where X_1 is the mole fraction of solvent.

Example 11-a. The vapor pressure of water at 20°C is 17.5 torr. Determine the vapor pressure of a solution prepared by dissolving 12.0 g of urea, $CO(NH_2)_2$, in 0.500 kg of H_2O.

Solution. To determine mole fraction of solvent we must first determine moles of both solute and solvent present.

$$\text{Moles urea} = 12.0\ \cancel{\text{g urea}} \times \frac{1\ \text{mole urea}}{60.0\ \cancel{\text{g urea}}} = 0.200\ \text{mole urea}$$

$$\text{Moles } H_2O = 5.00 \times 10^2\ \cancel{\text{g } H_2O} \times \frac{1\ \text{mole } H_2O}{18.0\ \cancel{\text{g } H_2O}} = 27.8\ \text{moles } H_2O$$

$$\text{Mole fraction solvent} = X_1 = \frac{\text{Moles solvent}}{\text{Moles solvent} + \text{Moles solute}}$$

$$= \frac{27.8}{27.8 + 0.200} = 0.993$$

Substituting into formula (2) above, $P = P_0 X_1$

$$\text{Vapor pressure of the solution} = (17.5\ \text{torr})(0.993)$$
$$= 17.4\ \text{torr}$$

Example 11-b. The vapor pressure of benzene, C_6H_6, at 30°C is 121.8 torr. If 10.0 g of a nonvolatile solute dissolved in 1.00×10^2 g of benzene lowers the vapor pressure 8.8 torr, what is the molecular weight of the solute?

Solution. Calculate the mole fraction of the solute by substituting in equation (1).

$$8.8\ \text{torr} = (121.8\ \text{torr})(X_2)$$

$$X_2 = \frac{8.8\ \cancel{\text{torr}}}{121.8\ \cancel{\text{torr}}} = 0.072$$

$$X_2 = \frac{\text{Moles solute}}{\text{Moles solute} + \text{Moles solvent}}$$

$$\text{Moles solvent} = 1.00 \times 10^2\ \cancel{\text{g benzene}} \times \frac{1\ \text{mole benzene}}{78.0\ \cancel{\text{g benzene}}} = 1.28\ \text{moles benzene}$$

$$X_2 = 0.072 = \frac{\text{Moles solute}}{\text{Moles solute} + 1.28} = \frac{n}{n + 1.28}$$

$$n = 0.072n + (0.072)(1.28)$$
$$0.928n = 0.092$$

$$\text{Moles solute} = n = 0.099\ \text{mole}$$

$$\text{Molecular weight} = \frac{10.0\ \text{g}}{0.099\ \text{mole}} = 101\ \text{g/mole}$$

Problems (Vapor Pressures)

11-1. The vapor pressure of H_2O at 25°C is 23.76 torr. Determine the vapor pressure of a solution of 84.0 g of glucose, $C_6H_{12}O_6$, in 250.0 g of H_2O.
11-2. What is the vapor pressure of a 23.5% aqueous solution of urea, $CO(NH_2)_2$, at 25°C if the vapor pressure of pure water is 23.76 torr at this temperature?
11-3. The vapor pressure of benzene at 75°C is 640.0 torr. A solution of 7.95 g of a solute in 110.5 g of benzene has a vapor pressure of 595.0 torr at 75°C. Calculate the molecular weight of the solute.
11-4. The vapor pressure of water at 40°C is 55.32 torr. What mass of sucrose (MW = 342.3) must be dissolved in 175.0 g of H_2O at 40°C to lower the vapor pressure to 53.95 torr?

11-B. Freezing Points and Boiling Points of Solutions of Nonelectrolytes

If 1 mole of a nonvolatile nonelectrolyte is dissolved in 1000 g of a solvent (this would be a 1 molal solution), the freezing point is depressed a constant amount below the freezing point of the pure solvent. It makes no difference what the nonelectrolyte is. The effect produced by 1 mole of solute in a kilogram of a given solvent is termed the molal freezing point depression constant for that particular solvent. For water the constant is 1.86°C per mole of solute per kilogram of solvent. This is written as

$$1.86\,\frac{°\text{C}}{\text{Mole solute/Kg solvent}} = 1.86\,\frac{°\text{C}}{m}$$

where m represents the units of molality. Consequently a 1 molal aqueous solution of a nonvolatile, nonionized solute freezes at −1.86°C and a 2 molal solution freezes at −3.72°C. In other words,

$$\Delta T = K_f \times \text{Molality}$$

where ΔT is the change in freezing point temperature and K_f is the freezing point depression constant.

In a similar manner, a solute elevates the boiling point of a solution. A solution of 1 mole of a nonvolatile nonelectrolyte in 1000 g of water boils at a temperature of 100.52°C at standard pressure. The boiling point has been elevated 0.52°C for a 1 molal solution. The molal boiling point elevation constant for water is therefore 0.52°C/(mole solute/kg solvent) or 0.52(°C/m). Other solvents have different numerical values for this constant. The boiling point elevation is a result of a lowering of the vapor pressure of the solvent by the dissolved substance. As a nonvolatile solute is added to a solvent, the vapor pressure of the solution decreases in accordance with Raoult's law. Therefore the temperature at which the vapor pressure equals the external pressure (boiling point) is increased.

What has been said about the effect of a nonvolatile solute on the boiling point does not hold for a volatile solute. It will be recalled that a liquid boils when the vapor pressure of the liquid becomes equal to the external pressure (usually atmospheric). For a solution of a volatile solute, both solute and solvent exert a pressure at a given temperature. Hence the solution boils when the combination of the two pressures equals the external pressure. The boiling point of such a solution may be either lower or higher than the boiling point of the solvent.

Since one molecular weight (expressed in grams) of a nonvolatile nonelectrolyte dissolved in 1 kg of solvent produces a known effect on the boiling and freezing point of any particular solvent, the molecular weight of the solute can be obtained by determining the boiling or freezing points of solutions with known quantities of solute and solvent. How this is done will be evident from a study of the following problem examples.

Example 11-c. If 20.0 g of the nonelectrolyte urea, $CO(NH_2)_2$, is dissolved in 2.50×10^2 g of H_2O, (a) what is the freezing point of the solution, and (b) what is the boiling point at standard pressure? The molecular weight of urea is 60.0.

Solution. (a) First, determine the molality; next, obtain the change in temperature of the freezing point by multiplying the molality by the molal freezing point depression constant. Subtract the change in temperature from the normal freezing point of the solvent to obtain the freezing point of the solution.

$$\text{Moles } CO(NH_2)_2 = 20.0\ \cancel{\text{g urea}} \times \frac{1 \text{ mole urea}}{60.0\ \cancel{\text{g urea}}} = 0.333 \text{ mole urea}$$

$$\text{Molality urea} = \frac{0.333 \text{ mole urea}}{0.250 \text{ kg } H_2O} = 1.33\ m$$

$$\Delta T = \frac{1.86°C}{\cancel{m}} \times 1.33\ \cancel{m} = 2.47°C$$

Freezing point $= 0.00 - 2.47°C = -2.47°C$

(b) The boiling point of the solution is obtained in a manner similar to the freezing point.

$$\Delta T = \frac{0.52°C}{\cancel{m}} \times 1.33\ \cancel{m} = 0.69°C$$

Boiling point $= 100.00 + 0.69 = 100.69°C$

Example 11-d. A solution of 10.0 g of A in 0.100 kg of H_2O boils at 100.78°C at standard pressure. What is the molecular weight of A if it is a nonvolatile nonelectrolyte?

Solution. First, calculate the molality from the boiling point elevation, and then, from the masses of solute and solvent, calculate the molecular weight.

$$\Delta T = 100.78 - 100.00 = 0.78°\text{C}$$

$$\Delta T = \frac{0.52°\text{C}}{m} \times \text{Molality}$$

$$\text{Molality} = \frac{\Delta T m}{0.52°\text{C}} = \frac{0.78°\cancel{\text{C}}m}{0.52°\cancel{\text{C}}} = 1.5\ m$$

$$\text{Moles solute} = 0.100\ \cancel{\text{kg H}_2\text{O}} \times \frac{1.5\ \text{moles solute}}{1\ \cancel{\text{kg H}_2\text{O}}} = 0.15\ \text{mole solute}$$

$$\text{MW} = \frac{\text{Grams solute}}{\text{Moles solute}} = \frac{10.0\ \text{g solute}}{0.15\ \text{mole solute}} = 67\,\frac{\text{g}}{\text{mole}}$$

Hence the molecular weight is 67.

Example 11-e. What mass of sugar (MW = 342) must be dissolved in 4.00 kg of H_2O to yield a solution that will freeze at −3.72°C?

Solution. First calculate the molality of a solution that will freeze at −3.72°C. Then calculate the mass of solute needed for 4.00 kg of solvent.

$$\Delta T = \frac{1.86°\text{C}}{m} \times \text{Molality}$$

$$\text{Molality} = \frac{\Delta T m}{1.86°\text{C}} = \frac{3.72°\cancel{\text{C}}m}{1.86°\cancel{\text{C}}} = 2.00\ m$$

$$\text{Grams sugar} = 4.00\ \cancel{\text{kg H}_2\text{O}} \times \frac{2.00\ \cancel{\text{moles sugar}}}{1\ \cancel{\text{kg H}_2\text{O}}} \times \frac{342\ \text{g sugar}}{1\ \cancel{\text{mole sugar}}}$$
$$= 2.74 \times 10^3\ \text{g sugar}$$

Problems (Solutions of Nonelectrolytes)

11-5. What is the freezing point of a solution prepared by dissolving 25.0 g of thioacetamide, CH_3CSNH_2, in 333 g of H_2O?

11-6. What is the boiling point at 760 torr of the solution in Problem 11-5?

11-7. Calculate the boiling point at standard pressure of a solution made by dissolving 75.0 g of urea, $CO(NH_2)_2$ in 0.500 kg of water.

11-8. Determine the freezing point of an automobile radiator solution if the radiator contains 50.0% (by weight) ethylene glycol (antifreeze), $C_2H_4(OH)_2$, in H_2O.

11-9. Exactly 21 grams of a nonelectrolyte are dissolved in 0.200 kg of water. The resulting solution is found to freeze at −3.72°C. What is the molecular weight of the solute?

11-10. Calculate the freezing point of a solution that contains 92.1 g of the nonelectrolyte glycerin, $C_3H_8O_3$, dissolved in 333 g of water.

11-11. A solution of 84.0 g of a nonelectrolyte in 0.400 kg H_2O freezes at −2.79°C. What is the molecular weight of the solute?

11-12. An aqueous solution of $CO(NH_2)_2$ freezes at −3.15°C. What is the molality of the solution?

11-C. Osmotic Pressure

If a solution is separated from pure solvent by a semipermeable membrane, solvent molecules may pass between the two sides of the membrane, but the solute molecules or ions may not (Fig. 11-1). Molecules from the pure solvent will diffuse through the membrane more rapidly than those from the solution can diffuse back. Because of this difference, there is a tendency for a net transfer of solvent molecules to the solution, thus diluting the solution.

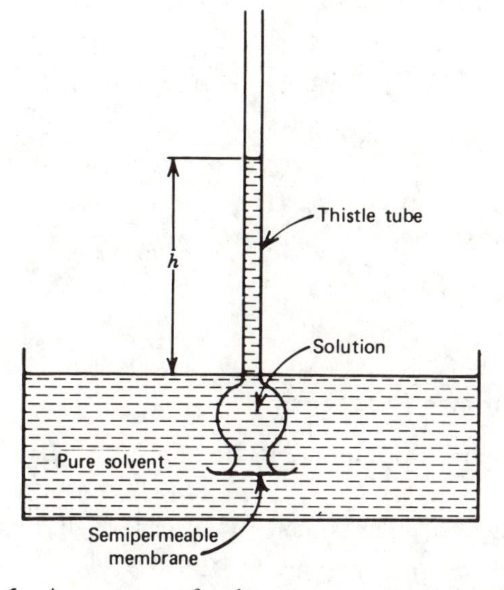

Figure 11-1 An apparatus for the measurement of osmotic pressure.

When the system reaches equilibrium, the pressure exerted downward by the column of solution of height h is just sufficient to prevent further dilution of the solution by the solvent. This pressure is termed the **osmotic pressure**, π, and the following relationships hold for **dilute** solutions:

$$\pi = \text{Molarity} \times RT \qquad \text{or} \qquad \pi = \text{Molality} \times RT$$

where R is the gas constant, 0.0821 liter atm/(mole K), and T is the absolute temperature. These relationships are particularly useful for the determination of molecular weights of large molecules, such as proteins.

Example 11-f. Determine the osmotic pressure of a 1.00×10^{-3} M aqueous solution at 0°C.

Solution.

$$\pi = \left(\frac{1.00 \times 10^{-3}\ \cancel{\text{mole}}}{\cancel{\text{liter}}}\right)\left(\frac{0.0821\ \cancel{\text{liter}}\text{-atm}}{\cancel{\text{mole}}\text{-}\cancel{K}}\right)(273\ \cancel{K})$$

$$= 0.0224 \text{ atm}$$

Example 11-g. An aqueous solution containing 5.0 g of a protein in 0.20 liter of solution exerts an osmotic pressure of 0.0237 atm at 27°C. Determine the molecular weight of the protein.

Solution.

$$\text{Molarity of solution} = \frac{\pi}{RT} = \frac{0.0237\ \cancel{\text{atm}}}{\left(\frac{0.0821\ \text{liter-}\cancel{\text{atm}}}{\text{mole-}\cancel{K}}\right)(300\ \cancel{K})}$$

$$= 9.6 \times 10^{-4} \text{ mole/liter}$$

$$\text{Grams protein per mole} = \frac{5.0 \text{ g}}{0.20\ \cancel{\text{liter}}} \times \frac{1\ \cancel{\text{liter}}}{9.6 \times 10^{-4} \text{ mole}} = 2.6 \times 10^{4}\ \frac{\text{g}}{\text{mole}}$$

Hence MW $= 2.6 \times 10^{4}$

Problems (Osmotic Pressure)

11-13. Determine the osmotic pressure of a 0.155 M solution of sucrose at 20°C.

11-14. Determine the approximate molecular weight of a protein, 1.60 g of which dissolved in 0.175 liter of an aqueous solution exert an osmotic pressure of 2.05 torr at 17°C.

11-15. The freezing point of human blood is $-0.58°C$. What is the osmotic pressure of blood at 98.6°F?

11-16. A 0.20% solution by weight of a certain protein exerts an osmotic pressure equivalent to a column of water 1.80 cm high at 20°C. Determine the approximate molecular weight of the protein.

11-D. Solutions of Electrolytes

Solutions of electrolytes in aqueous solution are, to a certain extent, ionized or dissociated into positive and negative ions. Some electrolytes, such as NaCl, $CaBr_2$, and K_2SO_4 approach 100% dissociation and are called strong electrolytes. Others, such as acetic acid, ammonia, and hydrofluoric acid, are only partially dissociated and are called weak electrolytes. As the concentrations of the solutions of both weak and strong electrolytes decrease, the extents of dissociation increase. At infinite dilution, both types are presumed to be completely dissociated.

In a solution of an electrolyte, more particles are present per mole of dissolved substance than is indicated by the molality of the solution, and hence abnormal colligative properties, such as freezing point depression, are observed. Since the freezing point of a solution is dependent on the number of dissolved particles, the extent to which a substance is ionized may be estimated from the freezing point.

Example 11-h. A solution of 20.0 g (1.00 mole) of HF in 1.00 kg of H_2O freezes at $-1.92°C$. What is the degree of ionization of HF? That is, what is the fraction of the molecules present that are ionized? The ionization of HF may be represented as $HF \rightleftharpoons H^+ + F^-$.

Solution.

$$HF \rightleftharpoons H^+ + F^-$$

Let

$$X = \text{Number of moles of } H^+ \text{ at equilibrium}$$

Because one HF molecule gives one H^+ and one F^-, then

$$X = \text{Number of moles of } F^- \text{ at equilibrium}$$

and

$$(1.00 - X) = \text{Number of moles of HF at equilibrium}$$

Total number of moles of all species in solution

$$= X + X + (1.00 - X) = (X + 1.00)$$

$$\text{Total molality} = (X + 1.00) \text{ moles/kg } H_2O$$

Since

$$\Delta T = (1.86°C/m) \times \text{Molality}$$

$$1.92°\cancel{C} = (1.86°\cancel{C}/\cancel{m}) \times (X + 1.00)\,\cancel{m}$$

$$\frac{1.92}{1.86} = X + 1.00$$

Hence $X + 1.00 = 1.03$

Therefore $X = 0.03$ and the degree (or fraction) of ionization is $0.03/1.00 = 0.03$ or 3%.

Example 11-i. Determine the freezing point of a 2.0 molal aqueous solution of an organic acid, HA, which is 10.0% ionized.

Solution.

$$HA \rightleftharpoons H^+ + A^-$$

$$\text{Moles of HA at equilibrium} = (0.90)(2.0) = 1.8$$
$$\text{Moles of } H^+ \text{ at equilibrium} = (0.10)(2.0) = 0.20$$
$$\text{Moles of } A^- \text{ at equilibrium} = (0.10)(2.0) = \underline{0.20}$$
$$\text{Total moles of all species} = 2.2$$

$$\text{Total molality of all species} = 2.2 \text{ moles/kg } H_2O$$

$$\Delta T = \frac{1.86°C}{\cancel{m}} \times 2.2\,\cancel{m} = 4.1°C$$

Freezing point $= -4.1°C$

Problems (Solutions of Electrolytes)

11-17. A 0.155 molal solution of NaCl may be considered to be completely ionized in aqueous solution. (a) What is the freezing point of the solution? (b) What is its boiling point?

11-18. A 0.118 molal solution of calcium perchlorate, $Ca(ClO_4)_2$, may be considered to be completely ionized in aqueous solution. (a) What is the freezing point of the solution? (b) What is its boiling point?

11-19. A solution of KNO_3 freezes at $-2.85°C$. What is the molality of the solution if the KNO_3 is considered to be completely ionized? The solvent is H_2O?

11-20. A 0.450 molal solution of $HC_2H_3O_2$ in H_2O freezes at $-0.855°C$. Calculate the apparent degree of ionization.

11-21. 2.40×10^2 g of acid HX, of molecular weight 120.0, are dissolved in 4.00 kg of H_2O. If the acid is 30.0% ionized, what is the freezing point of the solution?

11-22. A 0.100 molal solution of NH_3 freezes at $-0.192°C$. Calculate the apparent degree of ionization. Consider the reaction to be

$$NH_3 + H_2O \rightleftharpoons NH_4^+ + OH^-$$

Ignore the amount of water consumed.

11-23. What is the freezing point of an aqueous 1.50 molal solution of HX that is 65.0% ionized?

11-24. A 0.25 molal solution of the coordination compound (or complex compound) $Co(NH_3)_4Cl_3$ freezes at $-0.93°C$. In this compound, some of the chloride ions and all the ammonia molecules are bound to the cobalt and are not free to dissociate. The balance of the chloride ions are free to form ions, and a solution of this compound consists of a single positive ion and one or more free chloride ions. From the data given above, determine the number of ions produced per formula unit of the compound. The solvent is H_2O.

General Problems

11-25. A certain sample of paraffin has a freezing point depression constant of 4.50(°C/*m*). 20.0 grams of a solute are dissolved in 0.125 kg of paraffin. If the freezing point is found to be 6.00°C below the freezing point of pure paraffin, what is the molecular weight of the solute?

11-26. A solution of glucose, $C_6H_{12}O_6$, in water freezes at $-6.51°C$. What is the boiling point of this solution at standard pressure?

11-27. Ethylene glycol, $C_2H_4(OH)_2$, is used as an antifreeze in automobile radiators. How much $C_2H_4(OH)_2$ should be added to each quart of H_2O to give protection to $-25.0°C$? Assume that 1 quart of H_2O has a mass of 946 g.

11-28. How much ethyl alcohol, C_2H_5OH, should be added to 1.00 liter of H_2O to give a solution that freezes at $-20.0°C$?

11-29. A solution of sucrose, $C_{12}H_{22}O_{11}$, with a density of 1.19 g/ml freezes at $-4.09°C$. What is (a) the molality, and (b) the molarity of the solution? The solvent is H_2O.

11-30. An ethyl alcohol solution of a substance has a vapor pressure of 31.83 torr. If the vapor pressure of pure alcohol, C_2H_5OH, at this temperature is 32.16 torr, determine the molality of the solution.

11-31. A 0.315 kg portion of C_2H_5OH is dissolved in 1.00 quart of H_2O (946 ml). What is the freezing point of the resulting solution?

11-32. If 27.5 g of nonelectrolyte are dissolved in 100.0 g of H_2O, the solution freezes at -3.72°C. What is its molecular weight?

11-33. A 12.5 weight percent aqueous solution of a nonelectrolyte freezes at -7.44°C. What is the molecular weight of the solute?

11-34. Calculate the boiling point of a solution of 0.228 mole of a nonvolatile solute in 100.0 g of chloroform. Chloroform boils at 61.3°C at 1 atm pressure and has a boiling point constant of 3.63(°C/*m*).

11-35. A solution of 32.5 g of a nonvolatile solute in 325 g of water freezes at -2.79°C. What is the molecular weight of the solute?

11-36. A 6.36 weight percent aqueous solution freezes at -0.372°C. What is the molecular weight of the solute?

11-37. An organic compound consists of C = 40.7%, H = 5.1%, O = 54.2%. A solution of 23.6 g of this compound in 0.400 kg H_2O freezes at -0.93°C. What is the molecular formula of the compound?

11-38. Calculate the freezing point depression constant for benzene, C_6H_6, from the following data: 25.6 g of naphthalene, $C_{10}H_8$, in exactly 500 g of C_6H_6 depress the freezing point by 1.96°C.

11-39. (a) Calculate the vapor pressure of a 7.50 weight percentage aqueous solution of urea, $CO(NH_2)_2$, at 30°C. The vapor pressure of H_2O at 30° = 31.82 torr. (b) Calculate the osmotic pressure of the solution in (a).

11-40. A solution made by dissolving 21.0 g of an organic compound in 0.200 kg of chloroform, $CHCl_3$, freezes at a temperature 4.13°C below the freezing point of pure $CHCl_3$. If the freezing point depression constant for $CHCl_3$ is 4.70 (°C/*m*), what is the molecular weight of the solute?

11-41. (a) If 15.0 g of $CO(NH_2)_2$ in 0.200 kg of a solvent lower the freezing point of the latter 3.54°C, what is the freezing point depression constant for the solvent? (b) Calculate the freezing point depression for a 17.5 weight percent solution of $CO(NH_2)_2$ in this solvent.

11-42. When 6.84 g of an organic compound were dissolved in 120.0 ml of $CHCl_3$, it was found that the solution begins to solidify at a temperature 1.45 degrees below that of pure $CHCl_3$. The density of $CHCl_3$ is 1.50 g/ml and the freezing point depression constant for $CHCl_3$ is 4.70(°C/*m*). Analysis of the compound gave the following results: 58.5% C; 4.1% H; 11.4% N; 26.0% O. From these data, determine the true molecular formula of the compound.

11-43. The blood of cold-blooded marine animals and fish has about the same osmotic pressure (isotonic) as sea water. If sea water freezes at −2.30°C, calculate the osmotic pressure of the blood of marine life at 20°C.

11-44. The vapor pressure of an aqueous solution of sucrose, $C_{12}H_{22}O_{11}$, at 25°C is 99.5% that of pure H_2O. What is the osmotic pressure of this solution?

11-45. A method recently tested for the desalinization (removal of salt) of sea water is a reverse-osmosis process. With a membrane permeable by water but not by dissolved salts, fresh water can be obtained by exerting pressure on the salt-water side of the membrane. If sea water is considered to be composed of 3.0 weight percent NaCl for the purposes of this calculation, what minimum pressure (in addition to atmospheric pressure) must be applied to the sea-water side of a membrane to cause reverse osmosis to occur at 25°C?

11-46. In the following table, make the necessary calculations and fill in the blanks. Vapor pressure of H_2O at 20°C = 17.5 torr.

Solution	Vapor Pressure at 20°C (torr)	Boiling Point, °C	Freezing Point, °C	Osmotic Pressure (atm)
(a) 1.75 molal $C_6H_{12}O_6$	______	______	______	______
(b) 42.0 g $CO(NH_2)_2$ + 0.200 kg H_2O	______	______	______	______
(c) 34.2 g $C_{12}H_{22}O_{11}$ + 90.0 g H_2O	______	______	______	______

Chapter 12

Thermal Changes in Chemical Processes

Energy changes accompany all chemical and physical changes. The energy is usually manifested as heat energy. If heat is evolved to the surroundings, the process is said to be **exothermic**; if heat is absorbed from the surroundings, it is **endothermic**. When a change occurs in a system insulated from its surroundings, an increase in temperature indicates that the process is exothermic; conversely, a decrease in temperature indicates that the change is endothermic.

In considering heat changes, the following fundamental definitions are important.

The **joule** (J) is the principal unit of heat energy employed in chemical problems. It is a derived SI unit having the units kg m^2/sec^2. Until recently the unit calorie was employed when considering heat relationships in chemical systems. The calorie is now defined in terms of the joule—that is, 1 calorie $= 4.184$ J.

The **specific heat** is the number of joules required to raise the temperature of 1 g of substance 1°C. The specific heat of water is 4.184 J/(g °C), since 4.184 J are required to raise the temperature of one gram of water 1°C.

The **molar heat of vaporization** is the heat energy required to convert 1 mole of liquid at its boiling point to 1 mole of vapor at the same temperature. The molar heat of vaporization of water is $+40.7$ kJ/mole, with the plus sign indicating an endothermic process.

The **molar heat of condensation** is the energy evolved when 1 mole of vapor condenses to liquid at the boiling point. It is numerically equal, but opposite in sign, to the molar heat of vaporization.

The **molar heat of fusion** (melting) is the heat energy required to convert 1 mole of solid to 1 mole of liquid at the melting point. The heat of fusion for ice is about +6.0 kJ/mole.

The **molar heat of crystallization** is the heat evolved when 1 mole of liquid solidifies. It is numerically equal, but opposite in sign, to the molar heat of fusion.

The **molar heat of formation** is the energy change when 1 mole of a compound is formed from its elements. Standard heats of formation, ΔH_f°, for a few compounds are shown in Table 12-1. The term "standard" implies that one is considering substances at 25°C and 1 atm pressure. Under these conditions the substances are said to be in their standard state. **The heat of formation of an element is taken as zero.**

The **heat of reaction** (also called "enthalpy" of reaction; see Chapter 19) is the heat evolved or absorbed for a given chemical change when that number of moles shown in the balanced equation reacts. This heat may be considered as the difference in the heats of formation of products and reactants at the same temperature and pressure. Two conventions for showing heat changes in reactions are employed.

1. Energy changes may be shown as an adjunct to the chemical equation. For the reaction, $A + B \rightarrow C + D - \Delta H$, the symbol ΔH indicates the heat change accompanying the chemical reaction under a particular set of circumstances. ΔH has negative values for exothermic reactions and positive values for endothermic reactions. The exothermic reaction of CO and O_2 may be represented as

$$2\,CO + O_2 \longrightarrow 2\,CO_2 - \Delta H$$

or as

$$2\,CO + O_2 \longrightarrow 2\,CO_2 + 566\text{ kJ}$$

Table 12-1 Standard Heats of Formation

Compound	ΔH_f°, kJ/mole	Compound	ΔH_f°, kJ/mole
$H_2O(g)$	−241.8	$HCl(g)$	− 92.3
$H_2O(l)$	−285.8	$HF(g)$	−269.8
$CO(g)$	−110.5	$HBr(g)$	− 36.4
$CO_2(g)$	−393.5	HI(g)	+ 26.4
CH_3Cl(g)	− 82.0	$FCl(g)$	−113.0
$C_2H_2(g)$	+226.9	$H_2S(g)$	− 20.2
$C_2H_4(g)$	+ 52.6	$NO_2(g)$	+ 33.8
$C_2H_6(g)$	− 84.5	$NH_3(g)$	− 46.1

It should be obvious that $\Delta H = -566$ kJ. Fractional coefficients are permitted where it is desirable to show the energy change per mole of some reactant or product, as in the equation

$$CO + \tfrac{1}{2}O_2 \longrightarrow CO_2 + 283 \text{ kJ}$$

The endothermic process of formation of NO from its elements where $\Delta H = +180$ kJ may be represented

$$N_2 + O_2 \longrightarrow 2\,NO - 180 \text{ kJ}$$

2. Alternatively, energy changes may be shown following the equation to which they apply. Thus for the equations above,

$$2\,CO + O_2 \longrightarrow 2\,CO_2 \qquad \Delta H = -566 \text{ kJ}$$

$$CO + \tfrac{1}{2}O_2 \longrightarrow CO_2 \qquad \Delta H = -283 \text{ kJ}$$

$$N_2 + O_2 \longrightarrow 2\,NO \qquad \Delta H = +180 \text{ kJ}$$

This latter system is used in the following illustrations.

Hess' law of constant heat summation states that the heat change in a chemical reaction is the same whether the reaction proceeds in one or more steps. In other words, the heat change is independent of the path. Because of path independence, Hess' law allows us to calculate the heat of reaction for systems where this quantity is not easily measurable. Heats of reaction may be calculated from heats of formation of reactants and products. The heat of reaction is simply the sum of the heats of formation of all reactants subtracted from the sum of the heats of formation of all products, assuming all the heats have been measured at constant temperature. In equation form:

$$\Delta H = \sum \Delta H_f^\circ\text{ (products)} - \sum \Delta H_f^\circ\text{ (reactants)}$$

Further, Hess' law tells us that if two or more equations can be added algebraically to yield the desired reaction, ΔH for the desired reaction is equal to the sum of the ΔH's for the component equations.

The following examples illustrate the above principles.

Example 12-a. (a) ΔH_f° for NH_3 is -46.1 kJ/mole. Calculate this value in kcal/mole. (b) The heat of vaporization for HF is $+7.21$ kcal/mole. Determine this value in kJ/mole.

Solution. (a) Since 1.000 cal = 4.184 J or 1.000 kcal = 4.184 kJ, we have:

$$\Delta H_f^\circ(NH_3) = -\frac{46.1\,\cancel{\text{kJ}}}{\text{mole}} \times \frac{1.000\text{ kcal}}{4.184\,\cancel{\text{kJ}}} = -11.0\,\frac{\text{kcal}}{\text{mole}}$$

(b) Similar to (a) we have:

$$\Delta H_{vap}(HF) = +\frac{7.21\,\cancel{\text{kcal}}}{\text{mole}} \times \frac{4.184\text{ kJ}}{1.000\,\cancel{\text{kcal}}} = +30.2\,\frac{\text{kJ}}{\text{mole}}$$

Example 12-b. Calculate the quantity of heat required to convert 20.0 g of $H_2O(s)$ at 0°C to 20.0 g of $H_2O(g)$ at 100°C. Heat of fusion = 333 J/g; heat of vaporization = 2.26 kJ/g.

Solution. This requires three steps:

1. To melt 20.0 g of ice at 0°C requires:

$$20.0\,\text{g} \times \frac{333\text{ J}}{1\,\text{g}} = 6.66 \times 10^3\text{ J} = 6.66\text{ kJ}.$$

2. To heat 20.0 g of liquid from 0°C to 100°C requires:

$$20.0\,\text{g} \times 100°\text{C} \times \frac{4.18\text{ J}}{(\text{g}\,°\text{C})} = 8.36 \times 10^3\text{ J} = 8.36\text{ kJ}$$

3. To vaporize 20.0 g liquid at 100°C requires:

$$20.0\,\text{g} \times \frac{2.26\text{ kJ}}{1\,\text{g}} = 45.2\text{ kJ}$$

Total heat required = (6.7 + 8.4 + 45.2)kJ = 60.3 kJ

Example 12-c. 50.0 g of $H_2O(s)$ at 0°C is added to 100.0 g of $H_2O(l)$ at 80.0°C. What is the temperature of the $H_2O(l)$ when all the $H_2O(s)$ is melted? Assume that no heat is lost to the surroundings.

Solution. Let X = final temperature of liquid in °C.

The amount of heat needed to melt the ice and then to warm that liquid to the temperature X must be obtained by the cooling of the water originally at 80.0° to the temperature X. Therefore, the heat gained by water originally in ice = heat lost by water originally at 80.0°C.

$$50.0\,\text{g} \times \frac{333\text{ J}}{1\,\text{g}} + 50.0\,\text{g} \times (X - 0)°\text{C} \times \frac{4.18\text{ J}}{\text{g}\,°\text{C}} = 100.0\,\text{g} \times (80.0 - X)°\text{C} \times \frac{4.18\text{ J}}{\text{g}\,°\text{C}}$$

$$1.66 \times 10^4\text{ J} + 2.09 \times 10^2(X)\text{ J} = 3.34 \times 10^4\text{ J} - 4.18 \times 10^2(X)\text{ J}$$

$$6.27 \times 10^2(X)\text{ J} = 1.68 \times 10^4\text{ J}$$

$$X = \frac{1.68 \times 10^4\text{ J}}{6.27 \times 10^2\text{ J}} = 26.8$$

The final temperature therefore is 26.8°C.

Example 12-d. Calculate the heat of reaction for

$$CO(g) + \tfrac{1}{2}O_2(g) \longrightarrow CO_2(g)$$

from the molar heats of formation of CO and CO_2.

$$C(s) + O_2(g) \longrightarrow CO_2(g) \qquad \Delta H_1 = -393.5\text{ kJ} \qquad (1)$$

$$C(s) + \tfrac{1}{2}O_2(g) \longrightarrow CO(g) \qquad \Delta H_2 = -110.5\text{ kJ} \qquad (2)$$

Solution. The heat of reaction may be obtained by subtracting the heats of formation **per mole** of reactants from products, that is, heat of reaction = ΔH = heat of formation of CO_2 − (heat of formation of CO + 1/2 heat of formation of O_2) = −393.5 kJ − (−110.5 kJ + 0 kJ) = −283.0 kJ.

Or, we might arrange Equations (1) and (2) such that on addition we obtain the desired reaction; for example, turn Equation (2) around, as in Equation (2a), and add it to equation (1). ΔH of the desired reaction is then the sum of the individual heats of reaction. This is an application of Hess' Law.

$$C(s) + O_2(g) \longrightarrow CO_2(g) \qquad \Delta H_1 = -393.5\ \text{kJ} \qquad (1)$$

$$CO(g) \longrightarrow \tfrac{1}{2}\,O_2(g) + C(s) \qquad \Delta H_{2a} = +110.5\ \text{kJ} \qquad (2a)$$

$$CO(g) + \tfrac{1}{2}\,O_2(g) \longrightarrow CO_2(g) \qquad \Delta H = -283.0\ \text{kJ}$$

Example 12-e. Determine the standard heat of reaction for the process

$$Fe_2O_3(s) + 3\,CO(g) \longrightarrow 2\,Fe(s) + 3\,CO_2(g)$$

$\Delta H_f^\circ(Fe_2O_3) = -822.2$ kJ/mole.

Solution. Using values obtained in Table 12-1,

$$\begin{aligned}\Delta H^\circ &= \sum \Delta H_f^\circ\,(\text{products}) - \sum \Delta H_f^\circ\,(\text{reactants})\\ &= 3\,\Delta H_f^\circ(CO_2) + 2\Delta H_f^\circ(Fe) - [3\,\Delta H_f^\circ(CO) + \Delta H_f^\circ(Fe_2O_3)]\\ &= 3(-393.5)\ \text{kJ} + 2(0)\ \text{kJ} - [3(-110.5)\ \text{kJ} + (-822.2)\ \text{kJ}]\\ &= -1180.5\ \text{kJ} + 331.5\ \text{kJ} + 822.2\ \text{kJ}\\ &= -26.8\ \text{kJ}\end{aligned}$$

Example 12-f. C_2H_4 (ethylene) **cannot** be prepared directly by reaction of the elements C and H_2. However, ΔH for the reaction

$$2\,C(s) + 2\,H_2(g) \longrightarrow C_2H_4(g)$$

may be **calculated** using the following equations. (ΔH for these reactions can easily be measured.)

$$C(s) + O_2(g) \longrightarrow CO_2(g) \qquad \Delta H = -393.5\ \text{kJ} \qquad (1)$$

$$H_2(g) + \tfrac{1}{2}\,O_2(g) \longrightarrow H_2O(l) \qquad \Delta H = -285.8\ \text{kJ} \qquad (2)$$

$$C_2H_4(g) + 3\,O_2(g) \longrightarrow 2\,CO_2(g) + 2\,H_2O(l) \qquad \Delta H = -1411.2\ \text{kJ} \qquad (3)$$

Solution. Hess' law allows us to combine equations to obtain the desired result. We see that 2 × (1) + 2 × (2) − 1 × (3) yields the desired equation;

$$\begin{array}{lcll}
2\,C(s) + \cancel{2\,O_2(g)} & \longrightarrow & \cancel{2\,CO_2(g)} & \Delta H = -\ \ 787.0\text{ kJ} \\
2\,H_2(g) + \cancel{O_2(g)} & \longrightarrow & \cancel{2\,H_2O(l)} & \Delta H = -\ \ 571.6\text{ kJ} \\
\cancel{2\,CO_2(g)} + \cancel{2\,H_2O(l)} & \longrightarrow & \cancel{3\,O_2(g)} + C_2H_4(g) & \Delta H = +1411.2\text{ kJ} \\
\hline
2\,C(s) + 2\,H_2(g) & \longrightarrow & C_2H_4(g) & \Delta H = +\ \ 52.6\text{ kJ}
\end{array}$$

Bond Energies

Chemical bonds are the forces that hold atoms together. Energy is required to break these bonds and overcome the forces of attraction. This energy is termed the **bond** energy. It is usually expressed as kJ/mole—the energy required to break Avogadro's number of bonds.

Bond energies may be calculated from heats of formation. Since the heat of formation is the heat change in forming a compound from its elements, it follows that this is also a measure of the difference in energies of bonds in the compound and bonds in the element. If the latter are known, bond energy in compounds may be calculated.

Example 12-g. Determine the energy of the H—Cl bond. Given

$$H{-}H + Cl{-}Cl \longrightarrow 2\,H{-}Cl \qquad \Delta H = -184\text{ kJ}$$

$$H + H \longrightarrow H{-}H \qquad \Delta H = -435\text{ kJ}$$

and

$$Cl + Cl \longrightarrow Cl{-}Cl \qquad \Delta H = -242\text{ kJ}$$

Solution. Adding the above

$$2\,H + 2\,Cl \longrightarrow 2\,H{-}Cl \qquad \Delta H = -861\text{ kJ}$$

For the H—Cl bond $\Delta H = 861/2 = 430$ kJ/mole

A few bond energies are tabulated in Table 12-2. These tabulated values are actually **average** bond energies. The average term takes into account the fact that not all C—H (or N—H, etc.) bonds are **absolutely** identical in different environments or compounds.

Where bond energies are known, heats of formation of compounds may be calculated.

Table 12-2 Bond Energies

Bond	kJ/mole	Bond	kJ/mole
H—H	435	C=O	724
Cl—Cl	242	C—H	414
Br—Br	192	C—Br	276
F—F	159	N—H	389
C—C	347	F—Cl	254
O—O	146	I—Cl	209
O=O	498	F—H	565
O—H	463	Br—H	366
C—O	356	Cl—H	432
C=C	611	N≡N	946
C≡C	833	I—I	151
C—Cl	327	I—H	299

Example 12-h. Calculate the heat of formation of CH_4, that is:

$$C(s) + 2\,H_2(g) \longrightarrow CH_4(g) \qquad \Delta H = ?$$

using the table of bond energies and given

$$C(s) \longrightarrow C(g) \qquad \Delta H = +718 \text{ kJ}$$

Solution. From Table 12-2,

Energy released in forming 4 C—H bonds = 4 × (−414) = −1656 kJ

and

to break 2 H—H bonds requires 2 × 435 = +870 kJ

or

1. $$\cancel{C(g)} + \cancel{4\,H(g)} \longrightarrow CH_4(g) \qquad \Delta H = -1656 \text{ kJ}$$

2. $$2\,H_2(g) \longrightarrow \cancel{4\,H(g)} \qquad \Delta H = +870 \text{ kJ}$$

and

3. $$C(s) \longrightarrow \cancel{C(g)} \qquad \Delta H = +718 \text{ kJ}$$

On adding 1, 2, and 3

$$C(s) + 2\,H_2(g) \longrightarrow CH_4(g) \qquad \Delta H = -68 \text{ kJ}$$

Calorimetry

Heats of reaction are measured in a device called a calorimeter. Calorimeters generally contain a known amount of water and this water, as well as the calorimeter itself, absorbs the heat evolved during the reaction. If one knows the specific heat of water and the heat capacity of the calorimeter, the heat given off (or absorbed) during a reaction can easily be determined once the temperature change of the calorimeter is noted.

Example 12-i. A 0.600 g sample of carbon was burned in a calorimeter with an excess of $O_2(g)$. The calorimeter has a heat capacity of 923 J/°C and contained 815 g of H_2O. If the temperature increased from 25.00°C to 29.55°C, calculate ΔH for the reaction $C(s) + O_2(g) \rightarrow CO_2(g)$ in terms of kJ evolved per mole of carbon burned.

Solution. Temperature rise = (29.55 − 25.00) = 4.55°C. The evolved heat is absorbed both by the water and the calorimeter.

$$\text{Heat evolved} = 815\,\cancel{\text{g H}_2\text{O}} \times 4.55\,\cancel{^\circ\text{C}} \times 4.184\,\frac{\text{J}}{\cancel{\text{g H}_2\text{O}\,^\circ\text{C}}} + 4.55\,\cancel{^\circ\text{C}} \times 923\,\frac{\text{J}}{\cancel{^\circ\text{C}}}$$

$$= 15{,}500 \text{ J} + 4{,}200 \text{ J} = 19{,}700 \text{ J}$$

$$= 19.7 \text{ kJ}$$

Converting to kJ per mole of C burned and remembering the reaction is exothermic, we have

$$\Delta H = \frac{12.0\,\cancel{\text{g C}}}{1 \text{ mole C}} \times \frac{-19.7 \text{ kJ}}{0.600\,\cancel{\text{g C}}} = -394 \text{ kJ/mole C}$$

General Problems

12-1. The specific heat of Sb metal is 0.211 J/(g-°C). Calculate the quantity of heat required to raise the temperature of 0.500 kg of Sb from 20° to its melting point at 630°C.

12-2. Determine the quantity of heat, in kilojoules, necessary to convert 0.100 kg of $H_2O(s)$ at 0°C to $H_2O(g)$ at 100°C.

12-3. The heat of combustion of a bituminous coal is 25.1 kJ/g. What quantity of this coal on burning would supply sufficient heat to convert 17.5 kg of $H_2O(s)$ at 0°C to $H_2O(g)$ at 100°C?

12-4. A 1.00×10^2 g sample of $H_2O(s)$ at 0°C is added to 0.500 kg of $H_2O(l)$ at a temperature of 60.0°C. Assuming no transfer of heat to surroundings, what is the temperature of the $H_2O(l)$ after the ice is melted?

12-5. The following are physical properties of Hg: F.P. $-39°C$, B.P. 357°C, heat of fusion 0.554 kcal/mole, heat of vaporization 14.2 kcal/mole, specific heat 0.301 cal/(g °C). Calculate the number of kilocalories required to convert 100.3 g of Hg(*s*) at its freezing point to Hg(*g*) at the boiling point.

12-6. Neglect heat gained from or lost to the surroundings in this problem. 40.0 g of $H_2O(s)$ at 0°C are added to a quantity of $H_2O(l)$ that has a temperature of 60.0°C. The final temperature reached is 10.0°C. To what quantity of $H_2O(l)$ was the ice added?

12-7. Calculate the heat of reaction for

$$2\,PbS(s) + 3\,O_2(g) \longrightarrow 2\,PbO(s) + 2\,SO_2(g)$$

Heats of formation: PbS = -93.3 kJ/mole, PbO = -219.7 kJ/mole, $SO_2 = -296.6$ kJ/mole, $O_2 = 0$ kJ/mole.

12-8. Calculate the heat of reaction for

$$PbS(s) + PbSO_4(s) \longrightarrow 2\,Pb(s) + 2\,SO_2(g)$$

The heat of formation of $PbSO_4$ is -914.2 kJ/mole. See Problem 12-7 for other heats of formation.

12-9. Calculate the heat of reaction for

$$2\,SO_2(g) + O_2(g) \longrightarrow 2\,SO_3(g)$$

The heats of formation (ΔH) are: $SO_2 = -296.6$ kJ/mole, $SO_3 = -395.4$ kJ/mole, and $O_2 = 0$ kJ/mole.

12-10. From the following data, calculate the heat of formation of solid PCl_5 in kcal/mole.

$$P_4(s) + 6\,Cl_2(g) \longrightarrow 4\,PCl_3(l) \qquad \Delta H = -304.0 \text{ kcal}$$

$$PCl_3(l) + Cl_2(g) \longrightarrow PCl_5(s) \qquad \Delta H = -32.8 \text{ kcal}$$

12-11. Determine the number of kJ evolved in a calorimeter reaction where the temperature of the calorimeter increases from 19.50°C to 22.83°C. The calorimeter has a heat capacity of 645 J/°C and contains 318 g of water.

12-12. Calculate the molar heat of formation of methyl alcohol

$$C(s) + 2\,H_2(g) + \tfrac{1}{2}\,O_2(g) \longrightarrow CH_3OH(l)$$

from the following:

$$CH_3OH(l) + \tfrac{3}{2}\,O_2(g) \longrightarrow CO_2(g) + 2\,H_2O(l) \qquad \Delta H = -715.0 \text{ kJ}$$

$$C(s) + O_2(g) \longrightarrow CO_2(g) \qquad \Delta H = -393.3 \text{ kJ}$$

$$H_2(g) + \tfrac{1}{2}\,O_2(g) \longrightarrow H_2O(l) \qquad \Delta H = -285.8 \text{ kJ}$$

12-13. A piece of Fe with a mass of 0.350 kg at a temperature of 300°C is added to 1.00 kg of $H_2O(l)$ at a temperature of 20°C. What is the final equilibrium temperature of the $H_2O(l)$ and the Fe? The specific heat of Fe is 0.502 J/(*g* °C) and that of $H_2O(l)$ is 4.18 J/(g °C).

12-14. A 0.100 kg sample of $H_2O(s)$ at 0°C is added to 0.100 kg of $H_2O(l)$ at 60°C. When the temperature of the mixture is 0°C, what mass of $H_2O(s)$ is still present?

12-15. How many grams of ice at 0°C are necessary to cool 1 gallon of lemonade (about 3.8 kg) from 18°C to 3°C? Assume that the specific heat of lemonade is the same as that of H_2O.

12-16. Using Hess' law and heats of formation: $CaO(s) = -635.5$ kJ/mole, $CO_2(g) = -393.3$ kJ/mole, and $CaCO_3(s) = -1207.1$ kJ/mole, calculate ΔH for the reaction

$$CaO(s) + CO_2(g) \longrightarrow CaCO_3(s)$$

12-17. From the following

$$H_2(g) + Cl_2(g) \longrightarrow 2\,HCl(g) \qquad \Delta H = -184.9 \text{ kJ}$$

$$2\,H_2(g) + O_2(g) \longrightarrow 2\,H_2O(g) \qquad \Delta H = -483.7 \text{ kJ}$$

calculate ΔH for the reaction

$$4\,HCl(g) + O_2(g) \longrightarrow 2\,Cl_2(g) + 2\,H_2O(g)$$

12-18. Calculate ΔH for the reaction

(a) $$Na_2CO_3 \cdot 10\,H_2O(s) \longrightarrow Na_2CO_3(s) + 10\,H_2O(g)$$

from the following:

(b) $$Na_2CO_3 \cdot 10\,H_2O(s) \longrightarrow Na_2CO_3 \cdot 7\,H_2O(s) + 3\,H_2O(g) \qquad \Delta H = +155.2 \text{ kJ}$$

(c) $$Na_2CO_3 \cdot 7\,H_2O(s) \longrightarrow Na_2CO_3 \cdot H_2O(s) + 6\,H_2O(g) \qquad \Delta H = +320.1 \text{ kJ}$$

(d) $$Na_2CO_3 \cdot H_2O(s) \longrightarrow Na_2CO_3(s) + H_2O(g) \qquad \Delta H = +57.3 \text{ kJ}$$

12-19. Using the data in the previous problem, determine the heat absorbed per mole of water lost in each of the four reactions (a), (b), (c), and (d).

12-20. 2.56 g of S was placed in a calorimeter and burned in excess $O_2(g)$. During the process the calorimeter exhibited a temperature rise of 5.47°C. Determine ΔH (kJ/mole S) for the reaction $S(s) + O_2(g) \rightarrow SO_2(g)$. Assume use of the calorimeter employed in Example 12-i.

12-21. From the following equations, determine the molar heat of formation of $HNO_2(aq)$:

$$NH_4NO_2(aq) \longrightarrow N_2(g) + 2H_2O(l) \qquad \Delta H = -320.1 \text{ kJ}$$

$$NH_3(aq) + HNO_2(aq) \longrightarrow NH_4NO_2(aq) \qquad \Delta H = -\ 37.7 \text{ kJ}$$

$$2\,NH_3(aq) \longrightarrow N_2(g) + 3\,H_2(g) \qquad \Delta H = +169.9 \text{ kJ}$$

$$2\,H_2(g) + O_2(g) \longrightarrow 2\,H_2O(l) \qquad \Delta H = -571.5 \text{ kJ}$$

12-22. From the following equations, determine the heat of formation of HBr:

$$H_2(g) + Br_2(g) \longrightarrow 2\,HBr(g)$$

$$2\,KBr(aq) + Cl_2(g) \longrightarrow 2\,KCl(aq) + Br_2(aq) \qquad \Delta H = -96.2 \text{ kJ}$$

$$H_2(g) + Cl_2(g) \longrightarrow 2\,HCl(aq) \qquad \Delta H = -328.0 \text{ kJ}$$

$$KOH(aq) + HCl(aq) \longrightarrow KCl(aq) + H_2O \qquad \Delta H = -57.3 \text{ kJ}$$

$$KOH(aq) + HBr(aq) \longrightarrow KBr(aq) + H_2O \qquad \Delta H = -57.3 \text{ kJ}$$

$$Br_2(g) \longrightarrow Br_2(aq) \qquad \Delta H = -4.2 \text{ kJ}$$

$$HBr(g) \longrightarrow HBr(aq) \qquad \Delta H = -83.7 \text{ kJ}$$

12-23. The heat of formation of $H_2O(g)$ is -241.8 kJ/mole. From Table 12-2, which gives the bond energies of H_2 and O_2, determine the bond energy of O—H.

12-24. From thermal data in Tables 12-1 and 12-2 and in the example problems, determine ΔH for the following processes:

(a) $2\,C(g) + 4\,H(g) \rightarrow C_2H_4(g)$
(b) $3\,C(s) + 4\,H_2(g) \rightarrow C_3H_8(g)$
(c) $\frac{1}{2}\,I_2(g) + \frac{1}{2}\,Cl_2(g) \rightarrow ICl(g)$
(d) $C_2H_2(g) + 2\,H_2(g) \rightarrow C_2H_6(g)$
(e) $\frac{1}{2}\,N_2(g) + \frac{3}{2}\,H_2(g) \rightarrow NH_3(g)$
(f) $\frac{1}{2}\,H_2(g) + \frac{1}{2}\,Br_2(g) \rightarrow HBr(g)$
(g) $F_2(g) + Cl_2(g) \rightarrow 2\,FCl(g)$

12-25. From data in Tables 12-1 and 12-2 and example problems, calculate the following bond energies and compare with values recorded in Table 12-2.

(a) H—Br (b) H—I (c) F—Cl

12-26. 36.0 g of $H_2O(s)$ at 0.00°C were placed in a calorimeter containing 429 g of H_2O at 23.91°C. After melting, the entire calorimeter had dropped in temperature to 18.56°C. If the heat capacity of the calorimeter is 972 J/°C, calculate the molar heat of fusion for H_2O.

Chapter 13

Rates of Chemical Reactions

13-A. First-Order Chemical Reactions

The rates at which reactions occur are questions of utmost importance to chemists. The term "rate of a reaction" refers to the amount of substance reacting to form products per unit time; for example, in the simple reaction

$$A \rightarrow B + C$$

the rate of the reaction R is equal to the rate of **decrease** in the concentration of species A with time. The symbol $d[A]/dt$ is used to represent the rate of change of the concentration of A with time. Assuming it has been shown experimentally that the rate of the above reaction is proportional to the concentration of A, the **rate expression** or **rate law** may be written

$$\text{Rate of reaction} = \frac{-d[A]}{dt} = k[A] \quad (1)$$

where k is called the **rate constant**. This equation is a differential equation and any reaction that obeys this law is called a **first-order** chemical reaction. It is termed first-order because the sum of exponents of the concentrations for all species appearing in the rate law is one (1). In the above case the reaction is said to be first-order in A and first-order overall.

The rate constant, k, is a constant for a given reaction and does not change in value unless the temperature changes. Note this is not true for the rate of reaction. The rate of reaction varies with concentration change as described by the rate law.

By methods of calculus, it is possible to integrate equation (1) so that it takes the form

$$2.303 \log \frac{[A]}{[A]_0} = -kt \qquad (2)$$

where $[A]$ is the concentration of substance A remaining at time t, and $[A]_0$ is the amount originally present at zero time. The proportionality constant k is as defined in equation (1).

The time required for one-half of any given amount of material to decompose is known as the half-life of the substance in question. For a first-order reaction, the half-life is independent of the concentration or amount of starting material and is related to the rate constant by the equation

$$t_{1/2} = \frac{0.693}{k} \qquad (3)$$

Obviously, if either the rate constant or the half-life is known, the other may be calculated. A derivation of equation (3) is presented in Example 13-b.

It follows for first-order reactions that since one-half the amount of a given substance decomposes in one half-life, 1/2 times 1/2 or 1/4 of the original amount **remains** after two half-lives. In general the fraction of original substance **remaining** after n half-lives is $(1/2)^n$.

Example 13-a. The decomposition of SO_2Cl_2, according to the equation $SO_2Cl_2(g) \rightarrow SO_2(g) + Cl_2(g)$, has been found to be a first-order chemical reaction. At 602 K, 2.00% of the SO_2Cl_2 initially present had decomposed at the end of 6.72 min. What is the rate constant, k, at this temperature?

Solution. This problem may be solved by employing the integrated form of the first-order rate equation. Since only 2.00% of the original amount has decomposed, 98.00% remains, or $[A] = 0.9800\ [SO_2Cl_2]_0$ when $t = 6.72$ min. Substituting in Equation (2)

$$2.303 \log \frac{[SO_2Cl_2]}{[SO_2Cl_2]_0} = 2.303 \log \frac{0.9800[SO_2Cl_2]_0}{[SO_2Cl_2]_0} = -k(6.72 \text{ min})$$

$$k = \frac{-2.303 \log 0.9800}{6.72 \text{ min}} = \frac{-2.303(-0.00877)}{6.72 \text{ min}}$$

$$= 3.01 \times 10^{-3}/\text{min} = 3.01 \times 10^{-3} \text{ min}^{-1}$$

Example 13-b. From the data and solution given in Example 13-a, what is the half-life of the reaction for the decomposition of SO_2Cl_2?

Solution. This is analogous to asking the question "What time is required for one-half of any given amount of material to decompose?" This time is $t_{1/2}$ [Equation (3)], and is substituted for t in the integrated equation when the amount of reactant has decreased to half its initial value, that is, $t = t_{1/2}$ when $[SO_2Cl_2] = [SO_2Cl_2]_0/2$. Substituting into equation (2),

$$-2.303 \log \frac{[SO_2Cl_2]_0/2}{[SO_2Cl_2]_0} = kt_{1/2}$$

$$-2.303 \log \tfrac{1}{2} = kt_{1/2}$$

$$-2.303(-0.301) = 0.693 = kt_{1/2}$$

or

$$t_{1/2} = \frac{0.693}{k} \quad \text{[Equation (3)]}$$

Substituting the numerical value for k from Example 13-a, the half-life of SO_2Cl_2 at 602 K is

$$t_{1/2} = \frac{0.693}{k} = \frac{0.693}{3.01 \times 10^{-3}/\text{min}} = 2.30 \times 10^2 \text{ min}$$

(*Note*: Since the reaction rate constant is in the units of reciprocal minutes, the half-life of the reaction is obtained in minutes.)

Example 13-c. In 0.0400 *M* NaOH at 20°C, the decomposition of H_2O_2 by the reaction $2\,H_2O_2 \rightarrow 2\,H_2O + O_2$ has been shown to be first order in H_2O_2 only. (a) If the half-life is found to be 654 min, what fraction of the original H_2O_2 remains after exactly 100 min have elapsed? (b) What is the initial rate of reaction in a 0.0200 M solution of H_2O_2?

Solution. (a) To solve this problem, it is necessary to know k, the rate constant. This can be obtained by substituting the preceding data into the integrated form of the rate expression. When $t_{1/2} = 654$ min, $[H_2O_2] = [H_2O_2]_0/2$, and hence

$$-2.303 \log \frac{[H_2O_2]_0}{2[H_2O_2]_0} = k(654 \text{ min})$$

$$k = \frac{-2.303 \times (-0.301)}{654 \text{ min}} = 1.060 \times 10^{-3}/\text{min}$$

Since $t_{1/2}$ is known, the rate constant can also be evaluated from Equation (3)

$$k = \frac{0.693}{t_{1/2}} = \frac{0.693}{654 \text{ min}} = 1.060 \times 10^{-3}/\text{min}$$

Knowing k, it is now possible to solve for the fraction of hydrogen peroxide $[H_2O_2]/[H_2O_2]_0$ remaining after 100 min have elapsed. Making use of Equation (2),

$$2.303 \log \frac{[H_2O_2]}{[H_2O_2]_0} = -\frac{1.060 \times 10^{-3}}{\cancel{\text{min}}} \times (100\ \cancel{\text{min}})$$

$$\log \frac{[H_2O_2]}{[H_2O_2]_0} = \frac{-1.060 \times 10^{-1}}{2.303} = -4.60 \times 10^{-2}$$

$$= -0.0460$$

We obtain the antilog of -0.0460

$$\frac{[H_2O_2]}{[H_2O_2]_0} = 0.899$$

Hence 89.9% of the original H_2O_2 remains after the end of 100 min. (b) The initial rate of reaction is given by the equation Rate $= k[H_2O_2]_0$ and can easily be calculated if we know the rate constant and the concentration of H_2O_2. From the data given $[H_2O_2]_0 = 0.0200\ M$, and from the solution just presented, $k = 1.060 \times 10^{-3}$/min. Therefore, the Rate of reaction $= (1.060 \times 10^{-3}/\text{min}) \times (0.0200\ \text{mole/liter}) = 2.12 \times 10^{-5}$ (mole/liter)/min.

Problems (First-Order Chemical Reactions)

13-1. The reaction for the isomerization of cyclopropane gas to propene gas proceeds as illustrated by the equation

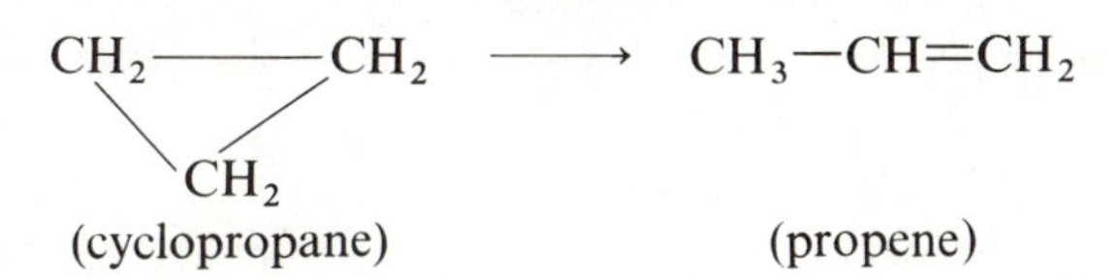

At 400°C and a pressure of 60.0 cm of Hg, the rate constant for this first-order reaction is 5.40×10^{-2}/hr. (a) What is the half-life of the cyclopropane under these conditions? (b) What fraction of cyclopropane remains at the end of 36.0 hr?

13-2. The rate constant for the *cis-trans* isomerization of liquid azobenzene is 2.22/hr at 77°C. The reaction is first order:

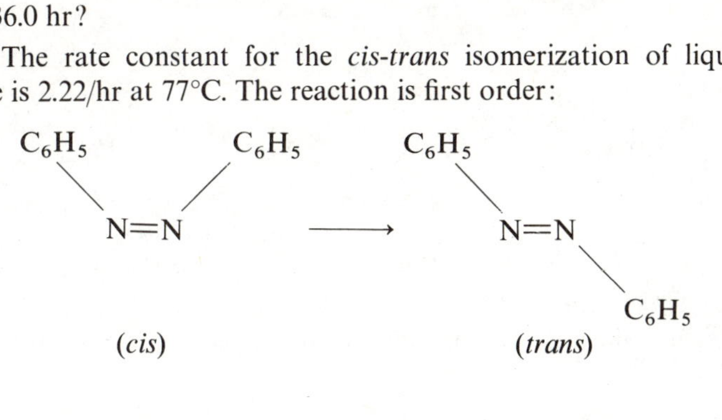

(a) What fraction of the *cis* has been converted at the end of 1.00 hr? (b) What is the half-life of the *cis* form under these conditions?

13-3. Acetic anhydride reacts with water to form acetic acid by the reaction

$$CH_3\overset{\overset{\displaystyle O}{\|}}{C}-O-\overset{\overset{\displaystyle O}{\|}}{C}CH_3 + H_2O \longrightarrow 2\,CH_3\overset{\overset{\displaystyle O}{\|}}{C}-OH$$

At 5°C, the reaction follows first-order kinetics and it has been found that a solution initially containing 0.802 M acetic anhydride is only 0.527 M at the end of 10.7 min. (a) Calculate the value of the rate constant, k. (b) What is the half-life of acetic anhydride under these conditions? (c) What is the initial rate of reaction (hydrolysis) of the acetic anhydride with water? (d) What is the rate of reaction at the end of one half-life?

13-4. The decomposition of phosphine at 719 K proceeds according to the reaction

$$4\,PH_3(g) \longrightarrow P_4(g) + 6\,H_2(g)$$

It is found that the reaction is first order and that the half-life is 37.9 sec. (a) Calculate the value of the rate constant. (b) What time is required for 87.5% of the material to decompose?

13-5. The inorganic complex ion *trans*-dioxalatodiaquochromate(III) undergoes isomerization in aqueous solution at 17°C to form the *cis* complex by a first-order reaction as shown by Professor Randall Hamm.

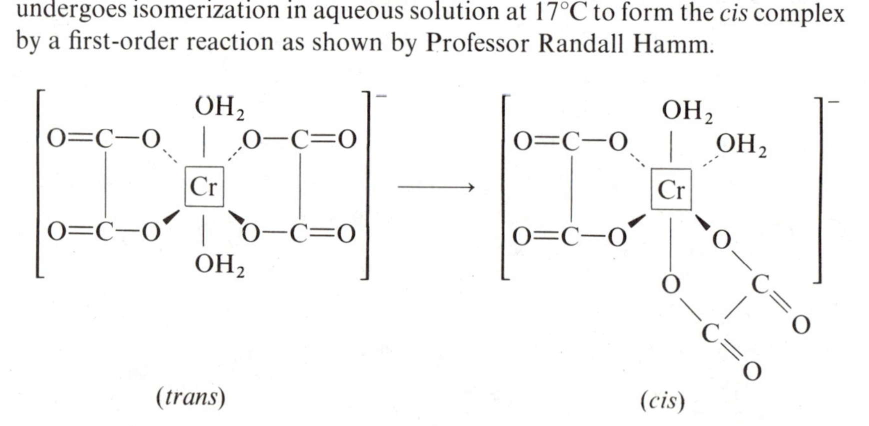

In an acidic aqueous solution, the half-life of the *trans* form is 1.07 hr. (a) Calculate the rate constant for the reaction. (b) What fraction of the *trans* form remains at the end of 3.21 hr? (c) What is the initial rate of reaction in a 0.125 M solution? (d) What is the rate of reaction in a solution at the end of one half-life if the initial concentration of complex ion was 0.096 M?

13-6. At 25°C, the hydrolysis of succinic anhydride in aqueous solution has been shown to be first order. The reaction is

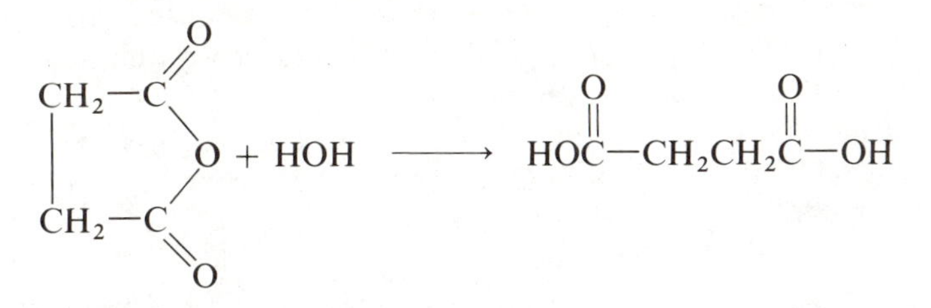

(a) If the half-life is 4.45 min, calculate the rate constant. (b) What fraction of material is left at the end of 12.0 min? (c) What fraction of material is left at the end of 4.0 half-lives? (d) What is the initial rate of the reaction in a solution containing 0.118 mole of succinic anhydride per liter?

13-7. The experimentally determined initial rates of the aquation reaction

$$[Co(NH_3)_5Cl]^{2+} + H_2O \longrightarrow [Co(NH_3)_5(H_2O)]^{3+} + Cl^-$$

are given below. From these data (a) prove that the reaction is first order. (b) Calculate the rate constant, k, for each set of data.

Experiment No.	Initial $[Co(NH_3)_5Cl]^{2+}$ (mole/liter)	Initial Rate [(mole/liter)/min]
1	1.0×10^{-3}	1.3×10^{-7}
2	2.0×10^{-3}	2.6×10^{-7}
3	3.0×10^{-3}	3.9×10^{-7}
4	1.0×10^{-2}	1.3×10^{-6}

13-B. Reactions other than First Order

For the general reaction

$$aA + bB \longrightarrow cC$$

it may be shown that

$$\text{Rate of reaction} = k[A]^x[B]^y$$

The sum of the exponents x and y gives the "overall order of the reaction." The numerical values of x and y **may** be equal to the coefficients a and b in the chemical equation, but **this must be demonstrated by experiment**. It **cannot** be assumed that x and y are equal to a and b, respectively, without some form of experimental proof. It has even been found for some reactions of the type $A \rightarrow B + C$ that the rate is independent of the concentration of A. Such a reaction is called a **zero-order** reaction.

In Section 13-A we did not mention how one might experimentally determine the order of a reaction. One relatively simple method of determination is outlined in the following examples.

Example 13-d. A reaction between gaseous substances A and B proceeds at a measurable rate. The following set of data has been accumulated.

Experiment No.	Initial $[A]$ (mole/liter)	Initial $[B]$ (mole/liter)	Initial Rate [(mole/liter)/min]
1	1.00×10^{-3}	0.25×10^{-3}	0.26×10^{-9}
2	1.00×10^{-3}	0.50×10^{-3}	0.52×10^{-9}
3	1.00×10^{-3}	1.00×10^{-3}	1.04×10^{-9}
4	2.00×10^{-3}	1.00×10^{-3}	4.16×10^{-9}
5	3.00×10^{-3}	1.00×10^{-3}	9.36×10^{-9}
6	4.00×10^{-3}	1.00×10^{-3}	16.64×10^{-9}

(a) What is the rate law for the reaction? What is the order of the reaction in A; B; overall? (b) Calculate the rate constant, k. (c) What is the initial rate of reaction in a solution where each of the reactants is at a concentration of $4.0 \times 10^{-4}\ M$?

Solution. (a) The questions that are really being asked are "What are the exponents x and y in the expression Rate $= k[A]^x[B]^y$?" and "What is the sum of $(x + y)$?" x represents the order in A, y the order in B, and $x + y$ the overall order.

Note in the data presented for Experiments 1–3 that the concentration of A is held constant and in Experiments 3–6, the concentration of B is held constant.

Consider first the variation of rate at constant $[A]$ and varying $[B]$. The rate of the reaction doubles from experiment 1 to 2 as the value of $[B]$ is doubled, and, likewise, in going from experiment 2 to 3, the rate doubles again as $[B]$ is doubled. Therefore, there is a direct proportionality between the rate and $[B]$, or rate $\propto [B]$.

Consider next the variation of rate at constant $[B]$ and varying $[A]$. The rate of reaction increases by a factor of 4 as the $[A]$ doubles, as is seen by comparing experiments 3 and 4 and experiments 4 and 6. In comparing experiments 3 and 5, the $[A]$ triples and the rate increases by a factor of 9. When comparing experiments 3 and 6, the concentration increases by a factor of 4 and the rate increases by a factor of 16. Thus it is seen that the rate is increasing by the square of the factor of the increase of $[A]$. Hence the rate is proportional to the square of the concentration of A, or rate $\alpha[A]^2$.

The rate law may now be written Rate $= k[A]^2[B]$. Hence the reaction is second order in A, first order in B, and third order overall since the sum of the exponents is 3.

(b) The rate constant may be evaluated by substituting any set of data from the table above into the rate law. For example, we may use the data from experiment number 5.

$$\text{Rate} = k[A]^2[B]$$

$$\text{Rate} = 9.36 \times 10^{-9} \text{ (mole/liter)/min}$$
$$[A] = 3.00 \times 10^{-3} \text{ mole/liter}$$
$$[B] = 1.00 \times 10^{-3} \text{ mole/liter}$$

Making the appropriate substitutions

$$9.36 \times 10^{-9} \text{ (mole/liter)/min} = k(3.00 \times 10^{-3} \text{ mole/liter})^2(1.00 \times 10^{-3} \text{ mole/liter})$$

$$9.36 \times 10^{-9} \text{ (mole/liter)/min} = k(9.00 \times 10^{-6})(1.00 \times 10^{-3}) \text{ (mole/liter)}^3$$

$$k = \frac{9.36 \times 10^{-9}}{9.00 \times 10^{-9}} = \frac{1.04}{\text{(mole/liter)}^2 \text{ min}} = 1.04\ M^{-2} \text{ min}^{-1}$$

The student should verify the fact that each set of data gives the same rate constant. This is required if the rate expression is valid.

(c) The initial rate when $[A] = [B] = 4.0 \times 10^{-4}\ M$ is obtained using the rate law and the constant evaluated in part (b).

$$\text{Rate} = k[A]^2[B]$$

$$= \frac{1.04}{\text{(mole/liter)}^2 \text{ min}} \times [4.0 \times 10^{-4} \text{ (mole/liter)}]^2 \times (4.0 \times 10^{-4} \text{ mole/liter})$$

$$= 1.04 \times 16 \times 10^{-8} \times 4.0 \times 10^{-4} \text{ (mole/liter)/min}$$

$$\text{Rate} = 6.7 \times 10^{-11} \text{ (mole/liter)/min}$$

Example 13-e. The substances X and Y react to form Z according to the following equation: $2X + Y \rightarrow 2Z$. At a particular temperature the following data were obtained during a series of experiments. (a) What is the order of the reaction in X; Y; overall? (b) Determine the rate law for the reaction.

Experiment No.	Initial $[X]$ (mole/liter)	Initial $[Y]$ (mole/liter)	Initial Rate [(mole/liter)/sec]
1	1.00×10^{-2}	4.00×10^{-4}	6.00×10^{-3}
2	2.00×10^{-2}	4.00×10^{-4}	1.20×10^{-2}
3	4.00×10^{-2}	4.00×10^{-4}	2.40×10^{-2}
4	1.00×10^{-2}	8.00×10^{-4}	6.00×10^{-3}

Solution. (a) Comparing Experiments 1 and 2, 1 and 3, and 2 and 3 we see there is a direct proportionality between the rate and $[X]$, or rate $\propto[X]$. Hence the reaction is first order in X.

Comparing Experiments 1 and 4 we see that changing $[Y]$ has *no* effect on the rate of reaction—that is, the reaction is zero order in Y.

The sum of the exponents is $1 + 0 = 1$, hence the reaction is first order overall.

(b) The rate law can be written in general form as Rate $= k[X]^a[Y]^b$. From (a) we determined $a = 1$ and $b = 0$, so the actual rate law is

$$\text{Rate} = k[X]^1[Y]^0 = k[X]$$

Note that experiment has shown that the coefficients in the balanced equation and the exponents in the rate law are *not* identical in this case.

Problems (Reactions other than First Order)

13-8. The reaction $CO(g) + NO_2(g) \rightarrow CO_2(g) + NO(g)$ has been studied at a temperature of 540 K. Some of the experimental data are listed in the table below.

Initial [CO] (mole/liter)	Initial [NO_2] (mole/liter)	Initial Rate [(mole/liter)/hr]
5.1×10^{-4}	0.35×10^{-4}	3.4×10^{-8}
5.1×10^{-4}	0.70×10^{-4}	6.8×10^{-8}
5.1×10^{-4}	0.18×10^{-4}	1.7×10^{-8}
10.2×10^{-4}	0.35×10^{-4}	6.8×10^{-8}
15.3×10^{-4}	0.35×10^{-4}	10.2×10^{-8}

(a) Derive an expression for the rate of reaction in terms of concentration of CO and NO_2. (b) What is the order of the reaction in CO; NO_2; overall? (c) Calculate the rate constant. (d) What is the initial rate of reaction if each of the species is present at a concentration of 2.8×10^{-4} *M*?

13-9. The compound N_2O_5 is found to decompose by the reaction $2\,N_2O_5(g) \rightarrow 4\,NO_2(g) + O_2(g)$. At a certain temperature, the following data were obtained.

Initial [N_2O_5] (mole/liter)	Initial Rate [(mole/liter)/sec]
2.00×10^{-4}	1.80×10^{-2}
3.00×10^{-4}	2.70×10^{-2}
4.00×10^{-4}	3.60×10^{-2}

(a) What is the rate law? (b) What is the order of the reaction for the decomposition of N_2O_5? (c) Determine the initial rate of the reaction when the initial concentration of N_2O_5 is 3.0×10^{-3} *M*.

13-10. Compound *A*, when enclosed in a glass vessel, spontaneously decomposes as shown by the equation $A \rightarrow B + C$. The following data were obtained for the rate of reaction as a function of initial concentration of *A*.

Initial [*A*] (mole/liter)	Initial Rate [(mole/liter)/sec]
4.0×10^{-2}	1.02×10^{-9}
5.0×10^{-2}	1.02×10^{-9}
8.0×10^{-2}	1.02×10^{-9}

(a) What is the rate law? (b) Find the order of the reaction.

13-11. Gaseous HI decomposes at 716 K by the reaction

$$HI(g) \longrightarrow \tfrac{1}{2} H_2(g) + \tfrac{1}{2} I_2(g)$$

The experimental rate data are shown below.

Initial [HI] (mole/liter)	Initial Rate [(mole/liter)/min]
1.0×10^{-3}	3.0×10^{-5}
2.0×10^{-3}	1.2×10^{-4}
3.0×10^{-3}	2.7×10^{-4}

(a) What is the rate law? (b) Determine the order of the reaction. (c) What is the initial rate of reaction if the concentration of HI is 2.5×10^{-4} *M*?

General Problems

13-12. Derive the relationship between $t_{1/2}$ and the rate constant for any first-order chemical reaction. Assume the equation $A \rightarrow B$ represents the reaction.

13-13. Acetyl fluoride hydrolyzes in a mixed water-acetone solvent according to the reaction

$$CH_3\overset{\overset{\displaystyle O}{\|}}{C}{-}F + HOH \longrightarrow CH_3\overset{\overset{\displaystyle O}{\|}}{C}{-}OH + HF$$

The rate expression is first order with respect to acetyl fluoride and is independent of water concentration. The first-order rate constant is 0.396/hr. Plot the fraction of acetyl fluoride remaining unreacted versus t for a period of three half-lives.

13-14. Gaseous *cis*-dichloroethylene at 185°C is converted to the *trans* form by a process that was experimentally found to be first order.

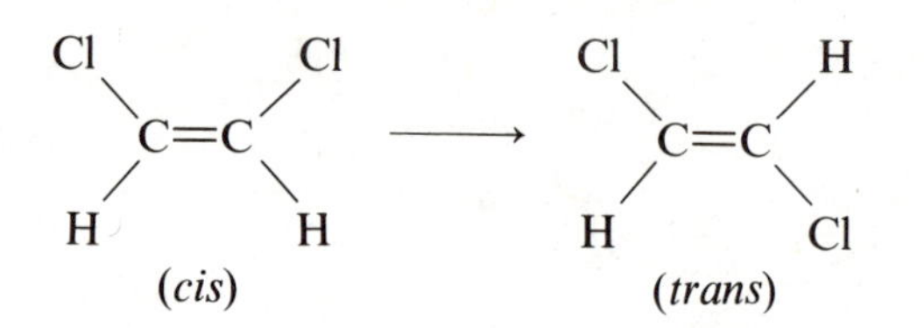

If after 3.5 hr the amount of the *cis* form is 95% of that originally present, calculate the value of the rate constant for the isomerization.

13-15. $H_2(g)$ and $I_2(g)$ combine by means of a reaction that is first order with respect to each species. The rate constant is 1.67×10^{-2}/[(mole/liter)-min]. If the concentration of H_2 is 1.08×10^{-2} mole/liter and the concentration of I_2 is 2.86×10^{-2} mole/liter, what is the initial rate of the reaction?

13-16. ${}^{226}_{88}Ra$ decays by alpha emission to ${}^{222}_{86}Rn$. This process is first order (as are all radioactive decay processes) with a half-life of 1620 yr. (a) How long will it take for 25% of the Ra to decay? (b) If the original sample was 6.40 g, how much Ra will be left after 16,200 yr?

13-17. The carbon found in a sample of charcoal from a cave in the western United States has a ${}^{14}C/{}^{12}C$ ratio roughly 19.0% as much as the ${}^{14}C/{}^{12}C$ ratio of presently live wood. Assume that the only change in the ${}^{14}C/{}^{12}C$ ratio results from the first-order decay process ${}^{14}_{6}C \rightarrow {}^{14}_{7}N + {}_{-1}^{0}e$, and that the ${}^{14}C/{}^{12}C$ ratio in new wood today is the same as when the charcoal sample was part of a living tree. If the half-life of the ${}^{14}C$ is 5760 yr, find the age of the wood from which the charcoal was prepared.

13-18. The compound dimethyl ether decomposes as shown by the reaction

$$CH_3OCH_3(g) \longrightarrow CH_4(g) + CO(g) + H_2(g)$$

by means of a first-order chemical reaction. At a certain temperature the rate constant was 0.12/min. Plot the fraction of dimethyl ether remaining in a sample versus the time since the reaction started. Plot the logarithm of this fraction versus the time since the reaction started. Which plot gives the most information?

13-19. The reaction of $2\,NO(g) + Cl_2(g) \rightarrow 2\,NOCl(g)$ appears to be a third-order reaction, and the rate determining step involves an intermediate with two molecules of NO and one of Cl_2. At a concentration of 0.20 *M* for each reactant, the rate constant is found to be approximately 1.0×10^{-3}/[(mole/liter)^2-min]. Determine (a) the rate law, and (b) the initial rate of reaction under these conditions. (c) What is the initial rate of reaction if the concentrations of reactants are doubled?

13-20. In the aquation reaction

$$[Cr(NH_3)_5I]^{2+} + H_2O \longrightarrow [Cr(NH_3)_5(H_2O)]^{3+} + I^-$$

1.5% of the initial concentration of the iodo complex is aquated in 10.4 min. (a) If this is a first-order reaction, what time is required for 15% of the material to react? (b) What is the time required for the fraction of material reacted to change from 0.50 to 0.60? (c) Calculate the initial rate of reaction in a 0.036 *M* solution of the complex.

Chapter 14

Molecular Equilibrium

Many chemical reactions are incomplete because as the reactants form products, the products of the reaction recombine to form the reactants. If the two processes are proceeding simultaneously, a state of equilibrium is eventually reached in which the rate of the forward reaction is equal to the rate of the reverse reaction. Consider the general equilibrium equation

$$mA + nB \rightleftharpoons pC + qD$$

where m, n, p, q are the coefficients of substances A, B, C, D, respectively, in the reaction. **When equilibrium is established**, the concentrations of reactants and products become fixed and are related through the mathematical relationship

$$\frac{[C]^p \times [D]^q}{[A]^m \times [B]^n} = K_c = \text{constant}$$

By convention, once a chemical equation is written, concentrations of the species on the **right** are put in the **numerator** of the equilibrium constant expression while the concentrations of the species on the **left** of the equation are put in the **denominator**.

K_c is termed the **equilibrium constant**, and has a particular value for each chemical reaction and each temperature. **The brackets denote concentrations expressed in moles per liter**, and it should be observed that each concentration is raised to that power that corresponds to the coefficient in the balanced equation. For example, in the reaction

$$2\,NOCl(g) \rightleftharpoons 2\,NO(g) + Cl_2(g)$$

$$K_c = \frac{[NO]^2[Cl_2]}{[NOCl]^2}$$

By various methods of chemical analysis, we can evaluate the concentrations of reactants and products at equilibrium. Once this information is available, the calculation of equilibrium constants is a simple matter.*

Example 14-a. Consider the reaction

$$2\,A(g) + B(g) \rightleftharpoons C(g) + 2\,D(g)$$

At a certain temperature the equilibrium concentrations were found to be $[A] = 0.50\ M$, $[B] = 0.40\ M$, $[C] = 2.0\ M$, and $[D] = 0.60\ M$. Calculate K_c.

Solution. By definition

$$K_c = \frac{[C][D]^2}{[A]^2[B]} = \frac{(2.0)(0.60)^2}{(0.50)^2(0.40)} = \frac{(2.0)(0.36)}{(0.25)(0.40)} = \frac{0.72}{0.10} = 7.2$$

Example 14-b. The equilibrium reaction

$$2\,NOCl(g) \rightleftharpoons 2\,NO(g) + Cl_2(g)$$

was studied at a temperature of 462°C and at a constant volume of 1.00 liter. Initially, 2.00 moles of NOCl were placed in the container, and when equilibrium was established, 0.66 mole of NO was found to be present. From these data, calculate the equilibrium constant.

Solution. Reaction

$$2\,NOCl(g) \rightleftharpoons 2\,NO(g) + Cl_2(g)$$

Initial amounts:

$$NOCl = 2.00\text{ moles};\ NO = 0\text{ mole};\ Cl_2 = 0\text{ mole}$$

* Equilibrium constants may or may not be unitless quantities. If the equilibrium reaction involves the same number of moles of reactants as products, K_c will be just a number without units; where there is a difference in numbers of moles of reactants versus products, K_c would carry units of concentration—that is, moles/liter, (moles/liter)^2, (moles/liter)$^{-1}$, and so on. In this presentation, the values of K_c will be expressed simply as **numbers**. The units in each case are easily obtainable from the equilibrium constant expression if desired.

Knowing that 0.66 mole of NO is present at equilibrium and using the balanced equation

$$\text{Moles NOCl reacted} = 0.66\ \cancel{\text{mole NO}} \times \frac{2\ \text{moles NOCl}}{2\ \cancel{\text{moles NO}}} = 0.66\ \text{mole NOCl}$$

$$\text{Moles } Cl_2 \text{ formed} = 0.66\ \cancel{\text{mole NO}} \times \frac{1\ \text{mole } Cl_2}{2\ \cancel{\text{moles NO}}} = 0.33\ \text{mole } Cl_2$$

Hence at equilibrium

$$\begin{aligned}\text{Moles NOCl remaining} &= (\text{Total} - \text{Moles dissociating})\\ &= 2.00 - 0.66 = 1.34\ \text{moles}\end{aligned}$$

$$\text{Moles NO} = 0.66;\ \text{moles } Cl_2 = 0.33$$

Since the volume is 1.00 liter the equilibrium concentrations are:

$$[NOCl] = 1.34\ \text{moles/liter};\ [NO] = 0.66\ \text{mole/liter};\ [Cl_2] \doteq 0.33\ \text{mole/liter}$$

Summarizing in tabular form:

	Initial Molarity	Change in Molarity	Equilibrium Molarity
NOCl	2.00	−0.66	1.34
NO	0	+0.66	0.66
Cl_2	0	+0.33	0.33

Now

$$K_c = \frac{[NO]^2[Cl_2]}{[NOCl]^2} = \frac{(0.66)^2(0.33)}{(1.34)^2}$$

$$K_c = \frac{0.436 \times 0.33}{1.80} = 8.0 \times 10^{-2}$$

Example 14-c. 1.00 mole of Br_2 gas is enclosed in a 1.00 liter container at 1483°C. It is found to be 1% dissociated under these conditions. Determine the equilibrium constant.

Solution. Reaction

$$Br_2(g) \rightleftharpoons 2\ Br(g)$$

$$K_c = \frac{[Br]^2}{[Br_2]}$$

Initial amounts:

$$Br_2 = 1.00 \text{ mole} \qquad Br = 0 \text{ mole}$$

Moles of Br_2 dissociating $= 1.00 \times 0.01 = 0.01$ mole

At equilibrium

$$\text{Moles } Br_2 = \text{Total moles } Br_2 - \text{Moles } Br_2 \text{ dissociated}$$
$$= 1.00 - 0.01 = 0.99 \text{ mole}$$

Using the balanced equation

$$\text{Moles Br} = 0.01 \cancel{\text{mole } Br_2} \times \frac{2 \text{ moles Br}}{1 \cancel{\text{mole } Br_2}} = 0.02 \text{ mole Br}$$

Since the volume is 1.00 liter

$$[Br_2] = 0.99 \text{ mole/liter}; [Br] = 0.02 \text{ mole/liter}$$

In tabular form:

	Initial Molarity	Change in Molarity	Equilibrium Molarity
Br_2	1.00	−0.01	0.99
Br	0	+0.02	0.02

Now

$$K_c = \frac{[Br]^2}{[Br_2]} = \frac{(2 \times 10^{-2})^2}{0.99} = 4 \times 10^{-4}$$

Example 14-d. A sample containing 2.00 moles of HI is enclosed in a 1.00-liter flask and heated to 628°C. The equilibrium reaction $2\,HI \rightleftharpoons H_2 + I_2$ is established for which the equilibrium constant is 3.80×10^{-2}. What is the percentage of dissociation under these conditions? What is the concentration of each species in the mixture?

Solution. Reaction

$$2\,HI(g) \rightleftharpoons H_2(g) + I_2(g)$$

$$K_c = \frac{[H_2][I_2]}{[HI]^2}$$

At equilibrium, let X be the number of moles of HI dissociated.

$$[HI] = \frac{\text{Total moles HI} - \text{Number of moles dissociated}}{\text{Number of liters}}$$

$$= \frac{2.00 - X}{1.00} = (2.00 - X) \text{ moles/liter}$$

Since 2 moles of HI form 1 mole of H_2 and 1 mole of I_2

$$[H_2] = X/2 \text{ moles/liter}$$

and

$$[I_2] = X/2 \text{ moles/liter}$$

In tabular form:

	Initial Molarity	Change in Molarity	Equilibrium Molarity
HI	2.00	$-X$	$2.00 - X$
H_2	0	$+X/2$	$X/2$
I_2	0	$+X/2$	$X/2$

Substituting in the equilibrium constant expression

$$0.0380 = \frac{(X/2)(X/2)}{(2.00 - X)^2} = \frac{(X/2)^2}{(2.00 - X)^2}$$

Taking the square root of both sides

$$\sqrt{0.0380} = \frac{X/2}{2.00 - X}$$

$$0.195 = \frac{X/2}{2.00 - X}$$

$$(0.195)(2.00) - 0.195X = 0.500X$$

$$0.390 = 0.695X$$

$$X = \frac{0.390}{0.695} = 0.561 \text{ mole/liter}$$

$$[HI] = 2.00 - 0.561 = 1.44 \text{ moles/liter}$$

$$[H_2] = [I_2] = 0.280 \text{ mole/liter}$$

$$\text{Percentage dissociation} = \frac{\text{Initial [HI]} - \text{Equil [HI]}}{\text{Initial [HI]}} \times 100$$

$$= \frac{2.00 - 1.44}{2.00} \times 100 = 28\%$$

Example 14-e. 2.50 moles of $COBr_2$ are introduced into a 2.00 liter flask at 0°C and heated to a temperature of 73°C. At equilibrium, the equilibrium constant K_c was found to be 0.190. What is the equilibrium concentration of each species? What is the percentage of dissociation? The chemical reaction involved is

$$COBr_2(g) \rightleftharpoons CO(g) + Br_2(g)$$

Solution. Initial concentrations

$$[COBr_2] = \frac{2.50 \text{ moles}}{2.00 \text{ liters}} = 1.25 \text{ moles/liter}$$

$$[CO] = [Br_2] = 0 \text{ mole/liter}$$

Let

$$X = \text{amount of } COBr_2 \text{ dissociated in moles/liter}$$

At equilibrium

$$[Br_2] = [CO] = X \text{ mole/liter}$$

and

$$[COBr_2] = (1.25 - X) \text{ mole/liter}$$

Summarizing:

	Initial Molarity	Change in Molarity	Equilibrium Molarity
	1.25	$-X$	$1.25 - X$
CO	0	$+X$	X
Br_2	0	$+X$	X

$$K_c = \frac{[CO][Br_2]}{[COBr_2]} = \frac{(X)(X)}{1.25 - X} = 0.190$$

$$X^2 = (0.190)(1.25 - X)$$

$$= 0.238 - 0.190X$$

or

$$X^2 + 0.190X - 0.238 = 0$$

To solve for X, we make use of the general solution to a quadratic equation (see Appendix I).

$$X = \frac{-b \pm \sqrt{b^2 - 4ac}}{2a}$$

where $a = 1 \qquad b = 0.190 \qquad c = -0.238$

$$X = \frac{-0.190 \pm \sqrt{(0.190)^2 - (4)(1)(-0.238)}}{(2)(1)}$$

$$X = \frac{-0.190 \pm \sqrt{0.036 + 0.952}}{2} = \frac{-0.190 \pm \sqrt{0.988}}{2}$$

$$X = \frac{-0.190 \pm 0.994}{2} = 0.402^*$$

Therefore, at equilibrium

$$[CO] = [Br_2] = 0.402 \text{ mole/liter}$$

and

$$[COBr_2] = 1.25 - 0.40 = 0.85 \text{ mole/liter}$$

$$\text{Percentage dissociation} = \frac{\text{Initial } [COBr_2] - \text{Equil } [COBr_2]}{\text{Initial } [COBr_2]} \times 100$$

$$= \frac{1.25 - 0.85}{1.25} \times 100 = 32\%$$

Example 14-f. If we add 2.00 moles of carbon monoxide to the above equilibrium system, determine the concentration of each species when equilibrium is reestablished.

Solution. From LeChatelier's principle, we can predict that the equilibrium would be shifted toward the formation of more $COBr_2$ and that the concentration of Br_2 would be decreased.

Let X be the concentration decrease of Br_2 because of reaction with the added CO. At the new equilibrium

$$[Br_2] = (0.40 - X) \text{ mole/liter}$$

$$[CO] = (0.40 - X) \text{ mole/liter} + \text{Concentration added}$$

$$= (0.40 - X) \text{ mole/liter} + \frac{2.00 \text{ moles}}{2.00 \text{ liters}}$$

$$= (0.40 + 1.00 - X) \text{ moles/liter}$$

$$= (1.40 - X) \text{ moles/liter}$$

$$[COBr_2] = \text{Initial concentration} + \text{Concentration increase (due to equilibrium shift)}$$

$$= (0.85 + X) \text{ mole/liter}$$

* For the ± 0.994 term, only $+0.994$ has any physical significance because it is not possible to have a negative concentration.

In tabular form:

	Initial Molarity	Change in Molarity	Equilibrium Molarity
$COBr_2$	0.85	$+X$	$0.85 + X$
CO	0.40	$+1.00 - X$	$1.40 - X$
Br_2	0.40	$-X$	$0.40 - X$

By substitution into the equilibrium constant expression

$$K_c = 0.190 = \frac{(1.40 - X)(0.40 - X)}{0.85 + X}$$

and using the quadratic expression

$$X = 0.23 \text{ mole/liter}$$

Hence

$$[Br_2] = 0.40 - 0.23 = 0.17 \text{ mole/liter}$$

$$[CO] = 1.40 - 0.23 = 1.17 \text{ moles/liter}$$

$$[COBr_2] = 0.85 + 0.23 = 1.08 \text{ moles/liter}$$

It is often convenient to express the equilibrium constant in terms of the partial pressures of the various gases present rather than as concentrations. In these cases K_p is used to designate the equilibrium constant to distinguish it from the concentration constant K_c. For example, considering the simple system $A(g) \rightleftharpoons B(g) + C(g)$

$$K_p = \frac{(p_B)(p_C)}{(p_A)}$$

where p represents the partial pressures of the respective substances.

In order to determine the relationship between K_p and K_c for the above example we proceed as follows.

Recall that $p_A V = n_A RT$ where A represents the substance above. Rearranging, we see that $n_A/V = p_A/RT$. But

$$\frac{n_A}{V} = \frac{\text{Moles } A}{\text{Liters}} = [A] = \frac{p_A}{RT} \quad \text{or} \quad [A] = p_A(RT)^{-1}.$$

Similarly $[B]$ would be expressed as $p_B(RT)^{-1}$ and $[C] = p_C(RT)^{-1}$.
Now

$$K_c = \frac{[B][C]}{[A]} = \frac{[p_B(RT)^{-1}][p_C(RT)^{-1}]}{[p_A(RT)^{-1}]} = \frac{(p_B)(p_C)}{(p_A)}(RT)^{-1}.$$

Hence for this example

$$K_c = K_p(RT)^{-1}$$

or

$$K_p = K_c(RT)^{+1}.$$

In general it can be shown that

$$K_p = K_c(RT)^{\Delta n}$$

where Δn is the difference in number of moles of gaseous products and reactants as expressed in the balanced equation. (Note in the above example that Δn = Moles products − Moles reactants = 2 − 1 = +1.)

Example 14-g. A sample of N_2 and H_2 is heated together at a temperature of 623°C and at a constant pressure of 30.0 atm. *If a stoichiometric ratio of N_2 and H_2 is used,* it is found that at equilibrium the pressure of N_2 is 4.0 atm. Using pressures instead of concentrations, calculate the equilibrium constant for the reaction

$$N_2(g) + 3\ H_2(g) \rightleftharpoons 2\ NH_3(g)$$

Solution. When equilibrium is finally established,

$$P_{\text{total}} = 30.0 \text{ atm} = p_{N_2} + p_{NH_3} + p_{H_2}$$

where p represents partial pressures. In a stoichiometric mixture, $p_{H_2} = 3p_{N_2}$ because in the balanced chemical equation, the ratio (moles of H_2/moles of N_2) = 3. If

$$p_{N_2} = 4.0 \text{ atm} \qquad \text{then} \qquad p_{H_2} = 12.0 \text{ atm}$$

$$p_{NH_3} = 30.0 - p_{N_2} - p_{H_2} = 30.0 - 4.0 - 12.0 = 14.0 \text{ atm}$$

$$K_p = \frac{(p_{NH_3})^2}{(p_{N_2})(p_{H_2})^3} = \frac{(14.0)^2}{(4.0)(12.0)^3} = 0.028$$

Example 14-h. In the case of equilibrium in heterogeneous systems where a gaseous phase and one or more condensed phases are present, the equilibrium constant is written similarly. In the equilibrium reaction

$$C(s) + CO_2(g) \rightleftharpoons 2\ CO(g)$$

$$K_p = \frac{(p_{CO})^2}{p_{CO_2} p_{C(s)}}$$

Since the vapor pressure of a pure solid (or a pure liquid) is constant at constant temperature, a new equilibrium constant K'_p may be defined as $K'_p = K_p \times p_{C(s)}$, and hence

$$K'_p = \frac{(p_{CO})^2}{p_{CO_2}}$$

At 1000 K when equilibrium between $CO_2(g)$, $CO(g)$, and $C(s)$ is established, the total pressure is found to be 4.70 atm. Calculate the equilibrium pressures of CO_2 and CO, if at 1000 K the value of K'_p is 1.72 and the vapor pressure of C(*s*) is negligible.

Solution. Let X = pressure of CO_2 in atm at equilibrium

$$P_{\text{total}} = 4.70 \text{ atm} = p_{CO_2} + p_{CO}$$

$$p_{CO} = (4.70 - X) \text{ atm}$$

On substitution into the equilibrium constant expression

$$1.72 = \frac{(4.70 - X)^2}{X}$$

On solution of this equation it is found that $X = 2.59$ atm, and hence

$$p_{CO_2} = 2.59 \text{ atm} \quad \text{and} \quad p_{CO} = 4.70 - 2.59 = 2.11 \text{ atm}$$

Example 14-i. If K_p for the equilibrium

$$PCl_5(g) \rightleftharpoons PCl_3(g) + Cl_2(g)$$

is 1.78 at 250°C, calculate K_c.

Solution.

$$K_p = K_c(RT)^{\Delta n} \quad \text{or} \quad K_c = \frac{K_p}{(RT)^{\Delta n}}$$

$$\Delta n = 2 \text{ moles products} - 1 \text{ mole reactant} = 1$$

Substituting

$$K_c = \frac{1.78}{(0.0821)(523)} = 4.15 \times 10^{-2}$$

General Problems*

14-1. A reaction vessel with a capacity of 1.0 liter, in which the reaction $SO_2(g) + NO_2(g) \rightleftharpoons SO_3(g) + NO(g)$ had reached a state of equilibrium, was found to contain 0.40 mole of SO_3, 0.30 mole of NO, 0.15 mole of NO_2, and 0.20 mole of SO_2. Calculate the equilibrium constant for the reaction.

14-2. Consider the equilibrium $H_2(g) + I_2(g) \rightleftharpoons 2\,HI(g)$. Set up the expression for the equilibrium constant and evaluate K_c if at 300 K the concentrations are $[H_2] = 0.40\ M$; $[I_2] = 0.45\ M$; $[HI] = 0.30\ M$.

14-3. Given the equilibrium $2\,A(g) + B(g) \rightleftharpoons C(g) + 3\,D(g)$ and the concentrations $[A] = 2.0\ M$, $[B] = 0.50\ M$, $[C] = 0.25\ M$, $[D] = 1.5\ M$, calculate the value of the equilibrium constant.

14-4. The equilibrium system $2\,SO_2(g) + O_2(g) \rightleftharpoons 2\,SO_3(g)$ was analyzed and found to contain the following concentrations: $[SO_2] = 0.40\ M$, $[O_2] = 0.13\ M$, $[SO_3] = 0.70\ M$. Calculate the equilibrium constant for this system.

14-5. The gaseous reaction $NO_2(g) + SO_2(g) \rightleftharpoons SO_3(g) + NO(g)$ is found to occur at elevated temperatures. In a given experiment, SO_3 and NO were mixed and allowed to come to equilibrium. Analysis of the mixture at equilibrium at a specific temperature demonstrated the presence of 0.60 mole of SO_3, 0.40 mole of NO, 0.30 mole of NO_2, and 0.20 mole of SO_2 in the 0.50 liter container. What is the equilibrium constant for the reaction?

14-6. The equilibrium constant for the formation of NH_3 by the reaction $N_2(g) + 3\,H_2(g) \rightleftharpoons 2\,NH_3(g)$ is 2.0 at a certain temperature. If the equilibrium concentration of N_2 in a mixture is 0.50 M and H_2 is 2.0 M, determine the concentration of ammonia.

14-7. At 2000 K a mixture of H_2, S_2, and H_2S rapidly reaches a state of equilibrium, represented by the equation

$$2\,H_2S(g) \rightleftharpoons 2\,H_2(g) + S_2(g)$$

An analysis of the equilibrium mixture shows that there are 2.00 moles of H_2S, 1.00 mole of H_2, and 0.130 mole of S_2 in the 2.00 liter flask. Calculate the concentration equilibrium constant.

14-8. The equilibrium constant is 4×10^{-3} for the dissociation of I_2 at elevated temperatures, as represented by the reaction $I_2(g) \rightleftharpoons 2\,I(g)$. If the equilibrium concentration of atomic iodine is equal to $4 \times 10^{-2}\ M$, which is the concentration of molecular iodine?

14-9. The equilibrium constant for the formation of CO_2 by the reaction $2\,CO(g) + O_2(g) \rightleftharpoons 2\,CO_2(g)$ is 1.0×10^4. If, at equilibrium, the concentra-

* All equilibrium constants for these problems are K_c unless specifically indicated as K_p.

tion of CO_2 is 3.0 M and the concentration of O_2 is 9.0 M, calculate the concentration of CO.

14-10. A mixture of 2.00 moles of Cl_2 gas and 2.00 moles of Br_2 gas is enclosed in a 4.0 liter flask. At a certain temperature the reaction $Br_2(g) + Cl_2(g) \rightleftharpoons 2\,BrCl(g)$, occurs. When equilibrium is established, it is found that 9.8% of the Br_2 has been used up. Determine the equilibrium constant for the reaction.

14-11. The equilibrium constant for the reaction

$$N_2(g) + 3\,H_2(g) \rightleftharpoons 2\,NH_3(g)$$

is 0.50 at a certain temperature. If 2.0 moles of N_2 and 4.0 moles of H_2 are in equilibrium with NH_3 in a 2.0 liter vessel at this temperature, (a) find the equilibrium concentration of NH_3. (b) What fraction of the **total** nitrogen in the system remains unreacted?

14-12. The equilibrium constant for the reaction

$$SO_2(g) + NO_2(g) \rightleftharpoons SO_3(g) + NO(g)$$

is 1.0 at a certain temperature. If 2.0 moles of SO_2 and 4.0 moles of NO_2 are placed together in a 1.0 liter vessel at this temperature, determine the concentration of SO_3 at equilibrium.

14-13. A sample containing 0.800 mole of $POCl_3$ is enclosed in a 0.500 liter vessel at a certain temperature. When the equilibrium for the dissociation reaction $POCl_3(g) \rightleftharpoons POCl(g) + Cl_2(g)$ is attained, it is found that the vessel contains 0.259 mole of Cl_2. Calculate the equilibrium constant.

14-14. In the equilibrium system $2\,NO(g) + O_2(g) \rightleftharpoons 2\,NO_2(g)$ at 1000 K, it is found that the concentration of NO is 2.50 M, and the concentration of O_2 is 1.25 M. If the equilibrium constant at this temperature is 1.20, (a) what is the concentration of NO_2? (b) What fraction of the original NO has been converted to NO_2?

14-15. Exactly 2.00 moles of HBr are enclosed in a 1.00 liter flask and heated to 1297 K, whereupon the equilibrium

$$2\,HBr(g) \rightleftharpoons H_2(g) + Br_2(g)$$

is rapidly established. The equilibrium constant is 7.32×10^{-6}. (a) Determine the percentage of dissociation under these conditions. (b) What is the concentration of each species?

14-16. The brown gas NO_2 on cooling is converted to the colorless gas dinitrogen tetroxide, N_2O_4, as described by the equation $2\,NO_2(g) \rightleftharpoons N_2O_4(g)$. If the original concentration of NO_2 is 0.90 M, and at equilibrium its concentration is only 0.26 M, what is the equilibrium constant for the reaction?

14-17. For the reaction $PCl_3(g) + PBr_3(g) \rightleftharpoons PCl_2Br(g) + PClBr_2(g)$, the equilibrium constant, K_p, is 1.5 at a certain temperature. The equilibrium pressure of PCl_2Br is equal to the pressure of $PClBr_2$ and is found to be 6.0 cm of Hg. If the pressure of PCl_3 is 5.0 cm of Hg, determine the pressure of PBr_3.

14-18. Consider the system $PCl_5(g) \rightleftharpoons PCl_3(g) + Cl_2(g)$. How many moles of PCl_5 must be added to a 4 liter flask at 250°C to obtain an equilibrium concentration of 0.3 mole/liter of Cl_2? $K_c = 4 \times 10^{-2}$.

14-19. At a certain temperature, the equilibrium constant, K_p, for the reaction $CO_2(g) + H_2(g) \rightleftharpoons CO(g) + H_2O(g)$ is 16.0. Originally, equal numbers of moles of H_2 and CO_2 were placed in a flask. At equilibrium, the pressure of H_2 is 1.20 atm. What is the pressure of each of the other gases?

14-20. A 0.750 mole sample of $COBr_2$ was placed in a 1.00 liter reaction vessel and allowed to come to equilibrium at a temperature of 454 K. After the equilibrium $COBr_2(g) \rightleftharpoons CO(g) + Br_2(g)$ was established, it was found that 57.1% of the initial material had dissociated. (a) Determine the concentration equilibrium constant of the reaction. (b) If 0.500 mole of Br_2 is added to the equilibrium mixture, what is the concentration of each species when equilibrium is reestablished?

14-21. At a certain temperature the equilibrium mixture

$$SO_2(g) + NO_2(g) \rightleftharpoons SO_3(g) + NO(g)$$

in a 1.0 liter flask is analyzed and found to contain 0.50 mole of SO_3, 0.45 mole of NO, 0.15 mole of SO_2, and 0.30 mole of NO_2. If 0.25 mole of SO_3 is added to the flask at constant temperature, calculate the new concentration of each gas in the mixture after establishment of equilibrium.

14-22. For the dissociation of $2\,HI(g) \rightleftharpoons H_2(g) + I_2(g)$, the equilibrium constant is 155 at 500 K. If 7.50 moles of HI are injected into an empty 3.00 liter flask at this temperature, (a) calculate the equilibrium concentrations of the various species. (b) What percentage of HI has dissociated?

14-23. The equation $CO_2(g) + H_2(g) \rightleftharpoons CO(g) + H_2O(g)$ represents the reaction between CO_2 and H_2 at elevated temperatures. (a) If at 1800°C there are 0.30 mole of CO_2, 0.30 mole of H_2, 0.60 mole of CO, and 0.60 mole of H_2O in a 2.00 liter flask, find the concentration equilibrium constant. (b) What number of moles of CO_2 must be added to increase the concentration of CO to 0.35 M?

14-24. At 650 K, the equilibrium $N_2(g) + 3\,H_2(g) \rightleftharpoons 2\,NH_3(g)$ is established. If a stoichiometric mixture of 3.00 moles of N_2 and 9.00 moles of H_2 is heated to 650 K in a 1.00 liter container, it is found that the reaction is 71.0%

complete when equilibrium is established. (a) Determine the equilibrium concentrations of the various species. (b) What is the equilibrium constant? (c) How many moles of NH_3 must be added to raise the equilibrium concentration of H_2 to 4.00 M?

14-25. The equilibrium constant for the formation of HBr by the reaction $H_2(g) + Br_2(g) \rightleftharpoons 2\,HBr(g)$ at 1000 K is 2.20×10^6. If 0.0600 mole of HBr is introduced into an empty 3.00 liter flask at 1000 K, determine the concentration of all species at equilibrium.

14-26. At 690 K, solid $TiCl_3$ is in equilibrium with HCl, $TiCl_4$, and H_2 as illustrated by the equation

$$2\,TiCl_3(s) + 2\,HCl(g) \rightleftharpoons 2\,TiCl_4(g) + H_2(g)$$

At equilibrium, the pressure of HCl is 1.153 atm, the pressure of $TiCl_4$ is 1.400 atm, and the total pressure of the system is 3.053 atm. (a) Calculate K_p, the equilibrium constant. (b) What is the equilibrium pressure of H_2 if the equilibrium pressure of HCl equals that of $TiCl_4(g)$? Assume the vapor pressure of $TiCl_3(s)$ is negligible.

14-27. At a certain temperature above 300°C, NOCl dissociates into NO and Cl_2, according to the equation $2\,NOCl(g) \rightleftharpoons 2\,NO(g) + Cl_2(g)$. The total pressure of the system is 708.1 torr, the pressure of NOCl is 241.0 torr, and the pressure of NO is 311.4 torr. (a) Calculate the equilibrium constant, K_p, using pressures in units of atmospheres. (b) What fraction of the total amount of NOCl initially present has dissociated?

14-28. Using the data in Example 14-b, calculate the equilibrium constant, K_p, employing the partial pressures of the various species present. *Hint:* Calculate the pressures using the ideal gas equation.

14-29. An industrial fuel gas is produced by reducing CO_2 with carbon according to the equation $CO_2(g) + C(s) \rightleftharpoons 2\,CO(g)$. At a total pressure of 10.00 atm and 1000°C, the mixture of gases in equilibrium is analyzed to be 7.00 mole % CO_2 and 93.0 mole % CO. (a) Determine the equilibrium constant, K_p, for the reaction. (b) Determine the composition of the gaseous mixture at a total pressure of 40.00 atm at 1000°C.

14-30. At elevated temperatures, limestone dissociates according to the equation $CaCO_3(s) \rightleftharpoons CaO(s) + CO_2(g)$. At 1000°C, the equilibrium pressure of CO_2 is 3.87 atm. If pure $CaCO_3$ is placed in an evacuated 2.50 liter container and heated to exactly 1000°, how much $CaCO_3$ will decompose to produce the equilibrium pressure? Assume the vapor pressures of the solids are negligible.

14-31. At 2000 K, the formation constant, K_p, of NO from its elements according to the equation $N_2(g) + O_2(g) \rightleftharpoons 2\,NO(g)$ is 4×10^{-4}. If the pressure of NO at equilibrium is found to be equal to 0.04 atm, and the pressure of N_2 is equal to the pressure of O_2, determine the equilibrium pressures of N_2 and O_2.

14-32. Derive the relationship between K_c and K_p, that is, $K_p = K_c(RT)^{\Delta n}$. *Hint*: Use the general gas law equation $PV = nRT$ where n/V = concentration in moles per liter. Δn is the difference between the number of moles of gaseous products and reactants.

14-33. 1.00 liter of an equilibrium mixture of NH_3, N_2, and H_2 at 750 K is composed of 1.20 moles of H_2, 1.00 mole of N_2, and 0.329 mole of NH_3. Consider the equilibrium to be $N_2 + 3\,H_2 \rightleftharpoons 2\,NH_3$. (a) Calculate K_c. (b) What is the total pressure of the mixture? (c) Determine the partial pressure of each gas. (d) Calculate K_p from partial pressures and also from the relationship $K_p = K_c(RT)^{\Delta n}$.

14-34. 5.00 moles of PCl_5 are placed in a 25.0 liter container and heated to 250°C. $K_c = 4.15 \times 10^{-2}$. Calculate the concentration of each species at equilibrium. Assume the reaction $PCl_5(g) \rightleftharpoons PCl_3(g) + Cl_2(g)$.

14-35. In the previous problem, if 2.50 moles of Cl_2 are added at constant temperature, what is the concentration of each species when equilibrium has been reestablished?

14-36. At 700 K, $K_p = K_c = 1.83$ for the equilibrium

$$2\,HI(g) \rightleftharpoons H_2(g) + I_2(g)$$

25.4 g of iodine and 0.505 g of hydrogen are heated to 700 K in a 4.00 liter flask. What are the equilibrium concentrations in moles/liter?

Chapter 15

Ionic Equilibrium

15-A. Ionization of Weak Electrolytes

Weak electrolytes are substances that ionize only partially in solution. Invariably in these solutions there are equilibria between the dissociated and undissociated species. Acetic acid typifies a weak electrolyte and its ionization may be represented by the equation

$$HC_2H_3O_2(aq) \rightleftharpoons H^+(aq) + C_2H_3O_2^-(aq)$$

The equilibrium constant for this reaction is written employing the same rules outlined in the previous section on molecular equilibrium—that is, the equilibrium constant is the product of the equilibrium concentrations of the substances on the right of the chemical equation divided by the product of the equilibrium concentrations of the substances on the left. The concentration of each substance is raised to the power given by the coefficient in the balanced chemical equation. For the acetic acid ionization,

$$K_a = \frac{[H^+][C_2H_3O_2^-]}{[HC_2H_3O_2]}$$

The constant K_a is usually referred to as the dissociation or ionization constant of the acid.*

For the ionization of the general weak base† BOH

$$BOH(aq) \rightleftharpoons B^+(aq) + OH^-(aq)$$

the dissociation or ionization constant of the base is expressed as

$$K_b = \frac{[B^+][OH^-]}{[BOH]}$$

A solution of ammonia in water is a typical example of such a weak base. There is a question as to the exact nature of the undissociated species present, and it appears that both NH_3 and hydrogen-bonded species ($HOH—NH_3$) may be present. This means the equilibrium may be complex as shown by the sequence

$$NH_3(aq) + H_2O \rightleftharpoons NH_4OH \rightleftharpoons NH_4^+(aq) + OH^-(aq)$$

However, the ionization equilibrium may, for all practical purposes, be adequately represented either as

$$NH_4OH \rightleftharpoons NH_4^+(aq) + OH^-(aq) \qquad K_b = \frac{[NH_4^+][OH^-]}{[NH_4OH]} \qquad (1)$$

or as

$$NH_3(aq) + H_2O \rightleftharpoons NH_4^+(aq) + OH^-(aq)$$

$$K_b = \frac{[NH_4^+][OH^-]}{[NH_3]} \qquad (2)$$

In the second case, the concentration of water has been omitted from the equilibrium constant expression because water is present in large excess, and its concentration does not change measurably during the course of the reaction. Since $[H_2O]$ is a constant, it is included in the value of K_b, so that the values of K_b for reactions 1 and 2 are numerically the same when

* Acids are molecules or ions that donate a proton to a solvent. For example,

$$HC_2H_3O_2 + H_2O \rightleftharpoons H_3O^+ + C_2H_3O_2^-$$

or

$$HC_2H_3O_2 + 4H_2O \rightleftharpoons H_9O_4^+ + C_2H_3O_2^-.$$

For simplicity, it is usual to show the ionization process as donation of the simple hydrogen ion H^+, but it should be remembered that H^+ is aquated, that is, bonded to H_2O molecules, in aqueous solution.

† A base is a substance that accepts H^+ or generates OH^- in solution.

expressed as above. In these equations, both $[NH_4OH]$ and $[NH_3]$ represent the equilibrium concentration of undissociated ammonia. Other nitrogen bases may be treated similarly.

Substances containing more than one acidic hydrogen ion in the molecule (polyprotic acids) undergo successive dissociation reactions. An equilibrium is eventually established involving all the species present in solution. A typical example is phosphoric acid, and the stepwise equilibria are shown below.

$$H_3PO_4(aq) \rightleftharpoons H^+(aq) + H_2PO_4^-(aq) \quad K_1 = \frac{[H^+][H_2PO_4^-]}{[H_3PO_4]}$$

$$H_2PO_4^-(aq) \rightleftharpoons H^+(aq) + HPO_4^{2-}(aq) \quad K_2 = \frac{[H^+][HPO_4^{2-}]}{[H_2PO_4^-]}$$

$$HPO_4^{2-}(aq) \rightleftharpoons H^+(aq) + PO_4^{3-}(aq) \quad K_3 = \frac{[H^+][PO_4^{3-}]}{[HPO_4^{2-}]}$$

Other common polyprotic acids are sulfuric acid, H_2SO_4; hydrosulfuric acid, H_2S; pyrophosphoric acid, $H_4P_2O_7$; sulfurous acid, H_2SO_3; and carbonic acid, H_2CO_3.

Example 15-a. Consider the ionization of the general weak acid HA:

$$HA(aq) \rightleftharpoons H^+(aq) + A^-(aq)$$

Determine the equilibrium concentrations, in moles/liter, of H^+, A^-, and HA in a solution prepared by adding 0.200 mole of HA to water and diluting to 2.00 liters. K_a for HA is 1.60×10^{-6}.

Solution. We see that the **total** amount of HA originally **added** (including both ionized and nonionized forms of the acid) is 0.200 mole HA. Hence the **total** concentration of HA is 0.200 mole HA/2.00 liters = 0.100 *M*.*

We call this **total** concentration the **stoichiometric concentration** and, using symbolism, we say $C_{HA} = 0.100\ M$. (This is read as "the stoichiometric concentration of HA is 0.100 *M*.")

The stoichiometric concentration represents the total concentration of all forms of the species **added** to the solution. If a substance has **not** been directly **added** (as a free chemical entity) to a solution we say its stoichiometric concentration is **zero**.

In this problem we are asked to find **equilibrium concentrations** and these are indicated by enclosing the appropriate formulas or symbols in brackets. For example $[H^+]$, $[A^-]$, and $[HA]$ represent equilibrium concentrations of H^+, A^-, and HA respectively.

* Some books use the word *initial* instead of *total* here. We choose to stress *total* in our presentation.

Note that C_{HA} and [HA] are **not** identical since part of the total HA ionizes into H^+ and A^-. Hence we see that $[HA] < C_{HA}$. ("$<$" means less than.)

Having introduced our symbolism we are now ready to attack the problem itself.

Since only HA was **added** to the solution we say that $C_{HA} = 0.100\ M$; $C_{H^+} = C_{A^-} = 0$. (C_{H^+} and $C_{A^-} = 0$ since no free H^+ and A^- have been added to the solution.)

When equilibrium is attained we see from the equation

$$HA(aq) \rightleftharpoons H^+(aq) + A^-(aq)$$

that $[H^+]$ and $[A^-]$ are **not** equal to zero and further, that

$$[H^+] = [A^-]$$

since each molecule of HA ionizing yields one H^+ and one A^-. Further,

$$[HA] = C_{HA} - [H^+]$$

since for each H^+ present at equilibrium one HA must have ionized.

The stoichiometric concentrations of the species involved, and the change each undergoes to reach its equilibrium concentration, are summarized in the table below.

Species	Stoichiometric Molarity	Change in *M* due to Ionization	Equilibrium Molarity
HA	C_{HA}	$-[H^+]$	$C_{HA} - [H^+]$
A^-	0	$+[H^+]$	$[H^+]$
H^+	0	$+[H^+]$	$[H^+]$

Using the definition of K_a and substituting we have:

$$K_a = \frac{[H^+][A^-]}{[HA]} = \frac{[H^+][H^+]}{C_{HA} - [H^+]} = \frac{[H^+]^2}{C_{HA} - [H^+]}$$

Since we know the values of K_a and C_{HA} we can solve the equation to determine $[H^+]$. Knowing $[H^+]$ allows us to calculate $[A^-]$ and [HA].

Replacing K_a and C_{HA} with their respective values we have

$$1.60 \times 10^{-6} = \frac{[H^+]^2}{0.100 - [H^+]} \tag{3}$$

or $[H^+]^2 + 1.60 \times 10^{-6}[H^+] - 1.60 \times 10^{-7} = 0$. On solution (by quadratic equation, see Appendix I)

$$[H^+] = 3.99 \times 10^{-4} \text{ mole/liter}$$

Hence

$$[A^-] = [H^+] = 3.99 \times 10^{-4} \text{ mole/liter}$$

$$[HA] = (0.100 - 3.99 \times 10^{-4}) \text{ mole/liter}$$

$$\cong 0.100\ M \text{ ("}\cong\text{" means approximately equal to)}$$

Note that [HA] is smaller than C_{HA} by 3.99×10^{-4} mole/liter. However, in terms of significant figures, we cannot realistically distinguish between C_{HA} and [HA]—that is, the amount of HA undergoing ionization is very, very small in this case. We make use of this fact below.

Alternate Solution of Equation (3). The solution to this problem can be considerably simplified by making the valid assumption that $[H^+]$ is small compared to 0.100. If we consider that $0.100 - [H^+] \cong 0.100$, equation (3) transforms to

$$[H^+]^2 = 1.60 \times 10^{-7}$$

On solution, $[H^+] = \sqrt{1.60 \times 10^{-7}} = 4.00 \times 10^{-4}$ mole/liter, which agrees well with the value of $[H^+]$ obtained above.

Whether this type of approximation procedure is acceptable depends entirely on the accuracy desired in the final answer. This is usually determined by the number of significant figures needed in each individual case. (*Note*: In obtaining the sum or difference of two numbers of different orders of magnitude, the smaller number is neglected. However, in multiplication and division, neglecting the smaller number is **not** permissible.)

Example 15-b. A 0.1000 *M* solution of $HC_2H_3O_2$ is found by conductivity measurements to be 1.33% ionized at 25°C. What is the ionization constant of this acid?

Solution. One and thirty-three hundredths percent of the total amount of $HC_2H_3O_2$ present has dissociated into ions, that is, $0.1000 \times 0.0133 = 0.00133$ mole/liter has dissociated. Therefore, $[H^+] = [C_2H_3O_2^-] = 0.00133$ mole/liter and the equilibrium concentration of unionized acetic acid

$$[HC_2H_3O_2] = C_{HC_2H_3O_2} - [H^+] = 0.1000 - 0.00133 = 0.0987 \text{ mole/liter.}$$

Summarizing:

Species	Stoichiometric Molarity	Change in *M* due to Ionization	Equilibrium Molarity
$HC_2H_3O_2$	0.1000	−0.00133	0.0987
$C_2H_3O_2^-$	0	+0.00133	0.00133
H^+	0	+0.00133	0.00133

Substituting in the equilibrium constant expression

$$K_a = \frac{[H^+][C_2H_3O_2^-]}{[HC_2H_3O_2]} = \frac{(1.33 \times 10^{-3})(1.33 \times 10^{-3})}{9.87 \times 10^{-2}}$$

$$= 1.79 \times 10^{-5}$$

Example 15-c. The numerical value of K_b for NH_3 is 1.8×10^{-5}. What percentage of a 0.0200 M solution of NH_3 is ionized?

Solution.

$$NH_3(aq) + H_2O \rightleftharpoons NH_4^+(aq) + OH^-(aq)$$

The stoichiometric concentrations are $C_{NH_3} = 0.0200$; $C_{OH^-} = C_{NH_4^+} = 0$. Since one OH^- is produced for every NH_4^+,

$$[OH^-] = [NH_4^+]$$

The concentration of NH_3 at equilibrium is equal to the stoichiometric amount present minus the amount ionized:

$$[NH_3] = C_{NH_3} - [OH^-] = 0.0200 - [OH^-]$$

In summary:

Species	Stoichiometric Molarity	Change in M due to Ionization	Equilibrium Molarity
NH_3	0.0200	$-[OH^-]$	$0.0200 - [OH^-]$
NH_4^+	0	$+[OH^-]$	$[OH^-]$
OH^-	0	$+[OH^-]$	$[OH^-]$

Substituting these quantities into the equilibrium constant expression

$$1.8 \times 10^{-5} = \frac{[NH_4^+][OH^-]}{[NH_3]} = \frac{[OH^-][OH^-]}{0.0200 - [OH^-]}$$

and assuming $[OH^-]$ is small compared to 0.0200 we have

$$[OH^-]^2 = (1.8 \times 10^{-5})(2.00 \times 10^{-2}) = 3.6 \times 10^{-7}$$

Hence

$$[OH^-] = 6.0 \times 10^{-4} \text{ mole/liter}$$

and

$$[NH_3] = 0.0200 - 0.0006 = 0.0194 \text{ mole/liter}$$

$$\text{Percentage dissociation} = \frac{6.0 \times 10^{-4}}{0.0200} \times 100 = 3.0\%$$

(Note that $[OH^-]$ is small compared to 0.0200 and hence our approximation is valid.)

Example 15-d. The ionization reactions for H_2S and their equilibrium constants are

$$H_2S(aq) \rightleftharpoons H^+(aq) + HS^-(aq) \quad K_1 = 1.10 \times 10^{-7}$$

$$HS^-(aq) \rightleftharpoons H^+(aq) + S^{2-}(aq) \quad K_2 = 1 \times 10^{-14}$$

(a) What is the equilibrium concentration of H^+ in a 0.100 M solution of H_2S? (b) What is the S^{2-} concentration in a 0.100 M H_2S solution?

Solution. (a) An approximate solution can be obtained as follows.

Since K_1 is very much larger than K_2, it may be assumed that the H^+ coming from the second reaction is negligible compared to that produced in the first reaction.* If this is the case, then the equilibrium concentration of H^+ is equal to the concentration of HS^-. The equilibrium concentration of H_2S is equal to the stoichiometric concentration minus the amount ionized:

$$[H^+] = [HS^-]$$

and

$$[H_2S] = C_{H_2S} - [H^+] = 0.100 - [H^+]$$

Tabulating:

Species	Stoichiometric Molarity	Change in M due to Ionization	Equilibrium Molarity
H_2S	0.100	$-[H^+]$	$0.100 - [H^+]$
HS^-	0	$+[H^+]$	$[H^+]$
H^+	0	$+[H^+]$	$[H^+]$

* *Note*: The hydrogen ion produced by the second ionization is equal to the sulfide ion concentration shown in part (b) to be 1×10^{-14} M. This value is negligible compared to the hydrogen ion produced from the first ionization, that is, 1.05×10^{-4} M.

Substituting these quantities into the equilibrium constant expression

$$K_1 = \frac{[H^+][HS^-]}{[H_2S]} = \frac{[H^+][H^+]}{0.100 - [H^+]} \cong \frac{[H^+]^2}{0.100}$$

if the assumption is made that $[H^+]$ is small compared to 0.100.

$$[H^+] = \sqrt{K_1 \times 0.100} = \sqrt{1.10 \times 10^{-7} \times 1.00 \times 10^{-1}}$$
$$= \sqrt{1.10 \times 10^{-8}} = 1.05 \times 10^{-4} \text{ mole/liter}$$

It is evident that $[H^+]$ is small compared to 0.100; hence the assumption that $0.100 - [H^+] \cong 0.100$ is justified.

(b) To calculate the S^{2-} concentration, the second equilibrium must be employed. From part (a) it is seen that $[HS^-] = 1.05 \times 10^{-4}$ *M* and $[H^+] = 1.05 \times 10^{-4}$ *M*. Substituting these values into the equilibrium expression for K_2, $[S^{2-}]$ can be obtained.

$$K_2 = \frac{[H^+][S^{2-}]}{[HS^-]} = \frac{(1.05 \times 10^{-4})[S^{2-}]}{1.05 \times 10^{-4}} = [S^{2-}]$$

Substituting the numerical value for K_2, $[S^{2-}] = 1 \times 10^{-14}$ mole/liter.

Problems (Ionization)

15-1. The ionization constant for the weak acid HA is 4.0×10^{-8}. Calculate the equilibrium concentration of H^+, A^-, and HA in a 0.050 *M* solution of the acid.

15-2. Calculate the ionization constant for the weak base MOH if a 0.20 *M* solution is 1.5% ionized.

15-3. K_a for amino-benzoic acid, $NH_2C_6H_4CO_2H$, is 1.2×10^{-5}. Calculate $[H^+]$ in a 0.080 *M* solution of the acid.

15-4. Find $[H^+]$ in a 0.15 *M* solution of $HC_2H_3O_2$. The ionization constant is 1.8×10^{-5}. What percentage of $HC_2H_3O_2$ is dissociated?

15-5. Calculate the equilibrium hydroxide ion concentration in a 0.25 *M* solution of methyl amine, CH_3NH_2. This substance reacts with H_2O as illustrated by the equation

$$CH_3NH_2(aq) + H_2O \rightleftharpoons CH_3NH_3^+(aq) + OH^-(aq),$$

for which the equilibrium constant is 3.7×10^{-4}. What percentage of CH_3NH_2 has reacted?

15-6. The aromatic base pyridine C_5H_5N is known to have an equilibrium constant $K_b = 1.7 \times 10^{-9}$ for the reaction

$$C_5H_5N(aq) + H_2O \rightleftharpoons C_5H_5NH^+(aq) + OH^-(aq).$$

(a) Determine $[OH^-]$ in a 0.30 *M* solution. (b) What is $[OH^-]$ in a 0.0060 *M* solution?

15-7. A 2.0 *M* solution of piperidine $C_5H_{10}NH$ is in equilibrium with $C_5H_{10}NH_2^+$ and OH^- as shown by the equation

$$C_5H_{10}NH(aq) + H_2O \rightleftharpoons C_5H_{10}NH_2^+(aq) + OH^-(aq)$$

At equilibrium, it is found that 1.4% of the $C_5H_{10}NH$ is in the ionized form. Calculate the equilibrium constant for the reaction.

15-8. The weak acid HCN is slightly ionized according to the reaction $HCN(aq) \rightleftharpoons H^+(aq) + CN^-(aq)$. A 0.10 *M* solution is found to be 7.0×10^{-3} percent ionized. What is the ionization constant of the acid?

15-9. The ionization constants for oxalic acid, $H_2C_2O_4$, are $K_1 = 5.6 \times 10^{-2}$ and $K_2 = 5.2 \times 10^{-5}$. (a) Determine the equilibrium concentration of H^+ in a 0.25 *M* solution of $H_2C_2O_4$. (b) What is $[C_2O_4^{2-}]$?

15-10. H_2SO_3 is a diprotic acid and undergoes stepwise ionization. The numerical value of K_1 is 1.6×10^{-2} and the value of K_2 is 6.3×10^{-8}. (a) Determine $[H^+]$ in a 0.45 *M* solution. (b) What is the equilibrium concentration of SO_3^{2-}?

15-B. Ionization of Water

The solvent water ionizes into H^+ and OH^- as shown by the equation $H_2O \rightleftharpoons H^+(aq) + OH^-(aq)$. By well-established procedures, it can be shown that the equilibrium concentration of each ion in pure H_2O is 1.00×10^{-7} *M* at 25°C. From this, the value of the ion product constant

$$K_w = [H^+][OH^-] \tag{4}$$

may be determined.* On substitution of the numerical values of the concentrations of H^+ and OH^- in (4), $K_w = 1.00 \times 10^{-7} \times 1.00 \times 10^{-7} = 1.00 \times 10^{-14}$. Under any conditions in dilute aqueous solution, the product of the concentrations of hydrogen and hydroxide ions must equal 1.00×10^{-14} at 25°C. Acidic solutions are those in which the hydrogen ion concentration is greater than the hydroxide ion concentration. For example,

* $[H_2O]$ may be omitted from the ion product expression for the same reason it was omitted in the K_b expression for NH_3 in Section 15-A.

in 0.10 *M* HCl, the acid is essentially completely dissociated so that the hydrogen ion concentration is 0.10 *M*. From the K_w relation

$$[OH^-] = \frac{K_w}{[H^+]} = \frac{1.00 \times 10^{-14}}{1.0 \times 10^{-1}} = 1.0 \times 10^{-13} \text{ mole/liter}$$

Similarly in basic solution, such as 0.010 *M* NaOH, the $[OH^-] = 1.0 \times 10^{-2}$ *M* and therefore $[H^+] = 1.0 \times 10^{-12}$ *M*. Note that in basic solution, the hydroxide ion concentration is greater than the hydrogen ion concentration.

Example 15-e. The $[H^+]$ in the 0.100 *M* solution of $HC_2H_3O_2$ described in Example 15-b was 0.00133 mole/liter. What is $[OH^-]$ in this solution?

Solution.

$$[OH^-] = \frac{K_w}{[H^+]} = \frac{1.00 \times 10^{-14}}{1.33 \times 10^{-3}} = 7.52 \times 10^{-12} \text{ mole/liter}$$

Problems (Ionization of Water)

15-11. Calculate the H^+ and OH^- concentrations of each of the following solutions of strong electrolytes. Assume 100% ionization.

(a) 0.015 *M* HCl
(b) 0.0020 *M* HNO_3
(c) 3.50×10^{-3} *M* $HClO_4$
(d) 0.085 *M* $HClO_4$
(e) 9.5×10^{-5} *M* HCl
(f) 7.75×10^{-4} *M* NaOH
(g) 0.025 *M* $Ba(OH)_2$
(h) 0.0380 *M* KOH
(i) 0.250 *M* KOH

15-C. The pH and pOH Values of Aqueous Solutions

The term pH is used to designate the acidity or basicity of a solution and is defined as $pH = -\log [H^+]$. This is a convenient scale on which the lower numbers indicate an acidic solution and the higher numbers indicate a basic solution. Similarly, the term pOH may be defined as $-\log [OH^-]$. A direct relationship between pH and pOH exists and is easily derived. It was stated in the preceding section that

$$[H^+][OH^-] = 1.0 \times 10^{-14}$$

Taking the logarithms of both sides of the equation

$$\log [H^+] + \log [OH^-] = \log (1.0 \times 10^{-14}) = -14.00$$

Substituting the definitions of pH and pOH, we find

$$pH + pOH = 14.00$$

Table 15-1 gives pH and pOH values for several solutions of strong acids and bases of varying concentrations. Note that the sum of pH and pOH is always 14 for dilute aqueous solutions at 25°C.

Table 15-1

Solution	$[H^+]$	pH	$[OH^-]$	pOH	Nature of Solution
1.0 *M* HCl	1.0 *M*	0	10^{-14} *M*	14	Acidic
0.1 *M* HCl	0.1	1	10^{-13}	13	Acidic
0.0001 *M* HCl	10^{-4}	4	10^{-10}	10	Acidic
Pure water	10^{-7}	7	10^{-7}	7	Neutral
0.001 *M* NaOH	10^{-11}	11	10^{-3}	3	Basic
0.1 *M* NaOH	10^{-13}	13	10^{-1}	1	Basic

Example 15-f. The hydrogen ion concentration of a dilute hydrochloric acid solution is 0.0076 mole/liter. Calculate the pH.

Solution.

$$[H^+] = 7.6 \times 10^{-3} \text{ mole/liter}$$

$$pH = -\log [H^+] = -\log (7.6 \times 10^{-3})^*$$

$$= -[\log 7.6 + \log 10^{-3}]$$

$$pH = -[0.88 - 3.00] = 2.12$$

Example 15-g. The hydroxide ion concentration of a dilute sodium hydroxide solution is 5.2×10^{-5} mole/liter. What is the pH?

Solution.

$$[OH^-] = 5.2 \times 10^{-5} \text{ mole/liter}$$

First calculate $[H^+]$.

$$[H^+] = \frac{1.0 \times 10^{-14}}{[OH^-]} = \frac{1.0 \times 10^{-14}}{5.2 \times 10^{-5}}$$

$$= 1.9 \times 10^{-10} \text{ mole/liter}$$

$$pH = -\log [H^+] = -\log (1.9 \times 10^{-10})$$

$$= -[\log 1.9 + \log 10^{-10}]$$

$$pH = -[0.28 - 10.00] = 9.72$$

* See Appendix III for a review of logarithms.

Example 15-h. The pH of a weakly basic solution is 10.30. Determine the concentration of hydrogen ion and hydroxide ion in the solution.

Solution.

$$\mathrm{pH} = -\log\,[\mathrm{H}^+]$$

$$10.30 = -\log\,[\mathrm{H}^+]$$

$$\log\,[\mathrm{H}^+] = -10.30 = 0.70 - 11.00$$

$$[\mathrm{H}^+] = 5.0 \times 10^{-11}\ \text{mole/liter}$$

$$[\mathrm{OH}^-] = \frac{1.0 \times 10^{-14}}{5.0 \times 10^{-11}} = 2.0 \times 10^{-4}\ \text{mole/liter}$$

Example 15-i. (a) Calculate the pOH of a 0.010 M NH_3 solution. (b) What is the pH? The equilibrium constant is 1.8×10^{-5} for the reaction $NH_3(aq) + H_2O \rightleftharpoons NH_4^+(aq) + OH^-(aq)$.

Solution. (a) Since only NH_3 was added to the solution we have

$$C_{\mathrm{NH_3}} = 0.010;\ C_{\mathrm{OH^-}} = C_{\mathrm{NH_4^+}} = 0.$$

At equilibrium

$$[\mathrm{OH}^-] = [\mathrm{NH_4^+}]$$

and

$$[\mathrm{NH_3}] = C_{\mathrm{NH_3}} - [\mathrm{OH}^-] = 0.010 - [\mathrm{OH}^-]$$

Substituting into the equilibrium constant expression and assuming $0.010 - [OH^-] \cong 0.010$ we have

$$1.8 \times 10^{-5} = \frac{[\mathrm{OH}^-][\mathrm{NH_4^+}]}{C_{\mathrm{NH_3}} - [\mathrm{OH}^-]} \cong \frac{[\mathrm{OH}^-]^2}{0.010}$$

Hence

$$[\mathrm{OH}^-]^2 = 1.8 \times 10^{-7}$$
$$[\mathrm{OH}^-] = 4.2 \times 10^{-4}\ \text{mole/liter}$$
$$\mathrm{pOH} = -\log\,(4.2 \times 10^{-4}) = 3.38$$

(b) The pH may be obtained in either of two ways. (1) From the K_w relationship,

$$[\mathrm{H}^+] = \frac{K_w}{[\mathrm{OH}^-]} = \frac{1.0 \times 10^{-14}}{4.2 \times 10^{-4}} = 2.4 \times 10^{-11}\ \text{mole/liter}$$

Since $pH = -\log [H^+]$,

$$pH = -\log (2.4 \times 10^{-11}) = 10.62$$

Or (2) since

$$pH + pOH = 14.00$$

$$pH = 14.00 - pOH = 14.00 - 3.38 = 10.62$$

Example 15-j. The pH of a 0.500 *M* solution of formic acid, HCOOH, is 2.03. Calculate the equilibrium constant.

Solution. Formic acid ionizes as

$$HCOOH(aq) \rightleftharpoons H^+(aq) + HCOO^-(aq) \quad K_a = \frac{[H^+][HCOO^-]}{[HCOOH]}$$

At equilibrium, the H^+ concentration may be calculated from the pH.

$$pH = 2.03 = -\log [H^+]; \quad [H^+] = 9.3 \times 10^{-3} \text{ mole/liter}$$

Since this is a solution containing only formic acid, then $[H^+] = [HCOO^-]$.

$$[HCOOH] = 0.500 - 0.009 = 0.491 \text{ mole/liter}$$

$$K_a = \frac{(9.3 \times 10^{-3})^2}{0.491} = 1.8 \times 10^{-4}$$

Problems (The pH and pOH Values of Aqueous Solutions)

15-12. Determine the pH of the following solutions. Assume that the electrolytes are 100% dissociated.

(a) 5.2×10^{-1} *M* HCl
(b) 7.6×10^{-3} *M* HNO_3
(c) 9.2×10^{-5} *M* HBr
(d) 8.0×10^{-6} *M* H_2SO_4
(e) 3.3×10^{-4} *M* HI
(f) 6.1×10^{-5} *M* NaOH
(g) 1.8×10^{-4} *M* KOH
(h) 5.0×10^{-3} *M* $Ba(OH)_2$
(i) 6.2×10^{-2} *M* LiOH
(j) 5.3×10^{-5} *M* NaOH

15-13. Determine the pH and pOH of each of the following solutions:
(a) 0.075 *M* NH_3, for which $K_b = 1.8 \times 10^{-5}$.
(b) 0.50 *M* NH_3, for which $K_b = 1.8 \times 10^{-5}$.
(c) 1.00 *M* $HC_2H_3O_2$, for which $K_a = 1.8 \times 10^{-5}$.

15-14. The pH of a 1.00 *M* solution of trichloroacetic acid, $HC_2Cl_3O_2$, is 0.44. What is the equilibrium constant for the acid dissociation reaction $HC_2Cl_3O_2(aq) \rightleftharpoons H^+(aq) + C_2Cl_3O_2^-(aq)$?

15-15. The base hydroxyethylamine $HOCH_2CH_2NH_2$ reacts with water to produce hydroxyethylammonium ion and hydroxide ion as illustrated by the equation

$$HOCH_2CH_2NH_2(aq) + H_2O \rightleftharpoons HOCH_2CH_2NH_3^+(aq) + OH^-(aq)$$

The pOH of a 0.500 *M* solution is 2.27. What is the base equilibrium constant for the reaction?

15-16. Hypochlorous acid dissociates as shown by the equation

$$HClO(aq) \rightleftharpoons H^+(aq) + ClO^-(aq)$$

If a 0.35 *M* solution has a pH = 3.73, what is the ionization constant of the acid?

15-17. A 0.50 *M* solution of hydrazine N_2H_4 has a pH of 11.00. If the equilibrium is represented by the equation

$$N_2H_4(aq) + H_2O \rightleftharpoons N_2H_5^+(aq) + OH^-(aq)$$

calculate the equilibrium constant for the reaction.

15-D. p*K*

Another term useful in considering equilibrium systems is pK, defined as follows:

$$pK = -\log K$$

where K is the equilibrium constant. In dealing with ionic systems the equilibrium constant may be the ionization constant such as K_a, K_b, K_w, and so on.

For the equilibrium

$$H_2O \rightleftharpoons H^+(aq) + OH^-(aq)$$

$$[H^+][OH^-] = K_w = 1.0 \times 10^{-14}$$

$$\log [H^+] + \log [OH^-] = \log K_w = \log (1.0 \times 10^{-14})$$

or

$$-\log [H^+] - \log [OH^-] = -\log K_w = -(-14.00) = 14.00 = pK_w$$

Hence

$$pH + pOH = 14.00 = pK_w$$

Example 15-k. Calculate pK_a for 0.0100 M benzoic acid ($HC_7H_5O_2$) that is 8.0% ionized.

Solution.

$$K_a = \frac{[H^+][C_7H_5O_2^-]}{[HC_7H_5O_2]}$$

$$[H^+] = [C_7H_5O_2^-] = (0.080)(0.0100) = 0.00080 \text{ mole/liter}$$

$$[HC_7H_5O_2] = (0.0100 - 0.0008) = 0.0092 \text{ mole/liter}$$

$$K_a = \frac{(8.0 \times 10^{-4})^2}{9.2 \times 10^{-3}} = 7.0 \times 10^{-5}$$

$$pK_a = -\log K_a = -\log(7.0 \times 10^{-5}) = 4.15$$

Example 15-l. What is the pH of a 0.10 molar solution of bromoacetic acid ($HC_2H_2O_2Br$) if $pK_a = 2.70$?

Solution.

$$pK_a = 2.70 = -\log K_a$$

$$K_a = \text{antilog}(-2.70) = 2.0 \times 10^{-3}$$

$$K_a = \frac{[H^+][C_2H_2O_2Br^-]}{[HC_2H_2O_2Br]} = 2.0 \times 10^{-3}$$

Since $[H^+] = [C_2H_2O_2Br^-]$,

$$\frac{[H^+]^2}{0.10 - [H]^+} = 2.0 \times 10^{-3}$$

and solving using the quadratic equation,

$$[H^+] = 1.3 \times 10^{-2}\ M$$

$$pH = -\log[H^+] = -\log(1.3 \times 10^{-2}) = 1.89$$

Problems (pK)

15-18. Calculate the pH of a 0.010 M solution of KOH at 30°C. Assume complete ionization. $pK_w = 13.83$ at 30°C.

15-19. An aqueous solution of $HC_2H_3O_2$ and $NaC_2H_3O_2$ has a pH of 5.75. What is the ratio of $[C_2H_3O_2^-]$ to $[HC_2H_3O_2]$ in this solution? $pK_a = 4.75$.

15-20. At 45°C $pK_w = 13.40$. Determine the pH of a neutral solution at this temperature.

15-21. pK_a for chloroacetic acid at 25°C = 2.85. Calculate the pH of a 0.0250 M solution of the acid.

15-22. A 0.50M solution of the acid HX has a pH of 4.52. Determine pK_a for this acid.

15-E. The Common Ion Effect

The addition of a strong electrolyte to a solution of a weak electrolyte has a pronounced effect on the equilibrium involving the weak electrolyte if the two electrolytes have a common ion; for example, the addition of the strong electrolyte NH_4Cl to a solution of NH_3 in H_2O disturbs the equilibrium

$$NH_3(aq) + H_2O \rightleftharpoons NH_4^+(aq) + OH^-(aq).$$

The introduction of NH_4Cl increases the NH_4^+ concentration. By LeChatelier's principle, the equilibrium shifts in such a way that the stress is minimized, that is, OH^- and NH_4^+ combine to form NH_3 and H_2O. Thus the equilibrium concentration of OH^- decreases and that of free NH_3 increases.

Example 15-m. Show that $[OH^-]$ in a solution of NH_3 is greater than that in a solution of NH_3 and NH_4Cl. In Example 15-i, it was demonstrated that the equilibrium OH^- concentration of a 0.010 M NH_3 solution was 4.2×10^{-4} M. What is $[OH^-]$ in a solution of 0.020 M NH_4Cl and 0.010 M NH_3?

Solution. Here $C_{NH_3} = 0.010$, $C_{NH_4Cl} = 0.020$, and $C_{OH^-} = 0$. The equilibrium equation is

$$NH_3(aq) + H_2O \rightleftharpoons NH_4^+(aq) + OH^-(aq)$$

At equilibrium we need to determine $[OH^-]$. Using the stoichiometry of the equation we have

$$[NH_3] = \text{Stoichiometric concentration of } NH_3 - \text{Amount ionized in mole/liter}$$
$$= C_{NH_3} - [OH^-] = 0.010 - [OH^-]$$

Similarly

$$[NH_4^+] = \text{Stoichiometric concentration of } NH_4Cl + \text{Amount obtained by ionization of } NH_3 \text{ in mole/liter}$$

$$= C_{NH_4Cl} + [OH^-] = 0.020 + [OH^-]$$

In summary:

Species	Stoichiometric Molarity	Change in M due to Ionization	Equilibrium Molarity
NH_3	0.010	$-[OH^-]$	$0.010 - [OH^-]$
NH_4^+	0.020	$+[OH^-]$	$0.020 + [OH^-]$
OH^-	0	$+[OH^-]$	$[OH^-]$

Substituting in the equilibrium constant expression for the ionization of NH_3

$$K_b = \frac{[NH_4^+][OH^-]}{[NH_3]}$$

$$1.8 \times 10^{-5} = \frac{(0.020 + [OH^-])[OH^-]}{0.010 - [OH^-]}$$

This equation may be solved by use of the quadratic or by an approximation procedure as follows.

If $[OH^-]$ is small compared to 0.010 and 0.020, then

$$[NH_3] = 0.010 - [OH^-] \cong 0.010 \text{ mole/liter}$$
$$[NH_4^+] = 0.020 + [OH^-] \cong 0.020 \text{ mole/liter}$$

Substituting into the equilibrium constant expression,

$$K_b = \frac{(0.020)[OH^-]}{0.010}$$

$$[OH^-] = \frac{(1.8 \times 10^{-5})(1.0 \times 10^{-2})}{2.0 \times 10^{-2}} = 9.0 \times 10^{-6} \text{ mole/liter}$$

To show that the assumption made concerning the relative magnitudes of $[OH^-]$ and 0.010 is valid:

$$[NH_3] = 0.010 - [OH^-] = 0.010 - 0.000009$$
$$= 0.009991 \text{ mole/liter}$$

For all practical purposes, and within the number of significant figures given, 0.009991 is the same as 0.010 and hence the assumption is justified.

In a solution containing only 0.010 M NH_3, $[OH^-] = 4.2 \times 10^{-4}$ M (Example 15-i, Section 15-C), while in a solution of 0.010 M NH_3 + 0.020 M NH_4Cl, $[OH^-] = 9.0 \times 10^{-6}$ M. Thus it is observed that the concentration of the noncommon ion (OH^-) of the weak electrolyte (NH_3) is reduced by the addition of a strong electrolyte (NH_4Cl) having an ion in common (NH_4^+) with the weak electrolyte.

Problems (The Common Ion Effect)

15-23. Show that $[H^+]$ is greater in (a) a solution of 0.20 *M* formic acid, HCO_2H, than in (b) a solution containing 0.20 *M* HCO_2H plus 0.10 *M* sodium formate, $NaHCO_2$. K_a for formic acid, HCO_2H, is 1.8×10^{-4}.

15-24. In which solution is $[OH^-]$ greater: (a) a solution containing 0.60 *M* hydrazinium chloride, N_2H_5Cl, and 0.60 *M* hydrazine, N_2H_4, or (b) in one containing 0.10 *M* N_2H_5Cl and 0.60 *M* N_2H_4? K_b for N_2H_4 is 4.2×10^{-6}. Establish your answer by calculation of $[OH^-]$ in each solution.

15-25. In which solution is $[C_2H_3O_2^-]$ greater: (a) one containing only 0.50 *M* $HC_2H_3O_2$, or (b) one containing 0.50 *M* $HC_2H_3O_2$ plus 0.10 *M* HCl? K_a for $HC_2H_3O_2$ is 1.8×10^{-5}. Establish your answer by calculation of $[C_2H_3O_2^-]$ in each solution.

15-26. In which solution is the equilibrium NH_4^+ concentration greater: (a) one containing 0.40 *M* NH_3, or (b) one containing 0.40 *M* NH_3 plus 0.20 *M* NaOH? K_b for NH_3 is 1.8×10^{-5}. Establish your answer by calculation of $[NH_4^+]$ in each solution.

15-F. Buffer Solutions

A buffer solution is one composed of substances of such a type that there is little or no change in pH on the addition of relatively small amounts of either acid or base. A solution containing both $NaC_2H_3O_2$ and $HC_2H_3O_2$ is such a solution, and, in general, a buffer solution is one composed of either a weak acid and a salt of that acid or a weak base and a salt of that base.

In the case of an $HC_2H_3O_2$–$NaC_2H_3O_2$ buffer, $NaC_2H_3O_2$ is completely ionized, $NaC_2H_3O_2(aq) \rightarrow Na^+(aq) + C_2H_3O_2^-(aq)$; molecular $HC_2H_3O_2$ is in equilibrium with its ions, $HC_2H_3O_2(aq) \rightleftharpoons H^+(aq) + C_2H_3O_2^-(aq)$. Such a solution resists a change in pH in the following ways. Addition of H^+ would ordinarily cause a large decrease in pH. As a result of the presence of $C_2H_3O_2^-$ in large amount, the H^+ is converted to nonionized acetic acid by the net reaction $H^+(aq) + C_2H_3O_2^-(aq) \rightarrow HC_2H_3O_2(aq)$. The decrease in pH is not nearly so large as it would have been in the absence of excess $C_2H_3O_2^-$. Similarly, the addition of OH^- should increase the pH markedly. But because of the presence of excess $HC_2H_3O_2$ molecules, the net reaction

$$HC_2H_3O_2(aq) + OH^-(aq) \rightarrow H_2O + C_2H_3O_2^-(aq)$$

nullifies most of the effect of the added hydroxide ion.

Example 15-n. (a) A solution contains both 1.00 mole of sodium acetate and 1.00 mole of acetic acid in 1.00 liter. What is the pH of the solution? The ionization constant of acetic acid is 1.8×10^{-5}.

Solution.

$$HC_2H_3O_2(aq) \rightleftharpoons H^+(aq) + C_2H_3O_2^-(aq) \quad K_a = \frac{[H^+][C_2H_3O_2^-]}{[HC_2H_3O_2]}$$

Initially we have $C_{HC_2H_3O_2} = 1.00$, $C_{NaC_2H_3O_2} = 1.00$, and $C_{H^+} = 0$. At equilibrium we are asked to find $[H^+]$. Further,

$$\begin{aligned}[HC_2H_3O_2] &= \text{Stoichiometric concentration of } HC_2H_3O_2 \\ &\quad - \text{Amount ionized in mole/liter} \\ &= C_{HC_2H_3O_2} - [H^+] = 1.00 - [H^+]\end{aligned}$$

and

$$\begin{aligned}[C_2H_3O_2^-] &= \text{Stoichiometric concentration of } NaC_2H_3O_2 \\ &\quad + \text{Amount obtained by ionization of } HC_2H_3O_2 \text{ in mole/liter} \\ &= C_{NaC_2H_3O_2} + [H^+] = 1.00 + [H^+]\end{aligned}$$

Summarizing:

Species	Stoichiometric Molarity	Change in M due to Ionization	Equilibrium Molarity
$HC_2H_3O_2$	1.00	$-[H^+]$	$1.00 - [H^+]$
$C_2H_3O_2^-$	1.00	$+[H^+]$	$1.00 + [H^+]$
H^+	0	$+[H^+]$	$[H^+]$

Substituting and assuming that $[H^+]$ is small compared to 1.00, we have

$$1.8 \times 10^{-5} = \frac{[H^+](1.00 + [H^+])}{1.00 - [H^+]}$$

$$\cong \frac{[H^+](1.00)}{(1.00)}$$

Therefore $[H^+] = 1.8 \times 10^{-5}$ mole/liter and the pH of the solution is 4.74.

(b) Assume that 0.20 mole of NaOH is added to 1.00 liter of the solution in part (a) in such a fashion that the volume does not change. What is the pH of the solution when equilibrium is reestablished?

Solution. There are actually two different factors we must consider in solving this problem. First, we must allow the stoichiometric reaction resulting from the addition of OH^- to occur:

$$OH^-(aq) + HC_2H_3O_2(aq) \longrightarrow C_2H_3O_2^-(aq) + H_2O$$

Hence 0.20 mole of NaOH uses up 0.20 mole of $HC_2H_3O_2$ and produces 0.20 mole of $NaC_2H_3O_2$. Second, after the above reactions, we consider changes in molarity due to ionization. These ionization changes have been detailed in previous examples in this chapter. Remember, we always allow the stoichiometric reaction to occur before considering ionization.

When the new equilibrium is established we are again asked to find $[H^+]$.

$$[H^+] = \text{Amount of } HC_2H_3O_2 \text{ ionized in mole/liter}$$

$$\begin{aligned}[C_2H_3O_2^-] &= (\text{Stoichiometric } NaC_2H_3O_2 \text{ concentration} \\ &\quad + NaC_2H_3O_2 \text{ concentration obtained from added NaOH} \\ &\quad + \text{Amount of } HC_2H_3O_2 \text{ ionized in mole/liter}) \\ &= 1.00 + 0.20 + [H^+] \\ &= 1.20 + [H^+] \cong 1.20 \text{ moles/liter}\end{aligned}$$

$$\begin{aligned}[HC_2H_3O_2] &= (\text{Stoichiometric } HC_2H_3O_2 \text{ concentration} \\ &\quad - HC_2H_3O_2 \text{ neutralized by NaOH in mole/liter} \\ &\quad - \text{Amount of } HC_2H_3O_2 \text{ ionized in mole/liter}) \\ &= 1.00 - 0.20 - [H^+] \\ &= 0.80 - [H^+] \cong 0.80 \text{ mole/liter}\end{aligned}$$

In tabular form:

Species	Stoichiometric Molarity	Change in *M* due to Stoichiometric Reaction	Change in *M* due to Ionization	Equilibrium Molarity
$HC_2H_3O_2$	1.00	-0.20	$-[H^+]$	$0.80 - [H^+] \cong 0.80$
$C_2H_3O_2^-$	1.00	$+0.20$	$+[H^+]$	$1.20 + [H^+] \cong 1.20$
H^+	0	0	$+[H^+]$	$[H^+]$

Substituting into the equilibrium constant expression

$$1.8 \times 10^{-5} = \frac{[H^+](1.20)}{0.80}$$

$$[H^+] = 1.2 \times 10^{-5} \text{ mole/liter}$$

$$pH = 4.92 \text{ for the solution.}$$

If the same amount of NaOH, 0.20 mole, had been added to a liter of water or other unbuffered solution, the pH would be 13.30. Thus it is observed that there is a tremendous effect of the buffer in resisting the change of pH.

(c) Assume that 0.50 mole of HCl is added to 1.00 liter of the solution in part (a) in such a fashion that the volume is not changed. What is the pH when equilibrium is reestablished?

Solution. The net stoichiometric reaction is the conversion of 0.50 mole of $C_2H_3O_2^-$ (from $NaC_2H_3O_2$) to 0.50 mole of $HC_2H_3O_2$. At the new equilibrium we must again determine $[H^+]$.

$$[H^+] = \text{Amount of } HC_2H_3O_2 \text{ dissociated in moles/liter}$$

$$\begin{aligned}[C_2H_3O_2^-] &= (\text{Stoichiometric } NaC_2H_3O_2 \text{ concentration} \\ &\quad - NaC_2H_3O_2 \text{ reacting with HCl in mole/liter} \\ &\quad + \text{Amount of } HC_2H_3O_2 \text{ ionized in mole/liter}) \\ &= 1.00 - 0.50 + [H^+] \\ &\cong 0.50 \text{ mole/liter}\end{aligned}$$

$$\begin{aligned}[HC_2H_3O_2] &= (\text{Stoichiometric } HC_2H_3O_2 \text{ concentration} \\ &\quad + HC_2H_3O_2 \text{ concentration obtained by protonation of } C_2H_3O_2^- \\ &\quad - \text{Amount of } HC_2H_3O_2 \text{ ionized in mole/liter}) \\ &= 1.00 + 0.50 - [H^+] \\ &\cong 1.50 \text{ moles/liter}\end{aligned}$$

Tabulating:

Species	Stoichiometric Molarity	Change in M due to Stoichiometric Reaction	Change in M due to Ionization	Equilibrium Molarity
$HC_2H_3O_2$	1.00	+0.50	$-[H^+]$	$1.50 - [H^+] \cong 1.50$
$C_2H_3O_2^-$	1.00	−0.50	$+[H^+]$	$0.50 + [H^+] \cong 0.50$
H^+	0	0	$+[H^+]$	$[H^+]$

Substituting into the equilibrium constant expression and solving for $[H^+]$

$$[H^+] = 1.8 \times 10^{-5} \times \frac{1.50}{0.50} = 5.4 \times 10^{-5} \text{ mole/liter}$$

The pH of the solution is 4.27.

If the same amount of HCl, 0.50 mole, had been added to 1.0 liter of water or other unbuffered solution, the pH of the solution would have been 0.30.

It should be clear from these examples that when substances (acids, bases, etc.) are added to systems, these substances must be allowed to react **before** any equilibrium calculations are carried out.

Problems (Buffer Solutions)

15-27. Find the pH of a buffer solution composed of 0.30 M $HC_2H_3O_2$ and 0.50 M $NaC_2H_3O_2$. $K_a = 1.8 \times 10^{-5}$.

15-28. What are the concentrations of all species (excepting H_2O) in a solution prepared by dissolving 0.50 mole of NH_4Cl in 1.00 liter of 0.50 M NH_3?

15-29. Calculate the pH of a buffer solution composed of 0.060 M NH_3 and 0.30 M NH_4NO_3. $K_b = 1.8 \times 10^{-5}$.

15-30. (a) A buffer solution is made up by dissolving 0.75 mole of $NaC_2H_3O_2$ in 1.00 liter of 0.50 molar $HC_2H_3O_2$. $K_a = 1.8 \times 10^{-5}$. Calculate the pH of the solution. (b) Calculate the pH after the addition of 0.25 mole of HCl to the solution in (a). Assume no volume change.

15-31. (a) What is the pH of a buffer solution composed of 1.00 M NH_3 and 1.00 M NH_4Cl? (b) What number of moles of HCl must be added to 1.00 liter of this buffer to decrease the pH by 1.00 unit, assuming no volume change? $K_b = 1.8 \times 10^{-5}$.

15-32. A buffer solution is composed of 0.500 M $HC_2H_3O_2$ and 0.300 M $NaC_2H_3O_2$. What number of moles of NaOH must be added to 1.00 liter of this buffer to increase the pH by 1.00 unit? Assume no volume change on addition of sodium hydroxide. $K_a = 1.8 \times 10^{-5}$.

General Problems

15-33. The fictitious weak acid HA has $K_a = 4.5 \times 10^{-5}$. Determine $[H^+]$, $[A^-]$, and [HA] in a 0.20 M solution of HA.

15-34. The fictitious weak base BOH has $K_b = 4.0 \times 10^{-4}$. Determine $[OH^-]$, $[B^+]$, $[H^+]$, and [BOH] in a 1.00 M BOH solution.

15-35. A 0.500 M solution of an acid, HA, is found by experiment to be 3.2% ionized. Calculate the ionization constant of the acid.

15-36. A 0.20 M solution of the weak base XOH is 15% ionized. Determine K_b.

15-37. Acetic acid has an ionization constant of 1.8×10^{-5}. What fraction of the $HC_2H_3O_2$ is ionized in (a) a 1.00 M solution, (b) a 0.100 M solution, and (c) a 0.0100 M solution?

15-38. Calculate the pH of a 0.080 M solution of the acid HX. $K_a = 2.5 \times 10^{-6}$.

15-39. The base butylamine has an equilibrium constant of 5.1×10^{-4} for the reaction $C_4H_9NH_2(aq) + H_2O \rightleftharpoons C_4H_9NH_3^+(aq) + OH^-(aq)$. (a) Calculate $[OH^-]$, $[C_4H_9NH_3^+]$, and $[C_4H_9NH_2]$ in a 0.25 M $C_4H_9NH_2$ solution. (b) Determine the pH of the solution. (c) What fraction of the $C_4H_9NH_2$ has dissociated in (i) a 1.0 M solution, (ii) a 0.010 M solution, and (iii) a 0.0010 M solution?

15-40. A 0.54 M solution of cyanoacetic acid, $NCCH_2COOH$, is approximately 50% ionized. Find the equilibrium constant for the reaction $NCCH_2COOH(aq) \rightleftharpoons (NCCH_2COO)^-(aq) + H^+(aq)$.

15-41. A solution of 0.100 M N_2H_4 and 0.300 M N_2H_5Cl was found to have an equilibrium OH^- concentration of 1.3×10^{-6} M. Determine the equilibrium constant for the reaction $N_2H_4(aq) + H_2O \rightleftharpoons N_2H_5^+(aq) + OH^-(aq)$.

15-42. Which of the following acids is **strongest**?

(a) Dichloroacetic acid $K_a = 3.3 \times 10^{-2}$
(b) Valeric acid $K_a = 1.5 \times 10^{-5}$
(c) Glutaric acid $K_a = 3.4 \times 10^{-4}$
(d) Formic acid $K_a = 1.8 \times 10^{-4}$
(e) Hypobromous acid $K_a = 2.1 \times 10^{-9}$

15-43. Calculate the H^+ concentrations and pH values of each of the following solutions of strong bases. Assume 100% ionization.

(a) 0.044 M $Ca(OH)_2$
(b) 0.0030 M KOH
(c) 7.5×10^{-3} M NaOH
(d) 2.4×10^{-4} M $Sr(OH)_2$
(e) 3.0×10^{-3} M $Ba(OH)_2$

15-44. Calculate the OH^- concentrations and pOH values of each of the following solutions of strong acids. Assume 100% ionization.

(a) 4.1×10^{-3} M HCl
(b) 2.3×10^{-2} M $HClO_4$
(c) 9.0×10^{-3} M HCl
(d) 0.20 M $HClO_4$
(e) 8.0×10^{-3} M HNO_3

15-45. A certain acid, HA, has an ionization constant of 1.6×10^{-5}. Determine the pH of a 0.30 M solution of the acid.

15-46. The ionization constant for

$$HC_2H_3O_2(aq) \rightleftharpoons H^+(aq) + C_2H_3O_2^-(aq)$$

is 1.8×10^{-5}. Calculate the quantity of $NaC_2H_3O_2$ in grams to be added to 1.00 liter of 0.250 M $HC_2H_3O_2$ to yield a solution with a pH of 4.50. Assume no volume change.

15-47. A buffer solution is made by dissolving 0.25 mole of NH_4Cl in exactly 500 ml of 0.20 M NH_3. What is the pH of the solution? K_b for $NH_3 = 1.8 \times 10^{-5}$.

15-48. A 0.75 M solution of a hypothetical base, BOH, is found to have a pOH of 3.00. What is the ionization constant of the base?

15-49. A 0.045 M solution of a hypothetical acid, HX, is found to have a pOH of 9.50. Determine the ionization constant of the acid.

15-50. A solution of a weak acid (HA), carefully prepared to be 0.15 M, exhibited a pH of 5.30. What is the ionization constant K_a of this weak acid?

15-51. A diprotic acid, H_2A, dissociates in a stepwise fashion with the equilibrium constants shown

$$H_2A(aq) \rightleftharpoons H^+(aq) + HA^-(aq) \quad K_1 = 3.6 \times 10^{-6}$$

$$HA^-(aq) \rightleftharpoons H^+(aq) + A^{2-}(aq) \quad K_2 = 2 \times 10^{-11}$$

(a) What is the pH of a 0.080 M solution of the acid?
(b) What is the equilibrium concentration of A^{2-}?
(c) What is the equilibrium concentration of HA^-?

15-52. A hypothetical base, $E(OH)_2$, ionizes in a stepwise manner with the equilibrium constants shown.

$$E(OH)_2(aq) \rightleftharpoons EOH^+(aq) + OH^-(aq) \quad K_1 = 4.0 \times 10^{-6}$$

$$EOH^+(aq) \rightleftharpoons E^{2+}(aq) + OH^-(aq) \quad K_2 = 2.0 \times 10^{-12}$$

(a) Calculate the pH of a 2.5×10^{-2} M solution of the base. (b) What is the equilibrium concentration of EOH^+? (c) What is the equilibrium concentration of E^{2+}?

15-53. The pH of a 2.5×10^{-2} M solution of H_2CO_3 is 4.02. Determine the value of K_1, the equilibrium constant for the reaction $H_2CO_3(aq) \rightleftharpoons HCO_3^-(aq) + H^+(aq)$. Assume ionization of HCO_3^- is negligible.

15-54. Find the pH of a solution prepared by mixing 1.0 liter of 0.60 M $HC_2H_3O_2$ and 1.0 liter of 0.30 M NaOH. $K_a = 1.8 \times 10^{-5}$.

15-55. Calculate the pH of a solution prepared by mixing 1.0 liter of 0.60 M NH_3 and 1.0 liter of 0.40 M HCl. $K_b = 1.8 \times 10^{-5}$.

15-56. A solution composed of 0.26 M hydrazoic acid, HN_3, and 0.26 M sodium azide, NaN_3, has a pH of 4.77. What is the ionization constant of HN_3?

15-57. Calculate the ionization constant for HF if a 0.10 M solution has a pH of 2.02.

15-58. The ionization constant of iodic acid, HIO_3, is 1.9×10^{-1}. What fraction of a 0.25 *M* solution is dissociated? Determine the fraction dissociated at several concentrations of the acid, and then plot the fraction dissociated versus the stoichiometric concentration of the acid.

15-59. A buffer is prepared containing 0.250 mole of $NaHCO_3$ and 0.250 mole of Na_2CO_3 in 0.500 liter of solution. The second ionization constant of H_2CO_3 is 4.4×10^{-11}, corresponding to the reaction

$$HCO_3^-(aq) \rightleftharpoons H^+(aq) + CO_3^{2-}(aq).$$

(a) What is the pH of the solution? (b) What is $[OH^-]$? (c) How many moles of HCl must be added to 1.00 liter of this solution to decrease the pH by 1.00 unit?

15-60. Determine the pH of a solution prepared by mixing 333 ml of 0.66 *M* NaOH and 1.0 liter of 0.40 *M* boric acid, H_3BO_3. Boric acid ionizes as shown by the equation $H_3BO_3(aq) \rightleftharpoons H^+(aq) + H_2BO_3^-(aq)$, and has an ionization constant equal to 6.4×10^{-10}. The reaction between NaOH and H_3BO_3 is $NaOH + H_3BO_3 \rightarrow NaH_2BO_3 + H_2O$.

15-61. A vitamin C (ascorbic acid) tablet containing 5.00×10^2 mg of $H_2C_6H_6O_6$ was dissolved in H_2O. The final volume of the solution was 28.4 ml. At equilibrium the pH of the solution was found to be 2.56. By further investigation $[C_6H_6O_6^{2-}]$ was determined to be 1.6×10^{-12} *M*. Calculate K_1 and K_2 for the diprotic acid $H_2C_6H_6O_6$.

Chapter 16
Equilibrium in Saturated Solutions

16-A. Solubility Product Constants

Salts on dissolving in water eventually reach a point of saturation at which an equilibrium is established between the solid salt and its solution. For many salts the amount dissolved is quite low, and the salts are said to be sparingly soluble or slightly soluble. The equilibrium of a sparingly soluble salt, such as $BaSO_4$ with its saturated solution, may be represented by an equation of the type

$$BaSO_4(s) \rightleftharpoons Ba^{2+}(aq) + SO_4^{2-}(aq)$$

An equilibrium constant may be written for this type of reaction just as for any other equilibrium reaction:

$$K = \frac{[Ba^{2+}][SO_4^{2-}]}{[BaSO_4(s)]}$$

Since $BaSO_4(s)$ is a pure substance, the **concentration** or **activity** of $BaSO_4$ in the solid is a constant. Therefore the product $K \times [BaSO_4(s)]$ is also a constant, termed the solubility product constant, and is designated as K_{sp}. Hence

$$K_{sp} = [Ba^{2+}][SO_4^{2-}]$$

For a salt of the general formula M_aX_b, the solubility equilibrium is $M_aX_b(s) \rightleftharpoons a\,M + b\,X$, and the solubility product constant is

$$K_{sp} = [M]^a[X]^b$$

The solubility product constant for any salt is thus the product of the concentrations of its ions in a saturated solution, each concentration raised to the power corresponding to its coefficient in the balanced chemical equation.* Concentrations are always expressed in moles per liter.

Example 16-a. Calculate K_{sp} for TlBr given that the experimentally determined solubility at 25°C is 0.0523 g per 0.100 liter of solution. The equation is

$$TlBr(s) \rightleftharpoons Tl^{+}(aq) + Br^{-}(aq) \quad \text{and} \quad K_{sp} = [Tl^{+}][Br^{-}]$$

Solution. The solubility of TlBr (FW = 284) is given in grams per 0.100 liter, and this must be converted to moles per liter, or molarity. The molarity of the thallous bromide solution is

$$\text{Molarity} = \frac{0.0523\ \text{g}}{0.100\ \text{liter}} \times \frac{1\ \text{mole}}{284\ \text{g}} = 1.84 \times 10^{-3}\ \text{mole/liter}$$

Since TlBr is completely ionized

$$[Tl^{+}] = 1.84 \times 10^{-3}\ \text{mole/liter}$$
$$[Br^{-}] = 1.84 \times 10^{-3}\ \text{mole/liter}$$
$$K_{sp} = [Tl^{+}][Br^{-}] = 3.4 \times 10^{-6}$$

Example 16-b. The solubility of $Cd_3(PO_4)_2$ in water at 20°C was found to be 1.15×10^{-7} mole per liter. What is the K_{sp}?

Solution.

$$Cd_3(PO_4)_2(s) \rightleftharpoons 3\,Cd^{2+}(aq) + 2\,PO_4^{3-}(aq)$$
$$K_{sp} = [Cd^{2+}]^3[PO_4^{3-}]^2$$

Since every formula weight of $Cd_3(PO_4)_2$ that dissolves produces three moles of Cd^{2+} and two of PO_4^{3-}

$$[Cd^{2+}] = \frac{1.15 \times 10^{-7}\ \cancel{\text{mole } Cd_3(PO_4)_2}}{\text{liter}} \times \frac{3\ \text{moles } Cd^{2+}}{1\ \cancel{\text{mole } Cd_3(PO_4)_2}}$$
$$= 3.45 \times 10^{-7}\ \text{mole } Cd^{2+}/\text{liter}$$

* The assumption is made in this treatment that the saturated solutions of slightly soluble solids are completely ionized and that no hydrolysis occurs. Although this is a good approximation in many instances, it does not apply to all electrolytes. In this introductory treatment, however, no attempt is made to include corrections for the association of electrolytes or hydrolysis.

$$[PO_4^{3-}] = \frac{1.15 \times 10^{-7} \text{ mole } Cd_3(PO_4)_2}{\text{liter}} \times \frac{2 \text{ moles } PO_4^{3-}}{1 \text{ mole } Cd_3(PO_4)_2}$$

$$= 2.30 \times 10^{-7} \text{ mole } PO_4^{3-}/\text{liter}$$

$$K_{sp} = (3.45 \times 10^{-7})^3(2.30 \times 10^{-7})^2$$

$$K_{sp} = (41.1 \times 10^{-21}) \times (5.29 \times 10^{-14})$$

$$= 217 \times 10^{-35}$$

$$K_{sp} = 2.2 \times 10^{-33}$$

Example 16-c. K_{sp} for AgI in water was found to be 8.3×10^{-17} at 25°C. What is the solubility in (a) moles per liter, and (b) grams per liter?

Solution.

$$AgI(s) \rightleftharpoons Ag^+(aq) + I^-(aq)$$

$$K_{sp} = [Ag^+][I^-]$$

Let S = solubility of AgI in moles/liter. The solubility of AgI will be numerically equal to the concentration of Ag^+ and the concentration of I^- since each AgI that dissolves produces one Ag^+ and one I^-, or $S = [Ag^+] = [I^-]$. In tabular form:

Species	Stoichiometric Molarity	Change in Molarity	Equilibrium Molarity
Ag^+	0	$+S$	S
I^-	0	$+S$	S

Substituting

$$K_{sp} = S^2$$

$$S = \sqrt{K_{sp}} = \sqrt{8.3 \times 10^{-17}} = \sqrt{83 \times 10^{-18}}$$

$$S = 9.1 \times 10^{-9} \text{ mole/liter}$$

The solubility in grams is obtained by multiplying S by the formula weight, 235. Thus the solubility in grams per liter is

$$\frac{9.1 \times 10^{-9} \text{ mole}}{\text{liter}} \times \frac{235 \text{ g}}{1 \text{ mole}} = 2.1 \times 10^{-6} \text{ g/liter}$$

Example 16-d. K_{sp} for CaF_2 is 4.0×10^{-11} at 25°C. Calculate the solubility of calcium fluoride (FW = 78.1) in moles/liter and in grams/liter.

Solution.

$$CaF_2(s) \rightleftharpoons Ca^{2+}(aq) + 2\,F^-(aq)$$

$$K_{sp} = [Ca^{2+}][F^-]^2$$

Let S = molar solubility of CaF_2 in H_2O

$$[Ca^{2+}] = S$$

$$[F^-] = 2S \text{ (since two } F^- \text{ result from each } CaF_2)$$

Tabulating:

Species	Stoichiometric Molarity	Change in Molarity	Equilibrium Molarity
Ca^{2+}	0	$+S$	S
F^-	0	$+2S$	$2S$

On substitution we have

$$K_{sp} = (S)(2S)^2 = 4S^3$$

$$S = \sqrt[3]{\frac{K_{sp}}{4}} = \sqrt[3]{\frac{4.0 \times 10^{-11}}{4}} = \sqrt[3]{1.0 \times 10^{-11}}$$

$$S = \sqrt[3]{10 \times 10^{-12}}$$

$$S = 2.2 \times 10^{-4} \text{ mole/liter}$$

The solubility in grams per liter is

$$\frac{2.2 \times 10^{-4} \cancel{\text{mole}}}{\text{liter}} \times \frac{78 \text{ g}}{1 \cancel{\text{mole}}} = 1.7 \times 10^{-2} \text{ g/liter}$$

Problems (Solubility Product Constants)

16-1. Which of the following salts is most (a) insoluble? (b) soluble?

(a) $CaCO_3$ $K_{sp} = 4.8 \times 10^{-9}$
(b) $BaCO_3$ $K_{sp} = 8.1 \times 10^{-9}$
(c) $MgCO_3$ $K_{sp} = 1 \times 10^{-5}$
(d) $SrCO_3$ $K_{sp} = 1.6 \times 10^{-9}$
(e) $PbCO_3$ $K_{sp} = 3.3 \times 10^{-14}$

16-2. The solubility of TlCl is 0.975 g in 0.31 liter of water. Determine K_{sp}.

16-3. The solubility of Ag_3PO_4 at 19°C in water is 4.7×10^{-6} M. Assuming the only reaction to be $Ag_3PO_4(s) \rightleftharpoons 3\,Ag^+(aq) + PO_4^{3-}(aq)$, calculate K_{sp}.

16-4. The solubility of Hg_2I_2 in water is 3.0×10^{-7} g/liter. What is K_{sp}? The equation representing the equilibrium involved is

$$Hg_2I_2(s) \rightleftharpoons Hg_2^{2+}(aq) + 2\,I^-(aq).$$

16-5. K_{sp} for TlI has been found to be 4.0×10^{-8}. (a) Calculate the solubility of TlI in moles per liter. (b) What is the solubility in grams per liter?

16-6. K_{sp} for PbI_2 is 7.1×10^{-9}. (a) Determine the solubility of PbI_2 in moles per liter. (b) What is the solubility in grams per liter?

16-7. K_{sp} for Hg_2Cl_2 is 1.3×10^{-18}. Calculate the solubility in water in moles per liter. The equilibrium may be represented by the equation $Hg_2Cl_2(s) \rightleftharpoons Hg_2^{2+}(aq) + 2\,Cl^-(aq)$.

16-B. Precipitation

A precipitate can form only when the solubility of a substance is exceeded. In other words, saturation must precede precipitation.

To predict whether precipitation will occur on mixing two solutions, we should first simply determine the product of the concentrations of ions in the resulting mixture **as if no precipitation occurs**. This product, which has the same form as the solubility product constant, is termed the **trial product**. Its value is then compared to the solubility product constant of the substance that might be expected to precipitate. If the trial product exceeds K_{sp}, precipitation should occur since in a saturated solution the product of cation and anion concentrations cannot exceed the value for a saturated solution—that is, the value of K_{sp}. If the trial product is less than K_{sp}, no precipitation occurs. The following examples clarify this point.

Example 16-e. K_{sp} for silver chloride is 1.8×10^{-10} at 25°C. Should precipitation occur if 1.0 liter of a solution of 1.0×10^{-4} *M* NaCl is mixed with (a) 1.0 liter of 6.0×10^{-7} *M* $AgNO_3$, (b) 2.0 liters of 9.0×10^{-3} *M* $AgNO_3$?

Solution. (a) After mixing, the volume is 2.0 liters; hence the concentration of each species will be halved:

$$C_{Cl^-} = \frac{1.0 \times 10^{-4}\ \text{mole}}{2.0\ \text{liters}} = 5.0 \times 10^{-5}\ \text{mole/liter}$$

$$C_{Ag^+} = \frac{6.0 \times 10^{-7}\ \text{mole}}{2.0\ \text{liters}} = 3.0 \times 10^{-7}\ \text{mole/liter}$$

$$\text{Trial product} = (C_{Ag^+})(C_{Cl^-}) = (3.0 \times 10^{-7})(5.0 \times 10^{-5})$$
$$= 15 \times 10^{-12} = 1.5 \times 10^{-11}$$

Since the trial product is less than the K_{sp} (1.8×10^{-10}), precipitation does **not** occur.

(b) After mixing, the final volume is 3.0 liters, and

$$C_{Cl^-} = \frac{1.0 \times 10^{-4} \text{ mole}}{3.0 \text{ liters}} = 3.3 \times 10^{-5} \text{ mole/liter}$$

$$C_{Ag^+} = \frac{2.0 \cancel{\text{liters}} \times \dfrac{9.0 \times 10^{-3} \text{ mole}}{\cancel{\text{liter}}}}{3.0 \text{ liters}} = \frac{1.8 \times 10^{-2} \text{ mole}}{3.0 \text{ liters}} = 6.0 \times 10^{-3} \text{ mole/liter}$$

$$\text{Trial product} = (6.0 \times 10^{-3})(3.3 \times 10^{-5})$$
$$= 2.0 \times 10^{-7}$$

The trial product of the final solution is greater than K_{sp}. Therefore precipitation **should occur** on mixing these solutions.

Example 16-f. What concentration of I^- is just sufficient to cause precipitation of PbI_2 at 20°C from a solution containing 1.0×10^{-2} *M* Pb^{2+}? That is, what concentration of iodide ion is just sufficient for the trial product to exceed the K_{sp}? K_{sp} for lead iodide is 7.1×10^{-9} at 20°C.

Solution.

$$PbI_2(s) \rightleftharpoons Pb^{2+}(aq) + 2I^-(aq); \quad K_{sp} = [Pb^{2+}][I^-]^2$$

$$[Pb^{2+}] = 1.0 \times 10^{-2} \text{ mole/liter}$$

Substituting into the K_{sp} expression,

$$7.1 \times 10^{-9} = (1.0 \times 10^{-2})[I^-]^2$$
$$[I^-]^2 = 7.1 \times 10^{-7}$$
$$[I^-] = 8.4 \times 10^{-4} \text{ mole/liter}$$

When the concentration of I^- is 8.4×10^{-4} *M*, the solution is just saturated. When the I^- concentration is slightly greater than 8.4×10^{-4} *M*, the trial product is slightly greater than 7.1×10^{-9} and, under these conditions, precipitation should begin.

Example 16-g. In the precipitation of PbI_2 by the addition of solid NaI to a solution of lead nitrate, as in Example 16-f, what concentration of Pb^{2+} remains in solution when sufficient NaI has been added so that the final equilibrium concentration of I^- is 2.0×10^{-3} *M*?

Solution.

$$K_{sp} = [Pb^{2+}][I^-]^2 = 7.1 \times 10^{-9}$$

$$[I^-] = 2.0 \times 10^{-3} \text{ mole/liter}$$

$$[Pb^{2+}] = \frac{K_{sp}}{[I^-]^2} = \frac{7.1 \times 10^{-9}}{(2.0 \times 10^{-3})^2} = \frac{7.1 \times 10^{-9}}{4.0 \times 10^{-6}}$$

$$[Pb^{2+}] = 1.8 \times 10^{-3} \text{ mole/liter}$$

$$= \text{Concentration in final solution}$$

Problems (Precipitation)

16-8. Exactly 0.50 liter of 2.8×10^{-4} *M* $TlNO_3$ solution is mixed with 0.50 liter of 2.8×10^{-4} *M* KI. K_{sp} for TlI is 4.0×10^{-8}. Should precipitation occur? Give the reason for your answer.

16-9. 50.0 ml of 4.5×10^{-6} *M* $Hg_2(NO_3)_2$ and 25.0 ml of 5.0×10^{-6} *M* NaCl are mixed. K_{sp} for Hg_2Cl_2 is 1.3×10^{-18}. Determine whether precipitation should occur and justify your answer.

16-10. K_{sp} for $CaCO_3$ is 4.8×10^{-9}. Should precipitation occur when 20.0 ml of 4.0×10^{-4} *M* $CaCl_2$ are mixed with 60.0 ml of 3.0×10^{-4} *M* Na_2CO_3? Justify your answer.

16-11. K_{sp} for $Be(OH)_2$ is 2.0×10^{-18}. (a) Determine the concentration of OH^- required to just start precipitation from a 5.0×10^{-4} *M* $Be(NO_3)_2$ solution. (b) What concentration of Be^{2+} remains in solution if the OH^- concentration is raised until the pH of the solution is 11.0?

16-12. K_{sp} for $PbBr_2$ is 9×10^{-6}. (a) Calculate the Br^- concentration required to begin precipitation of $PbBr_2$ from a 0.9 *M* solution of $Pb(NO_3)_2$. (b) What concentration of Pb^{2+} remains after the Br^- concentration has been raised to 3×10^{-2} *M*?

16-13. K_{sp} for silver bromate, $AgBrO_3$, is 4×10^{-5}. Calculate the Ag^+ concentration required to start precipitation of $AgBrO_3$ from a solution containing 0.5 g of $NaBrO_3$ per 0.3 liter.

16-C. Solubility and the Common Ion Effect

As in systems involving dissociation of weak electrolytes, the common ion effect is of importance in solubility equilibria; for example, consider the equilibrium

$$AgCl(s) \rightleftharpoons Ag^+(aq) + Cl^-(aq)$$

LeChatelier's principle predicts a decrease in solubility of silver chloride if either a soluble silver salt or a soluble chloride is added. Other equilibria behave similarly.

Example 16-h. Determine the solubility of AgCl in a 1.0×10^{-2} M NaCl solution. K_{sp} for AgCl is 1.8×10^{-10}.

Solution.

$$AgCl(s) \rightleftharpoons Ag^+(aq) + Cl^-(aq)$$

$$K_{sp} = [Ag^+][Cl^-]$$

Let S = Molar solubility of AgCl.

$$[Ag^+] = S \text{ mole/liter}$$

$$[Cl^-] = S + \text{Concentration of NaCl}$$
$$= (S + 0.010) \text{ mole/liter}$$

In tabular form:

Species	Stoichiometric Molarity	Change in Molarity	Equilibrium Molarity
Ag^+	0	$+S$	S
Cl^-	0.010	$+S$	$0.010 + S$

Substituting in the solubility product constant expression,

$$1.8 \times 10^{-10} = (S)(S + 0.010)$$

Assume S is small compared to 0.010, that is,

$$(0.010 + S) \cong 0.010$$
$$1.8 \times 10^{-10} = (S)0.010$$
$$S = 1.8 \times 10^{-8} \text{ mole/liter}$$

Note that S is very small compared to 0.010 and hence the approximation that $(0.010 + S) = 0.010$ is justified.*

The solubility of AgCl in pure water is 1.3×10^{-5} M, and thus the solubility of AgCl is reduced by a factor of about 1000 in 0.010 M NaCl compared to the solubility in pure water.

* The student is again cautioned to check such assumptions in the solution of problems to be certain that the use of an approximation is justified.

Example 16-i. K_{sp} for $PbBr_2$ is 9×10^{-6}. (a) Determine the molar solubility of $PbBr_2$ in a 0.30 *M* NaBr solution. (b) Determine the molar solubility of $PbBr_2$ in a 0.30 *M* $Pb(NO_3)_2$ solution.

Solution.

$$PbBr_2(s) \rightleftharpoons Pb^{2+}(aq) + 2\,Br^-(aq)$$

$$K_{sp} = [Pb^{2+}][Br^-]^2 = 9 \times 10^{-6}$$

(a) Let S = Molar solubility of $PbBr_2$

$$[Pb^{2+}] = S \text{ mole/liter}$$
$$[Br^-] = 2S + \text{Concentration of NaBr} = 2S + 0.30$$

Tabulating:

Species	Stoichiometric Molarity	Change in Molarity	Equilibrium Molarity
Pb^{2+}	0	$+S$	S
Br^-	0.30	$+2S$	$0.30 + 2S$

Substituting in the solubility product constant expression,

$$9 \times 10^{-6} = (S)(2S + 0.30)^2$$

Assume S is small compared to 0.30, that is,

$$2S + 0.30 \cong 0.30$$

Then $9 \times 10^{-6} = (S)(0.30)^2$

$$S = 1 \times 10^{-4} \text{ mole/liter}$$

(b) Let S = Molar solubility of $PbBr_2$

$$[Br^-] = 2S$$
$$[Pb^{2+}] = S + 0.30 \cong 0.30$$

In tabular form:

Species	Stoichiometric Molarity	Change in Molarity	Equilibrium Molarity
Pb^{2+}	0.30	$+S$	$0.30 + S \cong 0.30$
Br^-	0	$+2S$	$2S$

Then $9 \times 10^{-6} = (0.30)(2S)^2 = 1.2S^2$

$$S = \sqrt{7.5 \times 10^{-6}} = 3 \times 10^{-3} \text{ mole/liter}$$

Problems (Solubility and the Common Ion Effect)

16-14. K_{sp} for $PbBr_2$ is 9×10^{-6}. (a) What is the molar solubility of $PbBr_2$ in a 0.5 *M* NaBr solution? (b) What is the solubility of $PbBr_2$ (in grams/liter) in 0.25 *M* $Pb(NO_3)_2$?

16-15. K_{sp} for $BaSO_4$ is 1.1×10^{-10}. What is the molar solubility of $BaSO_4$ in a 3.6×10^{-3} *M* solution of Na_2SO_4?

16-16. K_{sp} for AgI is 8.3×10^{-17}. (a) Find the molar solubility of AgI in a 9.5×10^{-2} *M* $AgNO_3$ solution. (b) What is the solubility in grams per liter? (c) Demonstrate the fact that AgI should be more soluble in water than in the 9.5×10^{-2} *M* $AgNO_3$ solution.

16-17. K_{sp} for TlCl is 1.7×10^{-4}. Determine the molar solubility of TlCl in 5.0×10^{-2} *M* NaCl solution.

16-18. How many grams of CaF_2 will dissolve in 0.250 liter of 0.040 *M* NaF solution. K_{sp} for CaF_2 is 4.0×10^{-11}.

16-D. Simultaneous Equilibria

In many instances, the anion of a salt is the anion of a weak acid and, hence, the solubility of the salt is influenced by the pH of the solution. For example, calcium carbonate is relatively insoluble in water, but in the presence of dilute acid it is readily soluble. Qualitatively, this is illustrated by the following simultaneous equilibria.

$$CaCO_3(s) \rightleftharpoons Ca^{2+}(aq) + CO_3^{2-}(aq)$$

$$K_{sp} = [Ca^{2+}][CO_3^{2-}] = 4.8 \times 10^{-9}$$

$$H_2CO_3(aq) \rightleftharpoons H^+(aq) + HCO_3^-(aq)$$

$$K_1 = \frac{[H^+][HCO_3^-]}{[H_2CO_3]} = 4.3 \times 10^{-7}$$

$$HCO_3^-(aq) \rightleftharpoons H^+(aq) + CO_3^{2-}(aq)$$

$$K_2 = \frac{[H^+][CO_3^{2-}]}{[HCO_3^-]} = 5.6 \times 10^{-11}$$

LeChatelier's principle predicts that the addition of acid causes the latter two equilibria to be shifted to the left, decreasing the concentration of carbonate ion. This causes the first equilibrium to be shifted to the right, thus increasing the solubility of calcium carbonate.

Example 16-j. (a) Calculate the solubility of $CaCO_3$ in water, on the assumption that all carbonate dissolving remains as CO_3^{2-}. K_{sp} for $CaCO_3$ is 4.8×10^{-9}. (b) What is the solubility of calcium carbonate in a buffer at pH = 6.00?

Solution. (a) Let S = the molar solubility of $CaCO_3$ in H_2O.

$$S = [Ca^{2+}] = [CO_3^{2-}]$$

Substituting in the solubility product constant expression

$$K_{sp} = [Ca^{2+}][CO_3^{2-}] = S \times S = S^2$$

$$S = \sqrt{K_{sp}} = \sqrt{4.8 \times 10^{-9}} = 6.9 \times 10^{-5} \text{ mole/liter}$$

(b) Let S' be the molar solubility of $CaCO_3$ in the buffer. Then

$$S' = [Ca^{2+}]$$

S' is also the sum of the concentrations of all carbonate-containing species present:

$$S' = [CO_3^{2-}] + [HCO_3^-] + [H_2CO_3]$$

This is true because at pH 6.00 some of the CO_3^{2-} obtained from the solution of $CaCO_3$ has been protonated to form both HCO_3^- and H_2CO_3. The molar solubility of $CaCO_3$, S', is represented by the **total** amount of carbonate containing species present, whether protonated or not.

To calculate the solubility from the solubility product expression, the $[CO_3^{2-}]$ must be defined in terms of known quantities. From the two dissociation equilibria for carbonic acid, it can be shown that

$$[HCO_3^-] = \frac{[H^+][CO_3^{2-}]}{K_2}$$

and

$$[H_2CO_3] = \frac{[H^+]^2[CO_3^{2-}]}{K_1K_2}$$

By substituting these values into the equation for S',

$$S' = [CO_3^{2-}] + \frac{[H^+][CO_3^{2-}]}{K_2} + \frac{[H^+]^2[CO_3^{2-}]}{K_1K_2}$$

$$S' = [CO_3^{2-}]\left(1 + \frac{[H^+]}{K_2} + \frac{[H^+]^2}{K_1K_2}\right)$$

Solving for $[CO_3^{2-}]$ and substituting into the solubility product constant expression of calcium carbonate, the result is

$$K_{sp} = [Ca^{2+}][CO_3^{2-}] = (S') \times \frac{S'}{1 + \frac{[H^+]}{K_2} + \frac{[H^+]^2}{K_1K_2}}$$

The values of K_1, K_2, and K_{sp} are known, and $[H^+]$ may be calculated from the pH of 6.00 to be 1.0×10^{-6}. By substituting the numerical values into this equation, the solubility of $CaCO_3$ is

$$S' = 1.7 \times 10^{-2} \text{ mole/liter}$$

This represents an increase by a factor of 250 over the solubility in water under the conditions outlined in part (a).

Problems (Simultaneous Equilibria)

16-19. K_{sp} for MgF_2 is 7.1×10^{-9}. If the ionization constant for HF is 6.6×10^{-4}, calculate the molar solubility of MgF_2 in a solution whose pH is buffered at 3.00.

16-20. (a) What is the solubility of CaC_2O_4 in a solution whose pH is buffered at 3.00? K_{sp} for CaC_2O_4 is 2.6×10^{-9} and the acid dissociation constants for $H_2C_2O_4$ are $K_1 = 5.6 \times 10^{-2}$ and $K_2 = 5.2 \times 10^{-5}$. (b) Determine the solubility of CaC_2O_4 in a solution whose pH is buffered at 1.00. (Find S in mole/liter in each case.)

16-21. (a) Determine the solubility of ZnS in a solution whose pH is buffered at 4.00. (b) What is the solubility in a solution whose pH is buffered at 1.00? K_{sp} for ZnS is 1.2×10^{-23}, and the ionization constants for H_2S are $K_1 = 1.1 \times 10^{-7}$ and $K_2 = 1 \times 10^{-14}$. (Find S in mole/liter.)

16-E. Solubility and Separation Procedures

A knowledge of solubility product constants allows the prediction of separation procedures based on the differences in the solubility of the compounds in question. In many instances, considerable wasted time and effort can be avoided by using relatively simple and straightforward calculations. It is not always possible to determine that a process will work in practice, even though it should be successful in principle. However, an indication that a particular process is not possible in principle is almost always demonstrated by experiment to be a correct prediction. Thus, in this case, negative information is extremely valuable.

1. Separation of the Silver Group Cations as Insoluble Chlorides

In a list of solubility product constants of metal chlorides, we find only a few relatively small values—that is, only a few chlorides are relatively insoluble. The more insoluble ones are listed in Table 16-1. In working out a separation procedure, we ask several questions:

1. What concentration of Cl^- is necessary to begin precipitation of each

of the ions listed in Table 16-1 from a solution of particular concentration, for example, a 10^{-3} M solution?

This chloride concentration is easily calculated from the solubility product principles outlined earlier. For AgCl the minimum concentration of Cl^- required is given by the relationship

$$[Cl^-] = \frac{K_{sp}}{[Ag^+]} = \frac{1.8 \times 10^{-10}}{10^{-3}} = 1.8 \times 10^{-7} \text{ mole/liter}$$

Table 16-1 Relatively Insoluble Chlorides

Salt	K_{sp}	Theoretical Minimum $[Cl^-]$ Necessary to Start Precipitation from a 10^{-3} M Solution of Metal Ion
AgCl	1.8×10^{-10}	1.8×10^{-7} mole/liter
$PbCl_2$	1.6×10^{-5}	1.3×10^{-1} mole/liter
Hg_2Cl_2	1.3×10^{-18}	3.6×10^{-8} mole/liter
$HgCl_2$	9.5×10^{-3}	3.1 moles/liter

For $PbCl_2$ the concentration of chloride required is

$$[Cl^-] = \sqrt{\frac{K_{sp}}{[Pb^{2+}]}} = \sqrt{\frac{1.6 \times 10^{-5}}{10^{-3}}} = \sqrt{1.6 \times 10^{-2}}$$

$$= 1.3 \times 10^{-1} \text{ mole/liter}$$

Similarly, these quantities may be calculated for Hg_2^{2+}, Hg^{2+}, and other metal ions if the solubility product constants are known. The commonly encountered chlorides of other metals are all more soluble than mercuric chloride, and hence require even greater concentrations of chloride ion to cause precipitation.

2. The second question we ask is: What concentration of Cl^- is appropriate for a practical separation of Ag^+, Pb^{2+}, and Hg_2^{2+} from Hg^{2+} and other soluble ions? A convenient procedure is to select a Cl^- concentration large enough so that the trial products of those to be precipitated exceed the solubility product constants, but also one small enough that the solubility product constant of $HgCl_2$ is not exceeded. In this instance a Cl^- concentration of 1.0 M would suffice since this exceeds the $PbCl_2$ requirement but is less than that required for $HgCl_2$ precipitation. See Table 16-1.

3. Our third question is: What concentration of metal ion is left in solution after the $[Cl^-]$ is made 1.0 M?

This too can be obtained by use of the solubility product constant expression.

$$[Pb^{2+}] = \frac{K_{sp}}{[Cl^-]^2} = \frac{(1.6 \times 10^{-5})}{(1.0)^2} = 1.6 \times 10^{-5} \text{ mole/liter}$$

$$[Ag^+] = \frac{K_{sp}}{[Cl^-]} = \frac{1.8 \times 10^{-10}}{1.0} = 1.8 \times 10^{-10} \text{ mole/liter}$$

$$[Hg_2^{2+}] = \frac{K_{sp}}{[Cl^-]^2} = \frac{1.3 \times 10^{-18}}{(1.0)^2} = 1.3 \times 10^{-18} \text{ mole/liter}$$

Thus it is observed that Ag^+ and Hg_2^{2+} are quite effectively removed from solution, whereas the concentration of Pb^{2+} is reduced by a factor of 60 from its original concentration of 10^{-3} M.

2. Separation of the Copper Group as Insoluble Sulfides

Many metals form insoluble sulfides of quite varying degrees of solubility. In attempting to work out a scheme for the separation of certain sulfides into two groups, the same sorts of questions are raised as for the chlorides.

1. What minimum concentrations of S^{2-} are required to begin precipitation of each of the metal ions in Table 16-2 if the concentration of each of the metal ions is 10^{-3} M? (These have been calculated as indicated below and included in Table 16-2.)

In solutions of divalent metal sulfides, the equilibria may be represented as $MS(s) \rightleftharpoons M^{2+}(aq) + S^{2-}(aq)$, and K_{sp} may be written as

$$K_{sp} = [M^{2+}][S^{2-}].$$

The theoretical minimum S^{2-} concentration necessary to just start precipitation from a 10^{-3} M Hg^{2+} solution is

$$[S^{2-}] = \frac{K_{sp}}{[Hg^{2+}]} = \frac{10^{-54}}{10^{-3}} = 10^{-51} \text{ mole/liter}$$

For bismuth ion, $K_{sp} = [Bi^{3+}]^2[S^{2-}]^3$, and hence the theoretical minimum sulfide ion concentration necessary to cause precipitation from a 10^{-3} M solution of bismuth ion is

$$[S^{2-}] = \sqrt[3]{\frac{K_{sp}}{[Bi^{3+}]^2}} = \sqrt[3]{\frac{10^{-96}}{10^{-6}}} = 10^{-30} \text{ mole/liter}$$

Table 16-2 Some Insoluble Sulfides

Metal Sulfide	K_{sp}	Theoretical Minimum $[S^{2-}]$ Necessary to Start Precipitation from a 10^{-3} *M* Solution of Metal Ion
HgS	10^{-54}	10^{-51} mole/liter
CuS	10^{-36}	10^{-33} mole/liter
Bi_2S_3	10^{-96}	10^{-30} mole/liter
CdS	10^{-28}	10^{-25} mole/liter
ZnS	10^{-23}	10^{-20} mole/liter
CoS	10^{-21}	10^{-18} mole/liter
FeS	10^{-19}	10^{-16} mole/liter
MnS	10^{-15}	10^{-12} mole/liter

2. What is a convenient point for separation into two groups, and what $[S^{2-}]$ is required to make this separation?

The choice of separation of the ions into two groups depends on subsequent alternative procedures. From a chemical standpoint it is convenient to precipitate Cd^{2+} and the less soluble ones (ions above Cd^{2+} in Table 16-2) and leave Zn^{2+} and the more soluble ones (ions below Zn^{2+}) in solution. Therefore, a S^{2-} concentration just slightly less than 10^{-20} *M* would be ideal. Specifically, the concentration 10^{-21} *M* would work very nicely since this concentration would cause the solubility product constant of CdS to be exceeded, and yet it would be just short of that required to precipitate ZnS.

3. How is this concentration of S^{2-} obtained?

Earlier, H_2S in aqueous solution was described as a weak acid that ionizes in the stepwise fashion below.

$$H_2S(aq) \rightleftharpoons H^+(aq) + HS^-(aq)$$

$$K_1 = \frac{[H^+][HS^-]}{[H_2S]} = 1.1 \times 10^{-7} \quad (1)$$

$$HS^-(aq) \rightleftharpoons H^+(aq) + S^{2-}(aq)$$

$$K_2 = \frac{[H^+][S^{2-}]}{[HS^-]} = 1 \times 10^{-14} \quad (2)$$

Even though the actual dissociation of H_2S is a stepwise process, we can **mathematically** combine the two processes. For example, multiplying K_1 and K_2 we have

$$K_1 \times K_2 = \frac{[H^+][\cancel{HS^-}]}{[H_2S]} \times \frac{[H^+][S^{2-}]}{[\cancel{HS^-}]} = \frac{[H^+]^2[S^{2-}]}{[H_2S]} = 1.1 \times 10^{-21} \tag{3}$$

Hence in a solution of H_2S we see that knowing $[H_2S]$ and $[H^+]$ allows us to calculate $[S^{2-}]$, that is, mathematically

$$[S^{2-}] = K_1K_2 \frac{[H_2S]}{[H^+]^2}$$

From LeChatelier's principle, the common ion effect, and Equations (1) and (2) we can predict that addition of acid to a solution of H_2S would repress the ionization of H_2S and HS^-, thus decreasing the concentration of S^{2-} in solution. Conversely, the addition of OH^- would tend to increase the S^{2-} concentration. It was shown in Example 15-d that the concentration of S^{2-} in H_2S saturated water was 1×10^{-14} M. Therefore, acid must be added to decrease the S^{2-} concentration to 10^{-21} M. What concentration of H^+ produces this S^{2-} concentration?

Rearranging Equation (3) we have

$$[H^+]^2 = K_1K_2 \frac{[H_2S]}{[S^{2-}]}$$

The concentration of dissolved H_2S in a saturated aqueous solution at 1 atm pressure and room temperature is approximately 10^{-1} M. The desired concentration of S^{2-} is 10^{-21} M and $K_1K_2 = 1.1 \times 10^{-21}$. Making the appropriate substitutions,

$$[H^+]^2 = 1.1 \times 10^{-21} \frac{(10^{-1})}{(10^{-21})} = 0.11$$

$$[H^+] = 0.3 \text{ mole/liter}$$
$$pH = 0.5$$

Therefore, to precipitate the ions Hg^{2+}, Bi^{3+}, Cu^{2+}, Cd^{2+} as insoluble sulfides and to leave Zn^{2+}, Fe^{2+}, Co^{2+}, Mn^{2+} in solution, a pH = 0.5 is required to produce the necessary S^{2-} concentration. This pH is easily obtained by preparing a 0.3 M solution of HCl.

Problems (Separations)

16-22. A solution contains both I^- and Cl^-, each at a concentration of 3.0×10^{-2} *M*. What concentration of Pb^{2+} is required to precipitate the maximum amount of I^- without precipitating any Cl^- from the solution? K_{sp} for PbI_2 is 7.1×10^{-9} and that for $PbCl_2$ is 1.6×10^{-5}.

16-23. A solution contains both Tl^+ and Ag^+, each at a concentration of 5.4×10^{-3} *M*. What concentration of Cl^- is needed to precipitate the maximum amount of one of these ions without precipitating any of the other? Which ion precipitates first? K_{sp} for TlCl is 1.7×10^{-4} and that for AgCl is 1.8×10^{-10}.

16-24. A solution contains both Ba^{2+} and Ca^{2+}, each at a concentration of 2×10^{-3} *M*. (a) Calculate the concentration of $C_2O_4^{2-}$ required to precipitate the maximum amount of Ca^{2+} and little or no Ba^{2+}. (b) What pH must be maintained in the solution to achieve this $C_2O_4^{2-}$ concentration if the amount of undissociated $H_2C_2O_4$ is 2×10^{-3} *M*? K_{sp} for BaC_2O_4 is 2×10^{-8} and that for CaC_2O_4 is 3×10^{-9}. The value of K_1 for $H_2C_2O_4$ is 5.6×10^{-2} and K_2 is 5.2×10^{-5}.

16-25. A solution contains both Co^{2+} and Zn^{2+}, each at a concentration of 8×10^{-2} *M*. (a) Calculate the concentration of S^{2-} required to precipitate the maximum amount of Zn^{2+} while leaving as much Co^{2+} in solution as possible. K_{sp} for CoS is 1×10^{-21} and that for ZnS is 1×10^{-23}. (b) What pH is required in a saturated H_2S solution ($[H_2S] = 0.1$ *M*) in order to obtain the needed S^{2-} concentration?

General Problems

16-26. K_{sp} for TlCl is 1.7×10^{-4}. (a) What is the solubility in moles per liter? (b) What is the solubility in grams per liter?

16-27. Determine (a) the solubility of TlCl in a 0.42 *M* solution of NaCl. (b) What is the solubility in (a) in grams per liter? K_{sp} for TlCl is 1.7×10^{-4}.

16-28. K_{sp} for AgBr in water is 5.0×10^{-13}, and K_{sp} for AgBr in methyl alcohol is 6.3×10^{-16} at the same temperature. Calculate the molar solubility of AgBr in water and in methyl alcohol.

16-29. What is the molar solubility of AgBr in 0.125 *M* KBr? K_{sp} for AgBr is 5.0×10^{-13}.

16-30. Calculate K_{sp} for $PbCl_2$ at 25°C. The experimentally determined solubility at this temperature is 2.2 g per 0.500 liter of solution.

16-31. The solubility of Ag_2CrO_4 is 2.04×10^{-2} *grams* per 0.60 liter. Calculate K_{sp}.

16-32. The solubility of PbI_2 in water was found to be 0.083 g per 0.150 liter. What is K_{sp}?

16-33. K_{sp} for BiI_3 is 8×10^{-19}. What is the molar solubility of BiI_3 in 3×10^{-3} *M* sodium iodide?

16-34. How many grams of CaC_2O_4 will dissolve in 0.65 liter of 3.3×10^{-2} *M* $Na_2C_2O_4$ solution? K_{sp} for $CaC_2O_4 = 2.6 \times 10^{-9}$.

16-35. K_{sp} for TlI is 4.0×10^{-8}. (a) What concentration of I^- is required to just cause precipitation of Tl^+ from a 0.025 *M* solution of $TlNO_3$? (b) What concentration of I^- is necessary at equilibrium if $[Tl^+]$ has been reduced to 7.5×10^{-6} *M*?

16-36. K_{sp} for $Pb(IO_3)_2$ is 1.2×10^{-13}. (a) Determine the concentration of Pb^{2+} necessary to cause precipitation from a 0.040 *M* $NaIO_3$ solution. (b) What concentration of Pb^{2+} is necessary to reduce the equilibrium concentration of IO_3^- to 2.5×10^{-4} *M*?

16-37. Should precipitation occur on mixing 35 ml of 6.0×10^{-4} *M* $Pb(NO_3)_2$ and 35 ml of 1.1×10^{-4} *M* $NaIO_3$? K_{sp} for $Pb(IO_3)_2$ is 1.2×10^{-13}.

16-38. Should precipitation occur on mixing 0.25 liter of a solution containing 0.35 g of $NaIO_3$ and 0.25 liter of a solution containing 1.2 g of $AgNO_3$? K_{sp} for $AgIO_3$ is 3.0×10^{-8}.

16-39. Find the molar solubility of $AgBrO_3$ in a solution of 0.05 *M* $AgNO_3$. K_{sp} for $AgBrO_3$ is 4×10^{-5}.

16-40. What minimum concentration of Tl^+ is required to start precipitation of $TlBrO_3$ from a 2.1×10^{-3} *M* solution of $NaBrO_3$? K_{sp} for $TlBrO_3$ is 8.5×10^{-5}.

16-41. K_{sp} for PbS is 8×10^{-28}. Determine the solubility of PbS in a solution of 0.5 *M* HCl that is saturated with H_2S. The concentration of H_2S in a saturated solution is approximately 0.1 *M*. K_1 for H_2S is 1.1×10^{-7}; K_2 is 1×10^{-14}.

16-42. If a solution containing 0.6 *M* $MnCl_2$ and 0.01 *M* HCl is saturated with H_2S (0.1 *M*), does precipitation occur? K_{sp} for MnS is 1.4×10^{-15}.

16-43. The solubility of $TlIO_3$ in water was found to be 6.3×10^{-4} *M*. (a) What is K_{sp} for $TlIO_3$? (b) Determine the molar solubility of $TlIO_3$ in 0.085 *M* $NaIO_3$.

16-44. K_{sp} for $AgBrO_3$ is 4×10^{-5}. Calculate the molar solubility of $AgBrO_3$ in 0.5 *M* $NaBrO_3$.

16-45. How many grams of PbI_2 will dissolve in 0.75 liter of 0.021 *M* $Pb(NO_3)_2$ solution? K_{sp} for $PbI_2 = 7.1 \times 10^{-9}$.

16-46. How many grams of $CaSO_4$ will dissolve in 0.40 liter of 2.5×10^{-2} M Na_2SO_4 solution? K_{sp} for $CaSO_4 = 2.0 \times 10^{-4}$.

16-47. What concentration of PO_4^{3-} is required to just begin precipitation of Ag^+ from a 3.9×10^{-3} M solution of $AgNO_3$? K_{sp} for Ag_3PO_4 is 1.3×10^{-20}.

16-48. A solution contains the following metal ions at a concentration of 2.0×10^{-3} M. If the concentration of S^{2-} is slowly increased, in what order should the metal ions precipitate?

	K_{sp}		K_{sp}
Ag_2S	2.1×10^{-49}	ZnS	1.2×10^{-23}
Tl_2S	5.0×10^{-22}	Sb_2S_3	1.7×10^{-93}
CdS	1.6×10^{-28}	Bi_2S_3	3.0×10^{-96}
HgS	1.6×10^{-54}		

16-49. Sodium sulfate was added to a 0.015 M $BaCl_2$ solution until the solution concentration of SO_4^{2-} was 2.2×10^{-3} M. What concentration of Ba^{2+} (in moles per liter) remained in solution? $K_{sp} = 1.1 \times 10^{-10}$.

16-50. Exactly one liter of a solution originally contained 2.50×10^{-3} M $Pb(ClO_4)_2$ and 4.50×10^{-3} M $Sr(ClO_4)_2$. Sodium sulfate was added until the equilibrium SO_4^{2-} concentration was 5.00×10^{-3} M. (a) What are the concentrations of Pb^{2+} and Sr^{2+} in this equilibrium mixture? (b) What mass of $PbSO_4$ and what mass of $SrSO_4$ have been precipitated? K_{sp} for $PbSO_4 = 1.6 \times 10^{-8}$; K_{sp} for $SrSO_4 = 2.8 \times 10^{-7}$.

16-51. A solution is prepared by mixing $TlNO_3$, $AgNO_3$, and NaI. This solution is saturated with respect to both TlI and AgI. (a) If the concentration of Tl^+ is found to be 3.8×10^{-4} M, what is the concentration of Ag^+? K_{sp} for TlI is 4.0×10^{-8} and for AgI, K_{sp} is 8.3×10^{-17}. (b) What is the concentration of I^-?

16-52. The pH of a saturated $Mg(OH)_2$ solution in pure H_2O is 10.46. (a) Assuming the principal equilibrium to be

$$Mg(OH)_2(s) \rightleftharpoons Mg^{2+}(aq) + 2\,OH^-(aq),$$

determine K_{sp} for $Mg(OH)_2$. (b) What is the molar solubility of $Mg(OH)_2$ in 0.0150 M NaOH?

16-53. A 4.0×10^{-2} M solution of Al^{3+} begins to show precipitation of $Al(OH)_3$ at a pH of 3.90. (a) Calculate K_{sp}. Assume you can observe precipitation at the instant the solubility product is exceeded. (b) What is the molar solubility of $Al(OH)_3$ in a buffer solution of pH 10.70?

16-54. K_{sp} for $Fe(OH)_3$ is 4.0×10^{-38}. For a solution of 8.0×10^{-4} *M* $Fe(ClO_4)_3$, at what pH does precipitation begin? Assume the equation to be $Fe(OH)_3(s) \rightleftharpoons Fe^{3+}(aq) + 3\,OH^-(aq)$.

16-55. K_{sp} for AgCl is 1.8×10^{-10}. Calculate the concentrations of Ag^+ that can be in equilibrium with solid AgCl at each of the chloride ion concentrations given below. Plot $[Ag^+]$ versus $[Cl^-]$; plot points for each of these concentrations of Cl^-: $(0.1, 0.5, 0.7, 0.8, 0.9, 1.0, 2.0, 3.0, 5, 10) \times 10^{-5}$ *M* Cl^-. Is it possible to obtain complete precipitation?

Chapter 17

Hydrolysis and Complex Ion Formation

17-A. Hydrolysis

1. Equilibria and pH

Hydrolysis is the reaction of the anion of a weak acid with water to produce a basic solution or the reaction of the cation of a weak base with water to produce an acidic solution. Hydrolysis obviously affects the equilibrium between water and its ions. The production of H^+ by a hydrolysis reaction brings about a decrease in OH^- concentration in the solution, and similarly the production of OH^- decreases the H^+ concentration.

For example, an aqueous solution of $NaC_2H_3O_2$ has a pH greater than 7, indicating an excess of OH^- over H^+. $NaC_2H_3O_2$ is completely ionized as Na^+ and $C_2H_3O_2^-$, and because $HC_2H_3O_2$ is a weak acid, the reaction

$$C_2H_3O_2^-(aq) + H_2O \rightleftharpoons HC_2H_3O_2(aq) + OH^-(aq)$$

takes place to some extent with the consequent production of OH^-. There is no similar tendency for Na^+ to react with H_2O to form undissociated NaOH and H^+; hence there is a net increase of both the concentration of OH^- and the pH of the $NaC_2H_3O_2$ solution above those of pure water.

Just as other equilibria can be treated quantitatively, hydrolysis reactions lend themselves to mathematical description. For the above hydrolysis of acetate ion, the equilibrium constant may be written as

$$K_h' = \frac{[HC_2H_3O_2][OH^-]}{[C_2H_3O_2^-][H_2O]}$$

As in other cases of dilute solutions, $[H_2O]$ is essentially constant and $K'_h \times [H_2O]$ is a constant denoted as K_h and called a hydrolysis constant. Hence

$$K_h = \frac{[HC_2H_3O_2][OH^-]}{[C_2H_3O_2^-]}$$

Similarly, certain cations can be considered to undergo hydrolysis. The fact that a solution of $FeCl_3$ is acidic can be explained on the basis of the hydrolysis of Fe^{3+}. Because HCl is a strong acid, there is virtually no tendency of Cl^- to react with water to form HCl molecules and OH^-. The hydrolysis of Fe^{3+} may be represented by the equation*

$$Fe^{3+}(aq) + H_2O \rightleftharpoons FeOH^{2+}(aq) + H^+(aq)$$

2. Relationship Between Hydrolysis Constants and Dissociation Constants of Weak Acids and Bases

In the hydrolysis of $C_2H_3O_2^-$, the reaction can be considered to be the sum of two reactions: first, the dissociation of water,

$$H_2O \rightleftharpoons H^+(aq) + OH^-(aq) \quad K_w = [H^+][OH^-] = 1.00 \times 10^{-14} \tag{1}$$

and second, the combination of $C_2H_3O_2^-$ with H^+ to produce undissociated $HC_2H_3O_2$,

$$C_2H_3O_2^-(aq) + H^+(aq) \rightleftharpoons HC_2H_3O_2(aq)$$

$$\frac{1}{K_a} = \frac{[HC_2H_3O_2]}{[H^+][C_2H_3O_2^-]} = \frac{1}{1.8 \times 10^{-5}} \tag{2}$$

* Cations (and anions) in water are never "free," but are associated with numbers of water molecules. A more precise method of explaining the acidity of aquo cations is to consider that molecules of water coordinated to metal ions lose one or more protons to the solvent. The equations for successive ionizations by the aquated ferric ion may be written as

$$[Fe(H_2O)_6]^{3+}(aq) + H_2O \rightleftharpoons [Fe(H_2O)_5OH]^{2+}(aq) + H_3O^+$$

$$[Fe(H_2O)_5OH]^{2+}(aq) + H_2O \rightleftharpoons [Fe(H_2O)_4(OH)_2]^+(aq) + H_3O^+$$

$$[Fe(H_2O)_4(OH)_2]^+(aq) + H_2O \rightleftharpoons [Fe(H_2O)_3(OH)_3](s) + H_3O^+$$

It should be apparent that these equilibria may be treated similarly to those of polyprotic acids such as sulfuric and phosphoric acids. For the sake of simplicity, the coordinated water molecules of the metal ions and that of the hydronium ion have been omitted in the treatment presented here.

Adding the chemical Equations (1) and (2) we obtain

$$C_2H_3O_2^-(aq) + H_2O \rightleftharpoons HC_2H_3O_2(aq) + OH^-(aq)$$

$$K_h = \frac{[HC_2H_3O_2][OH^-]}{[C_2H_3O_2^-]} \qquad (3)$$

Now if we solve algebraic Equation (1) for $[OH^-]$

$$[OH^-] = \frac{K_w}{[H^+]}$$

and substitute in algebraic Equation (3), we obtain

$$K_h = \frac{[HC_2H_3O_2]K_w}{[H^+][C_2H_3O_2^-]}$$

But since

$$\frac{[HC_2H_3O_2]}{[H^+][C_2H_3O_2^-]} = \frac{1}{K_a}$$

$$K_h = \frac{K_w}{K_a}$$

For the hydrolysis of acetate ion, the equilibrium constant is

$$K_h = \frac{1.0 \times 10^{-14}}{1.8 \times 10^{-5}} = 5.6 \times 10^{-10}$$

For the hydrolysis of the cations of weak bases, it is possible to show that

$$K_h = \frac{K_w}{K_b}$$

The student should derive this in a manner similar to the above.

Example 17-a. What is the hydrolysis constant for the $HCOO^-$ ion? K_a for HCOOH is 1.8×10^{-4}.

Solution.

$$HCOO^-(aq) + H_2O \rightleftharpoons HCOOH(aq) + OH^-(aq)$$

$$K_h = \frac{[HCOOH][OH^-]}{[HCOO^-]} = \frac{K_w}{K_a} = \frac{1.0 \times 10^{-14}}{1.8 \times 10^{-4}} = 5.6 \times 10^{-11}$$

Example 17-b. What is the pH of a 1.0 *M* solution of $NaC_2H_3O_2$? The hydrolysis constant is 5.6×10^{-10}.

Solution.

$$C_2H_3O_2^-(aq) + H_2O \rightleftharpoons HC_2H_3O_2(aq) + OH^-(aq)$$

$$K_h = \frac{[HC_2H_3O_2][OH^-]}{[C_2H_3O_2^-]}$$

When equilibrium is attained we see from the hydrolysis equation that

$$[OH^-] = [HC_2H_3O_2]$$

since each $C_2H_3O_2^-$ hydrolyzing yields one OH^- and one $HC_2H_3O_2$. Further

$$\begin{aligned}[C_2H_3O_2^-] &= \text{Stoichiometric concentration of } NaC_2H_3O_2 \\ &\quad - \text{Amount hydrolyzed in mole/liter} \\ &= C_{NaC_2H_3O_2} - [OH^-] = 1.0 - [OH^-]\end{aligned}$$

Summarizing:

Species	Stoichiometric Molarity	Change in Molarity	Equilibrium Molarity
$C_2H_3O_2^-$	1.0	$-[OH^-]$	$1.0 - [OH^-]$
$HC_2H_3O_2$	0	$+[OH^-]$	$[OH^-]$
OH^-	0	$+[OH^-]$	$[OH^-]$

Substituting into the equilibrium constant expression,

$$5.6 \times 10^{-10} = \frac{[OH^-][OH^-]}{1.0 - [OH^-]}$$

On the assumption that $[OH^-]$ is small compared to 1.0, then $(1.0 - [OH^-]) \cong 1.0$, and $[OH^-]^2 = 5.6 \times 10^{-10}$.

$$[OH^-] = 2.4 \times 10^{-5} \text{ mole/liter}$$

$$[H^+] = \frac{K_w}{[OH^-]} = \frac{1.0 \times 10^{-14}}{2.4 \times 10^{-5}} = 4.2 \times 10^{-10} \text{ mole/liter}$$

$$pH = -\log(4.2 \times 10^{-10}) = 9.38$$

Example 17-c. The pH of a 0.100 *M* solution of $FeCl_3$ is 2.00. On the assumption that the only hydrolysis reaction of importance is

$$Fe^{3+}(aq) + H_2O \rightleftharpoons FeOH^{2+}(aq) + H^+(aq)$$

calculate the hydrolysis constant.

Solution.

$$K_h = \frac{[FeOH^{2+}][H^+]}{[Fe^{3+}]}$$

From the pH,

$$[H^+] = \text{antilog}(-2.00) = 1.0 \times 10^{-2} \text{ mole/liter}$$
$$[FeOH^{2+}] = [H^+]$$
$$[Fe^{3+}] = 0.100 - [FeOH^{2+}] = (0.100 - 0.010)$$
$$= 0.090 \text{ mole/liter}$$

Substituting in the equilibrium constant expression,

$$K_h = \frac{(1.0 \times 10^{-2})(1.0 \times 10^{-2})}{0.090} = 1.1 \times 10^{-3}$$

Problems (Hydrolysis)

17-1. (a) Calculate the hydrolysis constant for the hydrolysis reaction

$$NO_2^-(aq) + H_2O \rightleftharpoons HNO_2(aq) + OH^-(aq)$$

if K_a for HNO_2 is 4.6×10^{-4}.

(b) Determine the pH of a 0.55 M solution of $NaNO_2$.

17-2. What is the pH of a solution of 0.35 M $Zn(ClO_4)_2$ if K_h is 1.0×10^{-9} for the reaction $Zn^{2+}(aq) + H_2O \rightleftharpoons ZnOH^+(aq) + H^+(aq)$?

17-3. The hydrolysis of NH_4^+ proceeds according to the equation

$$NH_4^+(aq) + H_2O \rightleftharpoons NH_3(aq) + H_3O^+$$

and K_b for NH_3 is 1.8×10^{-5}. (a) Calculate the equilibrium constant for the hydrolysis reaction. (b) For a 0.22 M solution of NH_4NO_3 what is $[H^+]$? (c) What is the equilibrium OH^- concentration of the solution? (d) What is the pH of the solution? (e) Calculate the equilibrium concentration of NH_3.

17-4. A 0.250 M solution of NaCN is prepared. What is the pH of the solution if K_a for HCN is 4.9×10^{-10}?

17-5. The hydrolysis constant for the reaction

$$Be^{2+}(aq) + H_2O \rightleftharpoons BeOH^+(aq) + H^+(aq)$$

is 2.0×10^{-7}. (a) Determine the pH of a solution of 9.5×10^{-2} M $BeCl_2$. (b) What is the equilibrium concentration of $BeOH^+$?

17-6. The pH of a 0.1000 M solution of $Th(ClO_4)_4$ is 2.65. (a) Determine the hydrolysis constant for the reaction

$$Th^{4+}(aq) + H_2O \rightleftharpoons ThOH^{3+}(aq) + H^+(aq)$$

(b) What is the concentration of $ThOH^{3+}$ at equilibrium?

17-7. Calculate the pH at the end point of a titration of 0.36 M HNO_2 and 0.36 M KOH. K_a for HNO_2 is 4.6×10^{-4}.

17-8. Find the pH at the end point of a titration of 0.30 M NH_3 with 0.60 M HNO_3. K_b for ammonia is 1.8×10^{-5}.

17-B. Complex Ion Formation

Many positive ions associate in aqueous solution with negative ions or neutral molecules to form complex ions. An equilibrium exists between the various species present and may be written in the form of a dissociation reaction analogous to that for a weak electrolyte.

In a solution of $AgNO_3$ and NH_3, the diamminesilver complex ion, $Ag(NH_3)_2^+$, is formed and is in equilibrium with Ag^+ ions and NH_3 molecules. This is represented by the equation

$$Ag(NH_3)_2^+ \rightleftharpoons Ag^+ + 2\,NH_3$$

and the equilibrium or dissociation constant for the reaction is

$$K_d = \frac{[Ag^+][NH_3]^2}{[Ag(NH_3)_2^+]} = 6.2 \times 10^{-8}$$

Similarly, Ag^+ reacts with CN^- to form the complex $Ag(CN)_2^-$, whose equilibrium can be represented $Ag(CN)_2^- \rightleftharpoons Ag^+ + 2\,CN^-$. The dissociation constant is given by the equation

$$K_d = \frac{[Ag^+][CN^-]^2}{[Ag(CN)_2^-]} = 1.4 \times 10^{-20}$$

Complex ion formation is especially important in understanding the chemistry of ions in solution. The study of solubilities of normally insoluble substances in the presence of complexing agents is one of the more important areas of solution chemistry.

For example, AgCl(s) is soluble in excess NH_3 owing to the formation of the $Ag(NH_3)_2^+$ complex ion. This problem may be treated as an example of simultaneous equilibria analogous to those described previously. The equilibria involved are

$$AgCl(s) \rightleftharpoons Ag^+ + Cl^- \qquad K_{sp} = [Ag^+][Cl^-] \qquad (4)$$

$$Ag(NH_3)_2^+ \rightleftharpoons Ag^+ + 2\,NH_3 \qquad K_d = \frac{[Ag^+][NH_3]^2}{[Ag(NH_3)_2^+]} \qquad (5)$$

Addition of ammonia should, according to LeChatelier's principle, cause the conversion of Ag^+ to $Ag(NH_3)_2^+$. Therefore more $AgCl(s)$ should dissolve to replace part of the Ag^+ used up. The solubility equilibrium under these conditions is represented by the equation

$$AgCl(s) + 2\,NH_3 \rightleftharpoons Ag(NH_3)_2^+ + Cl^-.$$

The equilibrium constant for this reaction has the form

$$K_{eq} = \frac{[Ag(NH_3)_2^+][Cl^-]}{[NH_3]^2}$$

and can be related to the K_{sp} of AgCl and the K_d of the complex. Solving the K_{sp} and K_d equations for $[Ag^+]$ and equating them because the $[Ag^+]$ is the same for both equilibria, we have

$$[Ag^+] = \frac{K_{sp}}{[Cl^-]} = \frac{K_d[Ag(NH_3)_2^+]}{[NH_3]^2}$$

Solving for K_{sp}/K_d, we obtain

$$\frac{K_{sp}}{K_d} = \frac{[Ag(NH_3)_2^+][Cl^-]}{[NH_3]^2}$$

This equation is in the same form as that of the equilibrium constant for the net chemical reaction, and hence the ratio K_{sp}/K_d is numerically equal to K_{eq}.

Other systems may be treated similarly.

Example 17-d. A solution of 0.050 *M* $K[Ag(CN)_2]$ was prepared in which the equilibrium concentration of Ag^+ was found to be 5.6×10^{-8} *M*. Determine the value of K_d for the dissociation of $Ag(CN)_2^-$ ion.

Solution.

$$Ag(CN)_2^- \rightleftharpoons Ag^+ + 2\,CN^-$$

$$K_d = \frac{[Ag^+][CN^-]^2}{[Ag(CN)_2^-]}$$

At equilibrium $[Ag^+] = 5.6 \times 10^{-8}$ mole/liter. Since $2\,CN^-$ are obtained for each Ag^+

$$[CN^-] = \frac{5.6 \times 10^{-8}\ \cancel{\text{mole Ag}^+}}{1\ \text{liter}} \times \frac{2\ \text{mole CN}^-}{1\ \cancel{\text{mole Ag}^+}} = 1.12 \times 10^{-7}\ \frac{\text{mole CN}^-}{\text{liter}}$$

$$\begin{aligned}[Ag(CN)_2^-] &= \text{Stoichiometric concentration of } Ag(CN)_2^- \\ &\quad - \text{Amount of } Ag(CN)_2^- \text{ dissociated in mole/liter} \\ &= 0.050 - 5.6 \times 10^{-8} \cong 0.050 \text{ mole/liter}\end{aligned}$$

Summarizing:

Species	Stoichiometric Molarity	Change in Molarity	Equilibrium Molarity
$Ag(CN)_2^-$	0.050	-5.6×10^{-8}	$\cong 0.050$
Ag^+	0	$+5.6 \times 10^{-8}$	5.6×10^{-8}
CN^-	0	$+1.12 \times 10^{-7}$	1.12×10^{-7}

Substituting into the K_d expression

$$K_d = \frac{5.6 \times 10^{-8}(1.12 \times 10^{-7})^2}{5.0 \times 10^{-2}} = 1.4 \times 10^{-20}$$

Example 17-e. A 3.0×10^{-2} *M* solution of $[Ag(NH_3)_2]NO_3$ is prepared by dissolving the solid in water. If the dissociation constant of the complex ion is 6.2×10^{-8}, what is the concentration of Ag^+ and NH_3 in the equilibrium solution (assuming NH_3 does not dissociate as a base)? What fraction of the complex ion has dissociated?

Solution.

$$Ag(NH_3)_2^+ \rightleftharpoons Ag^+ + 2\,NH_3$$

$$K_d = \frac{[Ag^+][NH_3]^2}{[Ag(NH_3)_2^+]}$$

At equilibrium

$$[NH_3] = 2\,[Ag^+]$$

and

$$[Ag(NH_3)_2^+] = C_{Ag(NH_3)_2^+} - [Ag^+] \cong C_{Ag(NH_3)_2^+} = 3.0 \times 10^{-2} \text{ mole/liter}$$

Tabulating:

Species	Stoichiometric Molarity	Change in Molarity	Equilibrium Molarity
$Ag(NH_3)_2^+$	3.0×10^{-2}	$-[Ag^+]$	$\cong 3.0 \times 10^{-2}$
NH_3	0	$+2[Ag^+]$	$2[Ag^+]$
Ag^+	0	$+[Ag^+]$	$[Ag^+]$

Substituting into the equilibrium constant expression

$$6.2 \times 10^{-8} = \frac{[Ag^+](2[Ag^+])^2}{3.0 \times 10^{-2}} = \frac{4[Ag^+]^3}{3.0 \times 10^{-2}}$$

Solving for $[Ag^+]$ we obtain

$$[Ag^+] = 7.7 \times 10^{-4} \text{ mole/liter}$$

The *fraction* of complex dissociated is

$$\frac{[Ag^+]}{C_{Ag(NH_3)_2^+}} = \frac{7.7 \times 10^{-4}}{3.0 \times 10^{-2}} = 2.6 \times 10^{-2}$$

Example 17-f. Excess $AgCl(s)$ is treated with a 0.100 M solution of NH_3. The equation for the reaction is

$$AgCl(s) + 2\ NH_3 \rightleftharpoons Ag(NH_3)_2^+ + Cl^-$$

What are the equilibrium concentrations of NH_3, $Ag(NH_3)_2^+$, Cl^-, and Ag^+?

Solution. K_{eq} for the equilibrium above was shown to be K_{sp}/K_d. Substitution of the appropriate numerical values yields

$$K_{eq} = \frac{1.8 \times 10^{-10}}{6.2 \times 10^{-8}} = 2.9 \times 10^{-3}$$

Further

$$K_{eq} = \frac{[Ag(NH_3)_2^+][Cl^-]}{[NH_3]^2}$$

At equilibrium since one $Ag(NH_3)_2^+$ is obtained for each Cl^-

$$[Ag(NH_3)_2^+] = [Cl^-]$$

Further

$$[NH_3] = NH_3 \text{ total} - NH_3 \text{ reacted (both in mole/liter)}$$
$$= C_{NH_3} - 2[Cl^-] \text{ (since 2 } NH_3 \text{ react for each } Cl^- \text{ formed)}$$
$$= 0.100 - 2[Cl^-]$$

In tabular form:

Species	Stoichiometric Molarity	Change in Molarity	Equilibrium Molarity
NH_3	0.100	$-2[Cl^-]$	$0.100 - 2[Cl^-]$
$Ag(NH_3)_2^+$	0	$+[Cl^-]$	$[Cl^-]$
Cl^-	0	$+[Cl^-]$	$[Cl^-]$

Substituting into the expression for K_{eq}

$$2.9 \times 10^{-3} = \frac{[Cl^-][Cl^-]}{(0.100 - 2[Cl^-])^2} = \frac{[Cl^-]^2}{(0.100 - 2[Cl^-])^2}$$

Taking the square root of both sides we have

$$5.4 \times 10^{-2} = \frac{[Cl^-]}{0.100 - 2[Cl^-]}$$

Solving for $[Cl^-]$

$$[Cl^-] = 4.9 \times 10^{-3} \text{ mole/liter} = [Ag(NH_3)_2^+]$$

$$[NH_3] = 0.100 - 2[Cl^-] = 0.090 \text{ mole/liter}$$

$$[Ag^+] = \frac{K_{sp}}{[Cl^-]} = \frac{1.8 \times 10^{-10}}{4.9 \times 10^{-3}} = 3.7 \times 10^{-8} \text{ mole/liter}$$

Problems (Complex Ion Formation)

17-9. When a solution is prepared containing 0.100 M Cd^{2+} and 0.400 M NH_3, it is found that at equilibrium the concentration of free Cd^{2+} is 1.0×10^{-2} M. What is the equilibrium constant for the dissociation reaction $Cd(NH_3)_4^{2+} \rightleftharpoons Cd^{2+} + 4NH_3$ if this is the only reaction of significance occurring in solution?

17-10. The complex ion $CdBr^+$ forms on mixing solutions of Cd^{2+} and Br^-. In a particular solution containing 0.60 M total Cd^{2+}, it is found that only 20.0% of the total exists as free Cd^{2+} when the free Br^- concentration is 5.2×10^{-2} M. What is the dissociation constant for the reaction $CdBr^+ \rightleftharpoons Cd^{2+} + Br^-$?

17-11. A solution of 0.16 M $[Cu(NH_3)_4](NO_3)_2$ was prepared. On analysis it was found that the equilibrium concentration of NH_3 was 2.0×10^{-3} M. Assuming the only equilibrium to be

$$Cu(NH_3)_4^{2+} \rightleftharpoons Cu^{2+} + 4\,NH_3$$

determine the value of K_d.

17-12. A solution of $K[Au(CN)_2]$ was prepared by dissolving 63.0 g of the complex salt in water. The total volume was 0.600 liter. At equilibrium the concentration of CN^- was found to be 2.0×10^{-16} M. Assuming that the only equilibrium involved is the dissociation of the complex ion according to the equation $Au(CN)_2^- \rightleftharpoons Au^+ + 2\,CN^-$, what is the value of the dissociation constant?

17-13. The dissociation constant for $HgCl_4^{2-}$ in aqueous solution is 8.5×10^{-16} for the reaction $HgCl_4^{2-} \rightleftharpoons Hg^{2+} + 4\,Cl^-$. Assuming this to be the only equilibrium involved, what is the concentration of each of the above species in the equilibrium mixture prepared by dissolving 0.50 mole of $K_2[HgCl_4]$ in 1.00 liter of solution?

17-14. The reaction for the dissociation of $Hg(SCN)_4^{2-}$ may be represented by the equation $Hg(SCN)_4^{2-} \rightleftharpoons Hg^{2+} + 4\,SCN^-$, which has an equilibrium constant $K_d = 1.2 \times 10^{-21}$. (a) What percentage of a 0.100 M solution of $K_2[Hg(SCN)_4]$ dissociates, assuming that this is the only reaction occurring? (b) Calculate the concentrations of the various species involved when equilibrium is established.

17-15. $Pb(OH)_2$ dissolves to a certain extent in a 1.0 M solution of NaOH. Assuming an excess of solid $Pb(OH)_2$, (a) what are the equilibrium concentrations of Pb^{2+}, $Pb(OH)_3^-$, and OH^-? (b) What is the total solubility of $Pb(OH)_2$ in the 1.0 M solution of NaOH? Given the following equilibria,

$$Pb(OH)_2(s) \rightleftharpoons Pb^{2+} + 2\,OH^- \quad K_{sp} = 1.0 \times 10^{-16}$$

and

$$Pb(OH)_3^- \rightleftharpoons Pb^{2+} + 3\,OH^- \quad K_d = 1.3 \times 10^{-14}$$

17-16. (a) What is the equilibrium constant for the reaction

$$Co(NH_3)_6^{3+} + 6\,CN^- \rightleftharpoons Co(CN)_6^{3-} + 6\,NH_3$$

K_d for $Co(NH_3)_6^{3+} = 2.2 \times 10^{-34}$ and K_d for $Co(CN)_6^{3-}$ is 1.0×10^{-64}. (b) What is the concentration of NH_3 in an equilibrium solution when the same concentration of each cobalt complex is present and the concentration of uncomplexed CN^- is 1.0×10^{-6} mole/liter?

General Problems

17-17. Calculate the pH of a solution of 0.45 M $NaHCO_2$. K_a for HCOOH is 1.8×10^{-4}.

17-18. What is the pH of a 0.32 M solution of NH_4NO_3? K_b for NH_3 is 1.8×10^{-5}.

17-19. The pH of a 0.55 M solution of the salt NaX is found to be 9.15. Assuming that HX is a weak acid, determine the ionization constant of the acid.

17-20. The pH of a 0.250 M solution of the salt DCl is 4.70. Assuming that DOH is a weak base, calculate K_b, the ionization constant of the base.

17-21. A 25.0 ml sample of 0.360 M HZ (a weak acid) is titrated with 0.360 M NaOH. (a) If the pH of the solution is 9.30 at the equivalence point, calculate K_a for HZ. (b) Determine [HZ] at the equivalence point.

17-22. The equilibrium constant for the dissociation of CdI^+, by the reaction $CdI^+ \rightleftharpoons Cd^{2+} + I^-$, is 5.2×10^{-3}. In a solution containing 0.200 M Cd^{2+} total and 0.200 M I^- total, what fraction of the cadmium present exists in the form of the complex ion?

17-23. A 0.0100 M solution of $HgCl_2$ is found to be 0.029% dissociated in aqueous solution. Calculate K_d for the reaction

$$HgCl_2(aq) \rightleftharpoons Hg^{2+}(aq) + 2\,Cl^-(aq)$$

assuming that this is the only important equilibrium established in the solution.

17-24. The dissociation constant of $Ag(NH_3)_2^+$ has been established as 6.2×10^{-8}. If NH_3 is added to a solution containing 0.300 M $AgNO_3$, what is the concentration of free uncomplexed NH_3 when 85% of the silver has been converted to $Ag(NH_3)_2^+$?

17-25. A solution is prepared by mixing 0.500 liter of 0.0200 M $Fe(ClO_4)_3$ and 0.500 liter of 0.0200 M NaF. A reaction occurs that involves formation of FeF^{2+}. When equilibrium is established, the concentration of FeF^{2+} is found to be 9.8×10^{-3} M. (a) Calculate the equilibrium concentrations of Fe^{3+} and F^-. (b) What is K_d for FeF^{2+}?

17-26. The dissociation constant for $Cu(NH_3)_4^{2+}$ is 4.9×10^{-14}. If a solution is to be prepared in which 33.3% of the complex has dissociated as in the reaction $Cu(NH_3)_4^{2+} \rightleftharpoons Cu^{2+} + 4\,NH_3$, what concentration of uncomplexed NH_3 must there be in the solution if the above equilibrium is the only one to be considered?

17-27. Excess solid $AgIO_3$ ($K_{sp} = 3.0 \times 10^{-8}$) is shaken with a solution of NH_3 of initial concentration 0.50 *M*. (a) When equilibrium has been achieved, what mass of $AgIO_3$ has been dissolved? K_d for $Ag(NH_3)_2^+$ is 6.2×10^{-8}. (b) Find the equilibrium concentration of NH_3.

17-28. The hydroxide $Cr(OH)_3$ is slightly soluble in dilute NaOH. An excess of solid $Cr(OH)_3$ is equilibrated with a solution containing initially 0.60 *M* NaOH, and at equilibrium the concentration of OH^- is found to be 0.43 *M*. What is the equilibrium constant for the reaction

$$Cr(OH)_3(s) + OH^-(aq) \rightleftharpoons Cr(OH)_4^-(aq)?$$

If the solubility product constant of $Cr(OH)_3$ is 2.0×10^{-30}, calculate the dissociation constant for the reaction

$$Cr(OH)_4^-(aq) \rightleftharpoons Cr^{3+}(aq) + 4\,OH^-(aq).$$

17-29. Hydrolysis of a cation in aqueous solution is an alternative way of describing the formation of hydroxide ion complexes of metals. (a) Show that the hydrolysis constant of Be^{2+} is directly related to the equilibrium constant for the reaction

$$Be(OH)^+(aq) \rightleftharpoons Be^{2+}(aq) + OH^-(aq)$$

The hydrolysis constant for Be^{2+} to form $BeOH^+$ is 1×10^{-6}. (b) What is the dissociation constant for $BeOH^+$?

17-30. A solution contains 0.200 *M* NH_3, 0.200 *M* NaCN, and 0.020 *M* $AgNO_3$ · K_d for $Ag(NH_3)_2^+ = 6.2 \times 10^{-8}$, and K_d for $Ag(CN)_2^- = 1.4 \times 10^{-20}$. (a) What is the equilibrium constant for the reaction

$$Ag(NH_3)_2^+ + 2\,CN^- \rightleftharpoons Ag(CN)_2^- + 2\,NH_3?$$

(b) Determine the ratio of the concentrations of NH_3 and CN^- when equilibrium is established.

Chapter 18
Electrochemistry

18-A. Faraday's Laws

Electrochemistry is concerned with the interconversion of electrical and chemical energies. During electrolysis, electrical energy is expended in bringing about chemical change. In a voltaic or galvanic cell, a chemical change produces electrical energy. The relationships between electrical current and chemical change are given by **Faraday's laws**:

1. The mass of a given substance undergoing chemical change is directly proportional to the quantity of electricity passed.
2. The masses of different substances undergoing chemical change by passage of a given quantity of electricity are proportional to their equivalent weights.*

The quantity of electricity required to deposit 1 equivalent of a substance is 9.65×10^4 **coulombs** (coul).† This quantity of electricity consists of 6.02×10^{23} electrons and is often referred to as one **equivalent** or 1 **Faraday** of electrons. A coulomb is an ampere-second, that is, a current of 1 ampere flowing for 1 second.

* For review see Section 9-G.
† 9.6487×10^4 to five significant figures.

Example 18-a. What mass of Cu and Cl_2 will be produced at inert electrodes on passage of a current of 5.00 amp through a solution of $CuCl_2$ for 30.0 min? At the cathode, the reaction $Cu^{2+} + 2\,e^- \rightarrow Cu$ occurs, and at the anode, the reaction $2\,Cl^- \rightarrow Cl_2 + 2\,e^-$ occurs.

Solution.

$$\text{Quantity of electricity} = 5.00 \text{ amp} \times \left(30.0 \text{ min} \times \frac{60 \text{ sec}}{\text{min}}\right)$$

$$= 9.00 \times 10^3 \text{ amp-sec or } 9.00 \times 10^3 \text{ coulombs}$$

Equivalent weights:

$$Cu = \frac{63.6 \text{ g Cu}}{1 \text{ mole Cu}} \times \frac{1 \text{ mole Cu}}{2 \text{ equiv Cu}} = 31.8 \frac{\text{g Cu}}{\text{equiv}}$$

$$Cl_2 = \frac{70.9 \text{ g } Cl_2}{1 \text{ mole } Cl_2} \times \frac{1 \text{ mole } Cl_2}{2 \text{ equiv } Cl_2} = 35.5 \frac{\text{g } Cl_2}{\text{equiv}}$$

$$\text{Mass Cu} = 9.00 \times 10^3 \text{ coul} \times \frac{1 \text{ equiv}}{9.65 \times 10^4 \text{ coul}} \times \frac{31.8 \text{ g Cu}}{1 \text{ equiv}}$$

$$= 2.97 \text{ g Cu}$$

$$\text{Mass } Cl_2 = 9.00 \times 10^3 \text{ coul} \times \frac{1 \text{ equiv}}{9.65 \times 10^4 \text{ coul}} \times \frac{35.5 \text{ g } Cl_2}{1 \text{ equiv}}$$

$$= 3.31 \text{ g } Cl_2$$

Example 18-b. Calculate the quantity of electricity required to deposit 2.158 g of Ag from a solution of $AgNO_3$. $Ag^+ + e^- \rightarrow Ag$.

Solution. $$\text{Equivalent weight of Ag} = \frac{107.9 \text{ g Ag}}{1 \text{ mole Ag}} \times \frac{1 \text{ mole Ag}}{1 \text{ equiv Ag}} = 107.9 \frac{\text{g Ag}}{\text{equiv}}$$

$$\text{Equivalents of Ag} = 2.158 \text{ g Ag} \times \frac{1 \text{ equiv Ag}}{107.9 \text{ g Ag}} = 0.02000 \text{ equiv Ag}$$

$$\text{Quantity of electricity} = 0.02000 \text{ equiv} \times \frac{9.649 \times 10^4 \text{ coul}}{1 \text{ equiv}}$$

$$= 1.930 \times 10^3 \text{ coulombs}$$

Example 18-c. A quantity of electricity is passed through a series of solutions of $AgNO_3$, $CrCl_3$, $ZnSO_4$, and $CuSO_4$. (a) If 1.000 g of Ag is deposited from the first solution, what masses of the metallic ions in the other three solutions are simultaneously deposited? (b) Calculate the quantity of electricity in coulombs used.

Solution. (a) The equations for the half-reactions at the cathodes are:

$$Ag^+ + e^- \longrightarrow Ag; \qquad Zn^{2+} + 2\,e^- \longrightarrow Zn$$

$$Cr^{3+} + 3\,e^- \longrightarrow Cr; \qquad Cu^{2+} + 2\,e^- \longrightarrow Cu$$

Equivalent weights:

$$Ag = 107.9\,\frac{\text{g Ag}}{\text{equiv}} \quad \text{(See Ex. 18-b.)}$$

$$Zn = \frac{65.4\text{ g Zn}}{1\text{ mole Zn}} \times \frac{1\text{ mole Zn}}{2\text{ equiv Zn}} = 32.7\,\frac{\text{g Zn}}{\text{equiv}}$$

$$Cu = 31.8\,\frac{\text{g Cu}}{\text{equiv}} \quad \text{(See Ex. 18-a.)}$$

$$Cr = \frac{52.0\text{ g Cr}}{1\text{ mole Cr}} \times \frac{1\text{ mole Cr}}{3\text{ equiv Cr}} = 17.3\,\frac{\text{g Cr}}{\text{equiv}}$$

According to Faraday's laws the masses deposited will be proportional to their equivalent weights and to the number of equivalents of electricity passed.

$$\text{Equivalents of electricity passed} = 1.000\text{ g Ag} \times \frac{1\text{ equiv}}{107.9\text{ g Ag}} = 0.00927\text{ equiv}$$

$$\text{Mass Cr} = 0.00927\text{ equiv} \times \frac{17.3\text{ g Cr}}{1\text{ equiv}} = 0.160\text{ g Cr}$$

$$\text{Mass Zn} = 0.00927\text{ equiv} \times \frac{32.7\text{ g Zn}}{1\text{ equiv}} = 0.303\text{ g Zn}$$

$$\text{Mass Cu} = 0.00927\text{ equiv} \times \frac{31.8\text{ g Cu}}{1\text{ equiv}} = 0.295\text{ g Cu}$$

(b) Quantity of electricity

$$= 1.000\text{ g Ag} \times \frac{1\text{ equiv}}{107.9\text{ g Ag}} \times \frac{9.65 \times 10^4\text{ coul}}{1\text{ equiv}} = 894\text{ coulombs}$$

Problems (Faraday's Laws)

18-1. A steady current was passed through a solution of $CuSO_4$ until 9.54 g of metallic Cu were produced. How many coulombs of electricity were used?

18-2. A current of 3.25 amp is passed for 9.50 hr through a solution of $CoSO_4$ between Pt electrodes. What mass of cobalt will be deposited on the cathode?

18-3. A steady current is passed for 1.750 hr through the following series of solutions: $AgNO_3$; $FeCl_2$; $NiSO_4$; $CrCl_3$. 2.158 g of Ag are deposited from the first solution. (a) Calculate the masses of metals simultaneously produced in the remaining solutions. (b) What strength of current in amperes was used?

18-4. How long must a current of 2.50 amp be passed through a solution of $ZnCl_2$ to deposit 2.18 g of Zn?

18-5. In the electrolysis of an aqueous KCl solution the following reaction occurs:

$$2\,KCl + 2\,H_2O \longrightarrow 2\,KOH + H_2 + Cl_2$$

(a) Determine the volume of H_2 gas at 25°C and 1.00 atm obtained if 8.88 g of Cl_2 are produced. (b) If a current of 1.75 amps is used, how long must the electrolysis proceed?

18-6. Calculate the charge in coulombs of a single electron.

18-7. An aqueous solution containing both $ZnSO_4$ and $CdSO_4$ was electrolyzed until all the Zn and Cd was electrodeposited. If the mass of the mixture of metals was 47.5 g and 1.00 Faraday of electricity was used, calculate the mass of Zn in the mixture.

18-B. Reduction Potentials

1. Voltaic Cells

A voltaic cell converts chemical energy to electrical energy. To accomplish this, an oxidation half-reaction takes place at the anode of the cell and a reduction half-reaction occurs at the cathode. Simultaneously, a transfer of electrons from the anode to the cathode is effected by means of a metallic conductor connecting the electrodes outside the cell. The arrangement shown in Figure 18-1 constitutes a voltaic cell in which Zn metal is in contact with 1 *M* Zn^{2+} in the anode compartment and Cu metal is in contact with 1 *M* Cu^{2+} in the cathode compartment. This cell generates a voltage or potential of about 1.10 volts.

A system in which a metal electrode is in contact with its ions is called a **half-cell**; two half-cells connected by a salt bridge are required to complete the voltaic cell. The salt bridge, composed of a bent tube containing a solution of an electrolyte whose ions do not take part in the chemical reaction, serves to complete the circuit and allow movement of ions between the half-cells. During operation of the cell in Figure 18-1, the following chemical changes occur:

$$\text{At the anode:} \quad Zn \longrightarrow Zn^{2+} + 2\,e^- \quad \text{(oxidation)}$$

$$\text{At the cathode:} \quad Cu^{2+} + 2\,e^- \longrightarrow Cu \quad \text{(reduction)}$$

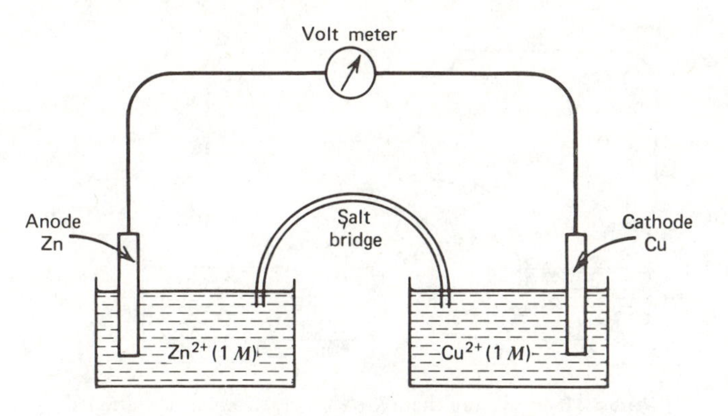

Figure 18-1 The Cu–Zn voltaic cell (Daniell cell).

Adding the two half-reactions, the overall **cell reaction** is obtained:

$$Zn + Cu^{2+} \longrightarrow Zn^{2+} + Cu$$

Any combination of two half-cells gives, in principle, a cell in which a potential difference exists between the two electrodes. By convention, such a cell is often represented by the notation

$$M \text{ (electrode)} | M^+ \text{ (solution)} \| N^+ \text{ (solution)} | N \text{ (electrode)}^*$$

when the cell reaction is $M + N^+ \rightarrow M^+ + N$. Thus the cell in Figure 18-1 would be represented as

$$Zn|Zn^{2+}(1\ M)\|Cu^{2+}(1\ M)|Cu$$

The cell reaction is $Zn + Cu^{2+} \rightarrow Zn^{2+} + Cu$ and electrons flow from Zn to Cu in the metallic circuit outside the cell.

The Hydrogen Electrode. Another voltaic cell is shown schematically in Figure 18-2, in which the $Zn|Zn^{2+}$ half-cell is similar to the one in Figure 18-1. The second-half cell is made up from a piece of platinum foil dipping into a solution of 1 *M* HCl or some other strong acid ($[H^+] = 1\ M$) with hydrogen gas (1 atm pressure) bubbling over the Pt electrode. The equilibrium, $2\,H^+ + 2\,e^- \rightleftharpoons H_2$, is established. This second electrode assembly is called a **standard hydrogen electrode**

The voltage generated by this cell (Figure 18-2) is 0.763 volt and the half-cell reactions are

At the anode: $Zn \longrightarrow Zn^{2+} + 2e^-$ (oxidation)

At the cathode: $2\,H^+ + 2\,e^- \longrightarrow H_2$ (reduction)

* $\|$ is the notation for a salt bridge connecting the two half-cells.

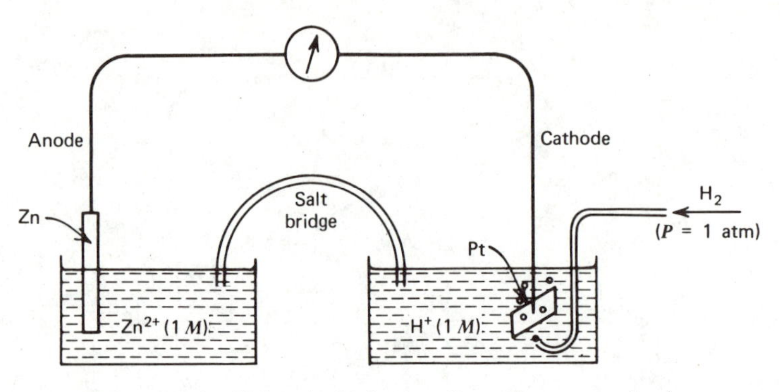

Figure 18-2 Using the hydrogen electrode as a standard.

The overall cell reaction is:

$$Zn + 2H^+ \longrightarrow Zn^{2+} + H_2$$

2. Standard Reduction Potentials

The observed voltage of a cell is simply the difference in potential between the two electrodes and is a measure of the relative tendency of substances to undergo reduction. In the cell shown in Figure 18-2 the voltage 0.763 measures the relative tendencies of the following two half-reactions to occur:

$$Zn^{2+}(1\ M) + 2e^- \rightleftharpoons Zn(s)$$

$$2H^+(1\ M) + 2e^- \rightleftharpoons H_2\ (1\ \text{atm})$$

By convention, the value 0.0000 volt has been assigned to the hydrogen electrode as its standard reduction potential (designated $\mathscr{E}^\circ$) when the pressure of H_2 gas is 1 atm and the hydrogen ion concentration is 1 M. Written as reductions, half-reactions that have a greater tendency to occur (as compared to the hydrogen half-cell) are assigned positive values of $\mathscr{E}^\circ$ while those with a lesser tendency to occur are assigned negative $\mathscr{E}^\circ$ values. Since it is known that hydrogen gas and zinc chloride form when a piece of zinc is placed in a 1 M HCl solution, then hydrogen ion has a greater tendency to be reduced than zinc ion. Hence the value of the standard reduction potential of zinc is -0.763 volt. By setting up other half-cells in conjunction with the standard hydrogen electrode and determining the voltage of the resulting voltaic cells, other standard reduction potentials may be assigned. Table 18-1 includes $\mathscr{E}^\circ$ values for a few half-reactions. All standard reduction potentials are for 1 M solutions of the ions (actually unit activity), a partial pressure of 1 atm for gases, and 25°C. Note that in Table 18-1 the half-reactions are written as reduction processes; hence the potentials are termed

Table 18-1 Reduction Potentials ($\mathscr{E}^\circ$)

Half-Reaction	$\mathscr{E}^\circ$ (Volts)
$F_2 + 2e^- \rightleftharpoons 2F^-$	+2.87
$H_2O_2 + 2H^+ + 2e^- \rightleftharpoons 2H_2O$	+1.78
$Ce^{4+} + e^- \rightleftharpoons Ce^{3+}$	+1.61
$MnO_4^- + 8H^+ + 5e^- \rightleftharpoons Mn^{2+} + 4H_2O$	+1.51
$PbO_2 + 4H^+ + 2e^- \rightleftharpoons Pb^{2+} + 2H_2O$	+1.46
$Cl_2 + 2e^- \rightleftharpoons 2Cl^-$	+1.36
$Cr_2O_7^{2-} + 14H^+ + 6e^- \rightleftharpoons 2Cr^{3+} + 7H_2O$	+1.33
$MnO_2 + 4H^+ + 2e^- \rightleftharpoons Mn^{2+} + 2H_2O$	+1.23
$Br_2 + 2e^- \rightleftharpoons 2Br^-$	+1.06
$NO_3^- + 4H^+ + 3e^- \rightleftharpoons NO + 2H_2O$	+0.96
$NO_3^- + 2H^+ + e^- \rightleftharpoons NO_2 + H_2O$	+0.80
$Ag^+ + e^- \rightleftharpoons Ag$	+0.799
$Hg_2^{2+} + 2e^- \rightleftharpoons 2Hg$	+0.79
$Fe^{3+} + e^- \rightleftharpoons Fe^{2+}$	+0.77
$O_2 + 2H^+ + 2e^- \rightleftharpoons H_2O_2$	+0.68
$H_3AsO_4 + 2H^+ + 2e^- \rightleftharpoons H_3AsO_3 + H_2O$	+0.56
$I_2 + 2e^- \rightleftharpoons 2I^-$	+0.53
$H_2SO_3 + 4H^+ + 4e^- \rightleftharpoons S + 3H_2O$	+0.45
$Cu^{2+} + 2e^- \rightleftharpoons Cu$	+0.337
$Hg_2Cl_2 + 2e^- \rightleftharpoons 2Hg + 2Cl^-$	+0.27
$AgCl + e^- \rightleftharpoons Ag + Cl^-$	+0.22
$SO_4^{2-} + 4H^+ + 2e^- \rightleftharpoons H_2SO_3 + H_2O$	+0.17
$Sn^{4+} + 2e^- \rightleftharpoons Sn^{2+}$	+0.15
$S + 2H^+ + 2e^- \rightleftharpoons H_2S$	+0.14
$2H^+ + 2e^- \rightleftharpoons H_2$	0.000
$Pb^{2+} + 2e^- \rightleftharpoons Pb$	−0.13
$Sn^{2+} + 2e^- \rightleftharpoons Sn$	−0.14
$Ni^{2+} + 2e^- \rightleftharpoons Ni$	−0.25
$Co^{2+} + 2e^- \rightleftharpoons Co$	−0.28
$In^{3+} + 3e^- \rightleftharpoons In$	−0.342
$PbSO_4 + 2e^- \rightleftharpoons Pb + SO_4^{2-}$	−0.36
$Cd^{2+} + 2e^- \rightleftharpoons Cd$	−0.40
$Cr^{3+} + e^- \rightleftharpoons Cr^{2+}$	−0.41
$2CO_2 + 2H^+ + 2e^- \rightleftharpoons H_2C_2O_4$	−0.49
$As + 3H^+ + 3e^- \rightleftharpoons AsH_3$	−0.60
$Zn^{2+} + 2e^- \rightleftharpoons Zn$	−0.763
$Al^{3+} + 3e^- \rightleftharpoons Al$	−1.66
$K^+ + e^- \rightleftharpoons K$	−2.92

reduction potentials. By rewriting the reaction to show oxidation and **changing the sign** of the potential, the corresponding **oxidation** potentials are obtained. There are also some reactions listed where an inert metal such as platinum must be used to remove electrons from the reducing agent or to supply electrons to the oxidizing agent. This is a situation analogous to that of the hydrogen electrode.

3. The Qualitative Use of Standard Reduction Potentials

By combining an oxidation half-reaction with a reduction half-reaction and adding the voltages of the two half-reactions, we may determine whether the reaction will proceed spontaneously as written. If the net voltage is positive, then the reaction as written is spontaneous and will proceed to a greater or lesser extent depending on the magnitude of the voltage. If the voltage is negative, the reaction as written occurs to only a very slight extent, and the reverse of the reaction written will take place to a greater extent.

Example 18-d. Will Ag metal reduce Sn^{2+}? Consider the two half-reactions

$$Ag \rightleftharpoons Ag^{+} + e^{-} \quad \text{(oxidation)} \quad \mathscr{E}^\circ = -0.80 \text{ volt}$$

$$Sn^{2+} + 2\,e^{-} \rightleftharpoons Sn \quad \text{(reduction)} \quad \mathscr{E}^\circ = -0.14 \text{ volt}$$

The sign for the first half-reaction is the reverse of that in Table 18-1 since this is oxidation.

Solution. To cancel electrons, the first equation is multiplied by 2 and the half-reactions are added. The voltage, however, is not changed.*

$$2\,Ag \rightleftharpoons 2\,Ag^{+} + 2\,e^{-} \qquad \mathscr{E}^\circ = -0.80 \text{ volt}$$

$$Sn^{2+} + 2\,e^{-} \rightleftharpoons Sn \qquad \mathscr{E}^\circ = -0.14 \text{ volt}$$

Cell reaction

$$2\,Ag + Sn^{2+} \rightleftharpoons 2\,Ag^{+} + Sn \qquad \mathscr{E}^\circ_{cell} = -0.94 \text{ volt}$$

Since the voltage is negative, the reaction as written will proceed to only a very slight extent; that is, Ag will not reduce Sn^{2+}. The reverse reaction will go to a much greater extent, however; that is, Sn will reduce Ag^{+}.

Example 18-e. Determine whether or not Fe^{2+} will reduce MnO_4^- in acid solution.

* The $\mathscr{E}^\circ$ value for Ag is not multiplied by two because an $\mathscr{E}^\circ$ value is an **intensive** property of a substance—that is, a property that depends only on the substance itself and **not** on the quantity of the substance present.

Solution.

$$5\,Fe^{2+} \rightleftharpoons 5\,Fe^{3+} + 5\,e^- \qquad \mathscr{E}^\circ = -0.77 \text{ volt}$$

$$MnO_4^- + 8\,H^+ + 5\,e^- \rightleftharpoons Mn^{2+} + 4\,H_2O \qquad \mathscr{E}^\circ = +1.51 \text{ volt}$$

$$5\,Fe^2 + MnO_4^- + 8\,H^+ \rightleftharpoons Mn^{2+} + 4\,H_2O + 5\,Fe^{3+} \qquad \mathscr{E}^\circ_{cell} = +0.74 \text{ volt}$$

The voltage is positive and therefore the reaction as written will proceed, that is, Fe^{2+} in acid solution will reduce MnO_4^-.

It may be noted that in the table of reduction potentials just given, the oxidized form will oxidize the reduced form of any substance that lies **below** it in the table; for example, Cu^{2+} will oxidize Zn but not Ag. Similarly, the reduced form will reduce the oxidized form of any substance that lies **above** it; for example, Ni will reduce Ag^+ but not Al^{3+}.

4. Oxidation-Reduction Potentials and Determination of Equilibrium Constants

It is possible to treat oxidation-reduction equilibria in exactly the same fashion as any other in that equilibrium constants for the reactions may be written. For the reaction

$$5\,Fe^{2+} + MnO_4^- + 8\,H^+ \rightleftharpoons Mn^{2+} + 4\,H_2O + 5\,Fe^{3+}$$

the equilibrium constant is written

$$K = \frac{[Mn^{2+}][Fe^{3+}]^5}{[Fe^{2+}]^5[MnO_4^-][H^+]^8}$$

The equilibrium constant for an oxidation-reduction reaction is easily obtained from the relation

$$\mathscr{E}^\circ_{cell} = 2.303\,\frac{RT}{nF}\log K$$

where n is the number of electrons transferred in the chemical equation for the cell reaction; $\mathscr{E}^\circ_{cell}$ is the standard cell potential obtained as in Section 18-3 above; F is the Faraday constant expressed as

$$9.649 \times 10^4\,\frac{\text{J}}{\text{volt} \times \text{equivalent}}.$$

At 25°C this relation becomes:

$$\mathscr{E}^\circ_{cell} = \frac{0.059}{n}\log K \qquad (1)$$

For the permanganate-ferrous ion reaction just cited, the cell voltage was 0.74 volt and n was 5. On substitution into the log K expression,

$$+0.74 = \frac{0.059}{5} \log K$$

$$\log K = \frac{5 \times 0.74}{0.059} = 62.7$$

$$K = 5 \times 10^{62}$$

From the magnitude of this equilibrium constant, it is obvious that the reaction should proceed virtually to completion.

Similarly, for the reaction $2\,Ag^+ + Sn \rightleftharpoons 2\,Ag + Sn^{2+}$, $\mathscr{E}^\circ_{cell}$ is $+0.94$ volt at 25°C, n is 2, and the equilibrium constant for the reaction is

$$K = \frac{[Sn^{2+}]}{[Ag^+]^2}$$

Substituting into expression (1) relating $\mathscr{E}^\circ_{cell}$ and log K above,

$$+0.94 = \frac{0.059}{2} \log K$$

$$\log K = \frac{2 \times 0.94}{0.059} = 31.9$$

$$K = 8 \times 10^{31}$$

It should be noted that if the value of $\mathscr{E}^\circ_{cell}$ is negative, the value of the equilibrium constant is less than 1. For example, the cell voltage in Example 18-d was -0.94 volt. This corresponds to a value of $K = 1.3 \times 10^{-32}$ for the reaction $2\,Ag + Sn^{2+} \rightleftharpoons 2\,Ag^+ + Sn$.

5. Solubility Product Constants from Oxidation-Reduction Potentials

It is difficult to measure the concentration of ions in a saturated solution of a slightly soluble salt in order to obtain the solubility product constant. However, this constant may be obtained from oxidation-reduction potentials.

Example 18-f. Determine K_{sp} for $PbSO_4$ at 25°C.

Solution. Add the two half-reactions (see Table 18-1 for $\mathscr{E}^\circ$ values).

$$\begin{array}{rcll} Pb & \rightleftharpoons & Pb^{2+} + 2\,e^- & \mathscr{E}^\circ = +0.13 \\ PbSO_4(s) + 2\,e^- & \rightleftharpoons & Pb + SO_4^{2-} & \mathscr{E}^\circ = -0.36 \\ \hline PbSO_4(s) & \rightleftharpoons & Pb^{2+} + SO_4^{2-} & \mathscr{E}^\circ = -0.23 \end{array}$$

The last equation represents the equilibrium for a solid with its ions, and the equilibrium constant in the process is the solubility product constant K_{sp}. Substituting in Equation (1), above,

$$\mathscr{E}^\circ = \frac{0.059}{n} \log K_{sp}$$

$$\log K_{sp} = \frac{n\mathscr{E}^\circ}{0.059} = \frac{(2)(-0.23)}{0.059} = \frac{-0.46}{0.059} = -7.8$$

$$K_{sp} = 1.6 \times 10^{-8}$$

6. Reduction Potential Changes with Concentration

The standard reduction potentials in Table 18-1 are for standard conditions, that is, all solids as pure substances, all gases at 1 atm pressure, and all solutions at 1 *M*. Often we are dealing with other concentrations and need to correct the standard potential for changes in concentration from 1 *M*.

Consider the half-reaction

$$aA + ne^- \rightleftharpoons bB$$

The relationship between concentration and reduction potential, at 25°C, is given by the Nernst equation

$$\mathscr{E} = \mathscr{E}^\circ - \frac{0.059}{n} \log \frac{[B]^b}{[A]^a} \quad (2)$$

where [ ] represents the molar concentration of the species in the half-reaction equation.

Example 18-g. Determine $\mathscr{E}$ for the $Zn^{2+} + 2e^- \rightleftharpoons Zn$ half-reaction at a concentration of $Zn^{2+} = 0.100\ M$.

Solution.

$$\mathscr{E} = \mathscr{E}^\circ - \frac{0.059}{n} \log \frac{[Zn]}{[Zn^{2+}]}$$

$$\mathscr{E} = -0.763 - \frac{0.059}{2} \log \frac{1}{0.100}$$

where the concentration or "activity" of pure metallic Zn is taken as unity.

$$\mathscr{E} = -0.763 - \frac{0.059}{2} \times (1.00)$$

$$= -0.763 - 0.030 = -0.793 \text{ volt}$$

Problems (Reduction Potentials)

18-8. For the cell $Ni|Ni^{2+}(1\ M)||Cu^{2+}(1\ M)|Cu$, (a) Determine the voltage of the cell. (b) Write the equation for the cell reaction that occurs. (c) In which direction do electrons flow outside the cell? (d) Calculate the equilibrium constant for the cell reaction.

18-9. From the table of standard reduction potentials, (a) determine whether or not the reaction $2\ Al^{3+} + 3\ Zn \rightleftharpoons 2\ Al + 3\ Zn^{2+}$ will occur to an appreciable extent. (b) Illustrate your answer by determining the equilibrium constant for the reaction.

18-10. (a) What is the value of $\mathscr{E}^\circ_{cell}$ for the cell $Sn|Sn^{2+}||Br_2, Br^-|Pt$? (b) Write the cell reaction. (c) In what direction do electrons flow outside the cell? (d) What is the equilibrium constant for the cell reaction?

18-11. Determine $\mathscr{E}$ (a) for an Ag electrode in a 0.035 M $AgNO_3$ solution, (b) for Cu in a 0.250 M $Cu(NO_3)_2$ solution, (c) for In in a 0.300 M $In(NO_3)_3$ solution, and (d) for In in a 0.300 M $In_2(SO_4)_3$ solution.

18-12. Calculate the voltage of a cell that is composed of a Ag electrode in a 0.020 M $AgNO_3$ solution and a Cd electrode in a 0.0010 M $Cd(NO_3)_2$ solution. The cell may be represented

$$Cd|Cd^{2+}(0.0010\ M)||Ag^+(0.020\ M)|Ag$$

18-13. Calculate the equilibrium constants for the following reactions:
(a) $NO_3^- + H^+ + Br^- \rightleftharpoons Br_2 + NO_2 + H_2O$
(b) $H_2O_2 + NO_3^- + H^+ \rightleftharpoons O_2 + NO + H_2O$
(c) $Zn + As + H^+ \rightleftharpoons Zn^{2+} + AsH_3$
(d) $H_2S + H_2SO_3 \rightleftharpoons S + H_2O$
(e) $Cl_2 + H_2SO_3 + H_2O \rightleftharpoons Cl^- + SO_4^{2-} + H^+$

18-14. What is the voltage of a cell composed of a Ni electrode in 3.0×10^{-5} M $NiCl_2$ solution and a platinum electrode in a 5.0×10^{-3} M $CrCl_2$ and 1.0×10^{-1} M $CrCl_3$ solution? The cell may be represented

$$Ni|Ni^{2+}(3.0 \times 10^{-5}\ M)||Cr^{3+}(1.0 \times 10^{-1}\ M), Cr^{2+}(5.0 \times 10^{-3}\ M)|Pt$$

General Problems

18-15. A current of 4.25 amp is passed through a solution of $ZnSO_4$ between Pt electrodes for 1.75 hr. Calculate the mass of Zn deposited at the cathode.

18-16. What steady current in amp will be needed to deposit 0.500 equiv of Cu from a $CuSO_4$ solution in 3.50 hr?

18-17. A quantity of electricity is passed through a series of electrolytic cells containing $AgNO_3$, $CuCl_2$, $Ni(NO_3)_2$, $Sc_2(SO_4)_3$, and $Au(CN)_2^-$ solutions. (a) If 0.275 equiv of Ag is deposited in the first cell, determine the mass in grams of each of the metals deposited from solution. (b) Determine the quantity of electricity employed.

18-18. Determine the equivalent weight of X if 3.50 g of X are deposited using a current of 3.00 amp for 45.0 min.

18-19. An acid solution is electrolyzed at a steady current of 0.725 amp. For how many minutes must the current flow to release 39.5 ml of hydrogen gas measured at 695 torr pressure and 27°C?

18-20. Water is electrolyzed at a current of 4.50 amp for 3.25 hr; H_2 and O_2 gases are liberated at the cathode and anode, respectively. Calculate the volume of each if dry and at STP.

18-21. Using the table of standard reduction potentials, determine the values of $\mathscr{E}^\circ_{cell}$ for the following cells (concentrations all 1 M):

(a) $Zn|Zn^{2+}\|Ag^+|Ag$
(b) $Zn|Zn^{2+}\|Ni^{2+}|Ni$
(c) $Cu|Cu^{2+}\|Fe^{3+}, Fe^{2+}|Pt$
(d) $Al|Al^{3+}\|Sn^{2+}|Sn$
(e) $Cu|Cu^{2+}\|Ag^+|Ag$
(f) $Pt|I^-, I_2\|Br_2, Br^-|Pt$

18-22. Write the cell reaction for each of the voltaic cells in Problem 18-21.

18-23. (a) Which of the following will be oxidized by Cu^{2+}?
(1) Ni (2) Ag (3) Al (4) Cr^{3+} (5) F^-
(b) Which of the following will be reduced by Sn?
(1) Ni^{2+} (2) Ag^+ (3) Br_2 (4) Al^{3+} (5) Cu^{2+}

18-24. Balance and then determine which of the following reactions will proceed spontaneously to an appreciable extent.

(a) $H_2O_2 + MnO_4^- + H^+ \rightleftharpoons O_2 + Mn^{2+} + H_2O$
(b) $Fe^{3+} + Cr^{3+} + H_2O \rightleftharpoons Fe^{2+} + Cr_2O_7^{2-} + H^+$
(c) $Ni + H^+ \rightleftharpoons Ni^{2+} + H_2$
(d) $PbO_2 + H^+ + Sn \rightleftharpoons Sn^{2+} + Pb^{2+} + H_2O$
(e) $Mn^{2+} + H_2O + AsO_4^{3-} \rightleftharpoons MnO_4^- + AsO_3^{3-} + H^+$

18-25. Determine $\mathscr{E}$ at 25°C for a copper electrode in contact with a 3.00×10^{-4} M solution of Cu^{2+}.

18-26. Using the table of standard reduction potentials and the principles outlined in Section 18-B-6, determine the values of $\mathscr{E}_{cell}$ for the following cells:

(a) $Zn|Zn^{2+}(1.0 \times 10^{-1}\ M)\|Ag^+(1.0 \times 10^{-3}\ M)|Ag$

(b) $Zn|Zn^{2+}(5.0 \times 10^{-2}\ M)\|Ni^{2+}(1.0 \times 10^{-2}\ M)|Ni$
(c) $Cu|Cu^{2+}(1.0 \times 10^{-1}\ M)\|Fe^{3+}(5.0 \times 10^{-3}\ M),$
$Fe^{2+}(2.5 \times 10^{-3}\ M)|Pt$
(d) $Cu|Cu^{2+}(1.0 \times 10^{-5}\ M)\|Cu^{2+}(1.0\ M)|Cu$
(e) $Cu|Cu^{2+}(1.0 \times 10^{-3}\ M)\|Sn^{2+}(2.0 \times 10^{-1}\ M)|Sn$
(f) $Al|Al^{3+}(5.0 \times 10^{-2}\ M)\|Sn^{2+}(1.0\ M)|Sn$

18-27. Calculate the equilibrium constant for each of the following reactions:
(a) $Cu^{2+} + Ni \rightleftharpoons Cu + Ni^{2+}$
(b) $Cu + NO_3^- + H^+ \rightleftharpoons Cu^{2+} + NO_2 + H_2O$
(c) $Ag + Ni^{2+} \rightleftharpoons Ag^+ + Ni$
(d) $I_2 + Br^- \rightleftharpoons Br_2 + I^-$
(e) $Br_2 + I^- \rightleftharpoons I_2 + Br^-$
(f) $F_2 + Br^- \rightleftharpoons Br_2 + F^-$

18-28. Determine the equilibrium constant for each of the following reactions:
(a) $Ce^{4+} + Cl^- \rightleftharpoons Cl_2 + Ce^{3+}$
(b) $Cl_2 + I^- \rightleftharpoons I_2 + Cl^-$
(c) $H_2O_2 + H_2SO_3 \rightleftharpoons SO_4^{2-} + H_2O + H^+$
(d) $H_2O_2 + Br^- + H^+ \rightleftharpoons H_2O + Br_2$
(e) $H_2O_2 + Cr_2O_7^{2-} + H^+ \rightleftharpoons Cr^{3+} + O_2 + H_2O$

18-29. The equilibrium constant for the reaction

$$Sn^{2+} + 2\,Hg^{2+} \rightleftharpoons Hg_2^{2+} + Sn^{4+}$$

is 5.0×10^{25}. (a) Calculate the voltage of the cell if each of the species in solution is present at a concentration of 1 *M*. (b) From the table of reduction potentials, obtain the value of $\mathscr{E}°$ for the half-reaction $Sn^{4+} + 2\,e^- \rightleftharpoons Sn^{2+}$, and calculate the value of $\mathscr{E}°$ for the half-reaction $2\,Hg^{2+} + 2\,e^- \rightleftharpoons Hg_2^{2+}$.

18-30. From the data in Table 18-1, determine K_{sp} at 25°C for (a) AgCl, and (b) Hg_2Cl_2.

Chapter 19

Chemical Thermodynamics

Energy changes accompany all chemical and physical changes. Chemical thermodynamics is a study of the relationships between all forms of energy as they pertain to chemical processes.

Some Thermodynamic Terms and Equations

It is assumed that the student's textbook will be a source of an elaboration and discussion of these terms in more detail. Therefore the definitions and relationships between terms are described only briefly here.

E = **Internal Energy**. This term represents the sum of all forms of kinetic and potential energy in a system.* The internal energy is a function of the temperature, pressure, and the chemical state of the system.

H = **Enthalpy**. H and E are related by the equation: $H = E + PV$ where P = pressure and V = volume. The enthalpy is equal to the internal energy plus the pressure-volume product.

S = **Entropy**. A measure of degree of disorder or randomness in a system. The entropy of a gas is greater than that of a liquid which in turn is greater than that of a solid, since a solid is the most ordered of the three physical states and a gas is the least ordered.

* A system is an isolated portion of the universe with physical boundaries separating it from other portions of the universe.

G = **Gibbs Free Energy**. $G = H - TS$ where T is absolute temperature. The free energy of a system is a measure of the energy that is available for useful work.

Thermodynamics is concerned with the **change** in these functions in passing from one state† to another. The mechanism by which the system gets from one state to a second is **not** a consideration. Therefore we may set up the following:

$\Delta E = E_2 - E_1$ where E_2 is energy of state 2 and E_1 is energy of state 1. (State 2 is the **final** state, state 1 the **initial** state.)

$\Delta H = H_2 - H_1$

$\Delta S = S_2 - S_1$

The change in enthalpy, ΔH, can be related to the change in internal energy, ΔE, by the equation $\Delta H = \Delta E + \Delta(PV)$. For a process at constant pressure, $\Delta H = \Delta E + P\,\Delta V$, where $P\,\Delta V$ represents the work done in increasing the volume of a system by ΔV against a constant pressure P.

19-A. The First Law of Thermodynamics

This is essentially a statement of the Law of Conservation of Energy, that is, "Energy may be converted from one form to another but can neither be created nor destroyed." The first law is concerned only with the conservation of energy during changes in a system and not with the conditions under which changes might or might not occur.

When a system changes from state 1 to state 2, the change in internal energy (ΔE) will be the quantity of heat absorbed (q) in the process less the amount of work (w) done by the system on the surroundings, that is, $\Delta E = q - w$. If the work done is pressure-volume work, that is, expansion of gas at constant pressure, $w = P\,\Delta V$ and thus $\Delta E = q_p - P\,\Delta V$ where q_p is used to denote the heat absorbed at constant pressure. Since $\Delta E = \Delta H - P\,\Delta V$ (constant P) then $\Delta H = q_p$. Also, for a process conducted at constant volume, no work is done, and the heat absorbed by the system, q_v, is equal to ΔE.

In considering $P\,\Delta V$ work the volume change in liquids and solids is relatively small and may be neglected—that is, where only liquids and solids are involved, it may be assumed that $P\,\Delta V = 0$ and therefore $q_v = q_p$. When gases are present in the reaction, there may be considerable volume

† The state of a system is defined by its T, P, V, and chemical composition.

change and the work may be calculated from the expression, work = $P\,\Delta V = (\Delta n)RT$, where Δn = number of moles of gaseous products minus the number of moles of gaseous reactants as shown by the balanced equation. R is the molar gas constant 8.314 J/(mole- K) and T is the absolute temperature.

Example 19-a. Given the value of ΔE, calculate ΔH for the following reaction at 25°C and one atmosphere pressure.

$$H_2(g) + \tfrac{1}{2}\,O_2(g) \longrightarrow H_2O(l) \qquad \Delta E = -282.1 \text{ kJ/mole } H_2O(l)$$

Solution.

$$\Delta H = \Delta E + P\,\Delta V$$

$$P\,\Delta V = (\Delta n)RT$$

The volume of the liquid phase is assumed to be negligible compared to the volume of the gaseous phase; therefore

$$\Delta n = 0 - (1.00 + 0.50) = -1.50 \text{ moles}$$

and

$$(\Delta n)RT = (-1.50\ \cancel{\text{moles}})\left(8.314\ \frac{\text{J}}{\cancel{\text{mole K}}}\right)(298\ \cancel{\text{K}})$$

$$= -3700 \text{ J} = -3.7 \text{ kJ}$$

$$\Delta H = -282.1 - 3.7 = -285.8 \text{ kJ/mole } H_2O(l)$$

For reactions at one atmosphere and 25°C, ΔH and ΔE usually differ only slightly.

1. Isothermal and Adiabatic Processes For Ideal Gases

An **isothermal** process is one conducted at constant temperature in a thermostat; heat flows in or out of the system to maintain the constant temperature. If a gas expands isothermally, work is done on the surroundings and energy is absorbed by the system equivalent to the work done. For the expansion of an ideal gas at constant temperature, the internal energy is constant, that is, $\Delta E = 0$. Since $\Delta E = q - w = 0$, then $q = w$. The work done is given by the equation, $w = nRT(2.303)\log\,(V_2/V_1)$ where V_2 is the final volume and V_1 is the initial volume.

If a process is carried out in an insulated system wherein no heat flows in or out of the system, the process is said to be **adiabatic**. In these cases,

$q = 0$ and $\Delta E = -w$. If a gas expands adiabatically, work is done on the surroundings and the temperature of the gas falls.

Example 19-b. Calculate the heat energy required to expand isothermally one mole of gas at 273 K from 1.00 atm to 0.500 atm.

Solution.

$$\text{Volume of 1 mole at 273 K and 1.00 atm} = 22.4 \text{ liters}$$

$$\text{Volume of 1 mole at 273 K and 0.500 atm} = 44.8 \text{ liters}$$

$$\Delta E = 0 = q - w \quad \text{so} \quad q = w$$

$$w = \text{work} = nRT(2.303) \log \frac{44.8}{22.4}$$

$$= (1.00 \cancel{\text{mole}}) \left(8.314 \frac{\text{J}}{\cancel{\text{mole K}}}\right) (273 \cancel{\text{K}})(2.303) \log 2.00$$

$$q = w = 1.57 \times 10^3 \text{ J}$$

2. Standard States and Changes

Of particular interest to chemists are changes of ΔH and ΔE for chemical reactions. Because ΔE and ΔH are both functions of temperature and pressure, it is necessary to establish certain conventions:

Standard State. The standard state of an element or compound is its most stable state at one atmosphere pressure and at some specified temperature, usually 25°C.

Standard Enthalpy of Formation. The standard enthalpy of formation, ΔH_f°, of any **compound** in its standard state is the enthalpy change when the compound in its standard state is formed from the elements in their standard states. The standard enthalpy of formation for any **element** in its standard state is zero.

Standard Enthalpy Changes. The standard enthalpy change, ΔH°, for a reaction where the reactants and products are in their standard states, is the difference between the sum of the standard enthalpies of formation of the products and the sum of the standard enthalpies of formation of the reactants.

Example 19-c. For the reaction shown by the equation

$$S(s) + O_2(g, 1 \text{ atm}) \longrightarrow SO_2(g, 1 \text{ atm})$$

ΔH° was found by experiment to be -296.1 kJ/mole. Determine the standard enthalpy of formation, $\Delta H^\circ_{SO_2}$, and the standard internal energy of formation, $\Delta E^\circ_{SO_2}$.

Solution.

$$\Delta H^\circ = \Delta H^\circ_{SO_2} - (\Delta H^\circ_S + \Delta H^\circ_{O_2})$$
$$-296.1 = \Delta H^\circ_{SO_2} - (0 + 0)$$
$$\Delta H^\circ_{SO_2} = -296.1 \text{ kJ/mole}$$
$$\Delta H^\circ_{SO_2} = \Delta E^\circ_{SO_2} + (\Delta n)RT$$

$\Delta n = 0$ assuming only the number of moles of gaseous products and reactants are important.

$$\Delta H^\circ_{SO_2} = \Delta E^\circ_{SO_2} = -296.1 \text{ kJ/mole}$$

Note that $\Delta H^\circ_{SO_2}$ corresponds to a rather unusual set of conditions—that is, the reactants in their standard states form the product in its standard state. If the conditions are different from standard, then ΔH will be different from ΔH°.

Example 19-d. Using the standard enthalpies of formation at 25°C in Table 19-1, determine the standard enthalpy change and also ΔE° for the reaction:

$$4\,NH_3(g, 1\text{ atm}) + 5\,O_2(g, 1\text{ atm}) \longrightarrow 4\,NO(g, 1\text{ atm}) + 6\,H_2O(g, 1\text{ atm})$$

Solution.

$$\Delta H^\circ = 4\Delta H^\circ_{NO} + 6\Delta H^\circ_{H_2O} - (4\,\Delta H^\circ_{NH_3} + 5\,\Delta H^\circ_{O_2})$$
$$= 4(90.4) + 6(-241.8) - [4(-46.2) + 5(0)]$$
$$= 361.6 - 1450.8 - (-184.8 + 0) = -904.4 \text{ kJ}$$

$$\Delta H^\circ = \Delta E^\circ + (\Delta n)RT$$

$$\Delta n = 10 - 9 = 1.00$$

$$\Delta E^\circ = -904.4 - \left(1.00\,\cancel{\text{mole}} \times 8.314\,\frac{\cancel{\text{J}}}{\cancel{\text{mole}}\,\cancel{\text{K}}} \times 298\,\cancel{\text{K}}\right) \times \left(10^{-3}\,\frac{\text{kJ}}{\cancel{\text{J}}}\right)$$

$$= -904.4 - 2.5 = -906.9 \text{ kJ}$$

Table 19-1 Thermodynamic Properties

Substance	ΔH_f°	ΔG_f°	S°
$NH_3(g)$	−46.19	−16.65	192.5
$HBr(g)$	−36.23	−53.1	198.3
$BrCl(g)$	14.69	−0.88	239.7
$CO_2(g)$	−393.5	−394.6	213.8
$CO(g)$	−110.5	−137.2	197.9
$HCl(g)$	−92.3	−95.4	186.6
$HF(g)$	−268.6	−270.7	173.6
$H_2O(g)$	−241.84	−228.61	188.7
$H_2O(l)$	−285.85	−237.19	69.9
$HI(g)$	25.94	1.30	206.3
$I_2(g)$	62.3	19.4	206.7
$Hg(g)$	60.7	31.80	174.9
$PBr_3(g)$	−150.2	−172.4	347.7
$PCl_3(g)$	−306.3	−286.2	311.7
$PCl_5(g)$	−399.2	−324.7	352.7
$NaCl(s)$	−411.3	−384.1	72.4
$CH_4(g)$	−74.9	−50.6	186.2
$CH_3OH(l)$	−238.5	−166.1	126.8
$COCl_2(g)$	−223.0	−210.5	289.1
$CH_3Cl(g)$	−82.0	−58.6	234.3
$C_2H_4(g)$	52.6	68.2	219.7
$C_2H_5OH(l)$	−277.8	−174.9	160.7
$C_2H_6(g)$	−84.5	−32.89	229.7
$NH_4NO_2(s)$	−264.0	—	—
$NO(g)$	90.37	86.69	210.5
$NOCl(g)$	52.59	66.36	263.6

Values are for substances at 25°C and 1 atm pressure. Enthalpies of formation (ΔH_f°) and free energies of formation (ΔG_f°) are expressed in kilojoules per mole; entropy (S°) values in joules per mole per degree absolute temperature.

Example 19-e. Calculate ΔH° and ΔE° for the following reaction at 25°C and 1 atm.

$$NH_4NO_2(s) \longrightarrow N_2(g) + 2\,H_2O(l)$$

Refer to Table 19-1 for enthalpies of formation at 25°C and 1 atm.

Solution.

$$\begin{aligned}\Delta H^\circ &= \Delta H_f^\circ(\text{products}) - \Delta H_f^\circ(\text{reactants})\\ &= [(0) + 2(-285.8)] - (-264.0)\\ &= -571.6 + 264.0 = -307.6 \text{ kJ}\\ \Delta E^\circ &= \Delta H^\circ - (\Delta n)RT \qquad \Delta n = (1) - (0) = 1.00\end{aligned}$$

assuming volumes of solid NH_4NO_2 and liquid H_2O are relatively small and may be neglected.

$$\begin{aligned}\Delta E^\circ &= -307.6 \text{ kJ} - (1.00\ \cancel{\text{mole}})\left(8.314\ \frac{\cancel{\text{J}}}{\cancel{\text{mole}}\ \cancel{\text{K}}}\right)(298\ \cancel{\text{K}})\left(10^{-3}\ \frac{\text{kJ}}{\cancel{\text{J}}}\right)\\ &= -307.6 \text{ kJ} - 2.5 \text{ kJ} = -310.1 \text{ kJ}\end{aligned}$$

19-B. The Second and Third Laws

1. Entropy and Entropy Change

As mentioned earlier, entropy is a measure of the disorder or randomness of a system. When a gas expands, at constant T, from a volume of 1.0 liter to a final volume of 2.0 liters, it possesses a greater degree of disorder or randomness in the larger volume and hence its entropy is said to have increased. In a chemical reaction, if a single molecule decomposes into two or more molecules or atoms, the products are therefore less ordered (more random) than the reactants and again the entropy for this system is said to have increased. When entropy increases, the entropy change is positive; if entropy decreases, the change is negative.

The **second law of thermodynamics** specifies that only certain processes will occur spontaneously. A simple statement of the law is as follows: **A spontaneous change takes place only with a net increase in total entropy of the system and its surroundings—that is, $\Delta S_{total} = \Delta S_{sys} + \Delta S_{surr} > 0$.** It must be emphasized that **both** the system and the surroundings must be considered when calculating the total entropy change!

Entropy changes for a process can be calculated in several ways. When a process occurs reversibly* at constant temperature and pressure, the change in entropy of the system, ΔS, is equal to the heat absorbed, q_p, divided by the absolute temperature at which the change occurs, that is, $\Delta S = q_p/T$. This indicates that in absorbing a fixed quantity of heat, a greater entropy change for the system will occur at lower as compared to higher temperatures.

* A reversible change is one carried out essentially under equilibrium conditions. For example, the freezing of water at its melting point may be carried out reversibly by the slow removal of heat at constant temperature and pressure; the process may be reversed by the addition of heat from the surroundings.

Example 19-f. Calculate ΔS for the melting of 1 mole of ice to liquid H_2O at 0°C. Heat of fusion = +6.00 kJ/mole.

Solution.

$$\Delta S = \frac{q_p}{T} = \frac{6.00 \times 10^3 \text{ J/mole}}{273 \text{ K}} = +22.0 \text{ J/(mole K)}$$

Since ΔS is positive, the entropy is increasing. Note that the disorder or randomness (entropy) is greater in the liquid state than in the solid state.

Calculations of entropy changes for chemical reactions can result from the **third law of thermodynamics** and the dependence of entropy on temperature. Hence we have a second method for calculation of ΔS. The third law states that **the entropy of a perfect crystalline substance at absolute zero (0 K) is zero**. The units in a crystal are arranged in a very definite way. This orderly arrangement possesses no disorder or randomness and hence may be regarded as having a minimum value for entropy: zero. Standard entropies for elements and compounds may be calculated in their standard states at 25°C and these may be used to calculate ΔS° for reactions by subtracting the sum of the entropy values of reactants from products. A few of these entropies are given in Table 19-2.

Table 19-2 Standard Entropies (S°) at 25°C in J/(mole K)

$H_2(g)$	130.5	$HCl(g)$	186.6	$CO_2(g)$	213.8
$O_2(g)$	205.0	$H_2O(l)$	69.94	$CH_4(g)$	186.2
$Cl_2(g)$	223.0	$H_2O(g)$	188.74	$ZnSO_4(aq, 1\ M)$	−89.1
$F_2(g)$	203.3	$Zn(s)$	160.9	$CO(g)$	197.9
$Na(s)$	51.0	$H_2SO_4(aq, 1\ M)$	17.2	$NO(g)$	210.5
		$C(s, \text{graphite})$	5.69	$NH_3(g)$	192.5

Example 19-g. Determine ΔS° for the reaction

$$H_2(g) + \tfrac{1}{2} O_2(g) \longrightarrow H_2O(l)$$

from the standard entropy values in Table 19-2.

Solution.

$$\begin{aligned}\Delta S^\circ &= S^\circ_{H_2O} - [S^\circ_{H_2} + \tfrac{1}{2} S^\circ_{O_2}] \\ &= 69.9 - [130.5 + \tfrac{1}{2}(205.0)] \\ &= -163.1 \text{ J/K [per mole } H_2O(l)]\end{aligned}$$

Example 19-h. From the values of $S°$ in Table 19-2, determine $\Delta S°$ for the reaction

$$4\,NH_3(g, 1\text{ atm}) + 5\,O_2(g, 1\text{ atm}) \longrightarrow 4\,NO(g, 1\text{ atm}) + 6\,H_2O(g, 1\text{ atm})$$

Solution.

$$\begin{aligned}\Delta S° &= \sum S°\,(\text{products}) - \sum S°\,(\text{reactants})\\ &= 4S°_{NO} + 6S°_{H_2O} - (4S°_{NH_3} + 5S°_{O_2})\\ &= 4(210.5) + 6(188.74) - [4(192.5) + 5(205.0)]\\ &= +\,179.4\text{ J/K}\end{aligned}$$

We have seen in Example 19-g that the standard entropy change, $\Delta S°$, for the **system** $H_2(g) + \frac{1}{2}O_2(g) \rightarrow H_2O(l)$ is negative. We also know, however, that the reaction is spontaneous and, once started, occurs with explosive force. Since the reaction is spontaneous, the **second law of thermodynamics** requires that $\Delta S_{total} > 0$. Because ΔS_{system} is negative, and since $\Delta S_{sys} + \Delta S_{surr}$ must be >0, this implies that ΔS_{surr} is not the same as ΔS_{sys}, that it must be greater in magnitude than ΔS_{sys}, and, further, that it must be positive.

Example 19-i. (a) Calculate ΔS_{surr} for the reaction $H_2(g) + \frac{1}{2}O_2(g) \rightarrow H_2O(l)$. Assume $T = 298.0$ K. (b) Determine whether this reaction should be spontaneous at this temperature.

Solution. (a) Recall that the heat lost by the system is gained by the surroundings, that is, $\Delta H_{surr} = -\Delta H_{sys}$. Since $\Delta S = q_p/T$ and $\Delta H = q_p$ for processes at constant pressure,

$$\Delta S_{surr} = \frac{\Delta H_{surr}}{T} = \frac{-\Delta H_{sys}}{T}$$

$$\begin{aligned}\Delta H_{sys} &= \Delta H_{f(H_2O(l))} - [\Delta H_{f(H_2)} + \tfrac{1}{2}\Delta H_{f(O_2)}]\\ &= -\,285.8 - [0 + \tfrac{1}{2}(0)] = -285.8\text{ kJ/mole}\end{aligned}$$

$$\Delta S_{surr} = \frac{+285.8\ \cancel{\text{kJ}}/\text{mole}}{298.0\text{ K}} \times \frac{10^3\text{ J}}{1\ \cancel{\text{kJ}}} = +959.1\text{ J/(mole K)}$$

$$\begin{aligned}\text{(b) } \Delta S_{total} &= \Delta S_{sys} + \Delta S_{surr}\\ &= (-163.1 + 959.1)\text{ J/(mole K)}\\ &= +796.0\text{ J/(mole K)}\end{aligned}$$

Since $\Delta S_{total} > 0$ the reaction can occur spontaneously.

2. Free Energy and Free Energy Change

Since ΔS_{surr} is related to ΔH_{sys} we can say the following:

$$\Delta S_{total} = \Delta S_{surr} + \Delta S_{sys} = \frac{-\Delta H_{sys}}{T} + \Delta S_{sys}$$

A reaction can proceed spontaneously therefore, if:

$$\frac{-\Delta H_{sys}}{T} + \Delta S_{sys} > 0$$

or

$$-\Delta H_{sys} + T\Delta S_{sys} > 0$$

or

$$\Delta H_{sys} - T\Delta S_{sys} < 0$$

This last expression is most important to us for it allows us to define "**free energy**," ***G***, a property of the state of the **system**, as

$$G = H - TS$$

or

$$\Delta G = \Delta H - T\Delta S \text{ at constant temperature and pressure}$$

Hence we see, according to the second law, that a reaction can proceed spontaneously if $\Delta G < 0$, **assuming constant pressure and temperature**. Note that one can predict spontaneity for a reaction (system) by determining ΔG for the **system** only. One need not be concerned with the surroundings as was the case for ΔS_{total}. The Gibbs free energy of a system, G, may be thought of as the energy possessed by the system that may be converted to useful work—that is, the net **available** energy.

ΔG is a combination of terms depending on the changes in enthalpy and entropy in the system and serves to determine whether a chemical reaction can occur under these conditions. If ΔG is negative for the equation as written, then that process can occur spontaneously; if ΔG is positive the reverse reaction can occur spontaneously. If ΔG is 0, the system is at equilibrium and no net reaction occurs. The same standard state conventions have been adopted for ΔG as for ΔH. The free energy of formation, ΔG_f°, of the free element in its most stable state at 1 atmosphere is taken as zero. Free energies of formation for a number of compounds are given in Table 19-1.

Example 19-j. Calculate ΔG° for the reaction:

$$NOCl(g) \longrightarrow NO(g) + \tfrac{1}{2}\,Cl_2(g)$$

and predict whether the reaction can occur spontaneously under these conditions. Refer to Table 19-1 for standard free energies of formation.

Solution.

$$\begin{aligned}\Delta G^\circ &= \Delta G^\circ_{NO} + \tfrac{1}{2}\,\Delta G^\circ_{Cl_2} - \Delta G^\circ_{NOCl}\\ &= (86.69 + 0) - 66.36 = +20.33 \text{ kJ}\end{aligned}$$

Since ΔG° is positive, the reaction cannot proceed spontaneously as written.

Example 19-k. Calculate $\Delta G^\circ_{NH_3}$ from data in Table 19-1 and the experimental fact that $\Delta G^\circ = -958.30$ kJ for the reaction:

$$4\,NH_3(g, 1 \text{ atm}) + 5\,O_2(g, 1 \text{ atm}) \longrightarrow 4\,NO(g, 1 \text{ atm}) + 6\,H_2O(g, 1 \text{ atm})$$

Solution.

$$\Delta G^\circ = (4\,\Delta G^\circ_{NO} + 6\,\Delta G^\circ_{H_2O}) - (4\,\Delta G^\circ_{NH_3} + 5\,\Delta G^\circ_{O_2})$$

$$-958.30 = 4(86.69) + 6(-228.61) - 4\,\Delta G^\circ_{NH_3}$$

$$-958.30 = 346.76 - 1371.66 - 4\,\Delta G^\circ_{NH_3}$$

$$\Delta G^\circ_{NH_3} = \frac{-66.60}{4} = -16.65 \text{ kJ/mole}$$

Once ΔG° and ΔH° are known for a particular reaction at a given T, then calculation of ΔS° at that temperature may be obtained from the relationship:

$$\Delta G^\circ = \Delta H^\circ - \mathrm{T}\,\Delta S^\circ \quad \text{or} \quad \Delta S^\circ = \frac{\Delta H^\circ - \Delta G^\circ}{T}$$

Example 19-l. For the reaction

$$4\,NH_3(g, 1 \text{ atm}) + 5\,O_2(g, 1 \text{ atm}) \longrightarrow 4\,NO(g, 1 \text{ atm}) + 6\,H_2O(g, 1 \text{ atm})$$

calculate ΔS°. Use values of ΔH° and ΔG° from Examples 19-d and 19-k. Note that ΔS° is expressed in joules; hence ΔH° and ΔG° must be converted to joules.

Solution.

$$\Delta H^\circ = -904.4 \text{ kJ} \qquad \Delta G^\circ = -958.3 \text{ kJ}$$

$$\Delta S^\circ = \frac{[-904{,}400 - (-958{,}300)]}{298} = +180.9 \text{ J/K}$$

19-C. Chemical Equilibrium and Free Energy

The standard free energy changes, ΔG°, calculated above are limited to reactions with both reactants and products in their standard states. Only rarely are these circumstances encountered in actual situations. Under conditions different from those of the standard states, the sign of ΔG is the controlling factor in predicting spontaneity of reactions at constant temperature and pressure. The equation

$$\Delta G = \Delta G^\circ + 2.303RT \log Q_P$$

allows values of ΔG to be evaluated for conditions other than standard state conditions. The second term on the right-hand side corrects for the differences between the actual initial conditions and the standard state conditions of the reactants and products. Q_P has the same form as the equilibrium constant, and for gaseous reactions, the partial pressures (in atmospheres) of the constituents are used as a measure of the initial amounts of the reacting species. For the general reaction

$$mA + nB \rightleftharpoons rC + sD$$

$$Q_P = \left[\frac{(p_C)^r(p_D)^s}{(p_A)^m(p_B)^n}\right]_{\text{initial}}$$

If any of the reactants or products are pure liquids or solids, their vapor pressures at constant temperature are constant and do not change during the course of the reaction. Such terms are not included in expressions for K_p or Q_p (see Examples 14-g and 14-h).

Example 19-m. Calculate ΔG at 298 K for the reaction at the conditions listed:

$$4\,NH_3(g, 0.10\text{ atm}) + 5\,O_2(g, 10.0\text{ atm}) \rightleftharpoons 4\,NO(g, 2.0\text{ atm}) + 6\,H_2O(g, 1.0\text{ atm})$$

where ΔG° is -958.3 kJ.

Solution.

$$Q_p = \frac{(p_{NO})^4(p_{H_2O})^6}{(p_{NH_3})^4(p_{O_2})^5}$$

$$\Delta G = \Delta G^\circ + (2.303)(8.314)\left(\frac{10^{-3}\text{ kJ}}{\text{J}}\right)(298)\log\frac{(2.0)^4(1.0)^6}{(0.10)^4(10.0)^5}$$

$$\Delta G = -958.3\text{ kJ} + [5.706 \log 1.6]\text{ kJ}$$

$$= -957.1\text{ kJ}$$

Because ΔG is **negative** 957.1 kJ, NH_3 at a pressure of 0.10 atm and O_2 at a pressure of 10.0 atm **will** react to produce NO at a pressure of 2.0 atm and H_2O at a pressure of 1.0 atm.

For a system in chemical equilibrium, there is no tendency for any further net chemical change. Under these conditions, $\Delta G = 0$ and Q_p becomes numerically equal to K_p. $\Delta G°$ then is given by the equation $\Delta G° = -2.303RT \log K_p$. Values of the equilibrium constant may be calculated from $\Delta G°$ and vice versa.

Example 19-n. Calculate K_p for the equilibrium

$$COCl_2(g) \rightleftharpoons CO(g) + Cl_2(g)$$

See Table 19-1 for free energies of formation.

Solution.

$$K_p = \frac{p_{CO}p_{Cl_2}}{p_{COCl_2}}$$

$$\Delta G° = -(2.303)(RT \log K_p)$$

$\Delta G°$ may be calculated from free energies of formation.

$$\Delta G° = \Delta G_f° \text{(products)} - \Delta G_f° \text{(reactants)}$$

$$\Delta G° = (-137.2 + 0) - (-210.5) = +73.3 \text{ kJ}$$

$$\Delta G° = -(2.303)(8.314)(298) \log K_p$$

$$= +73{,}300 \text{ J}$$

$$\log K_p = \frac{+73{,}300 \not{J}}{-5{,}706 \not{J}} = -12.8$$

$$K_p = 1.4 \times 10^{-13}$$

Example 19-o. Calculate K_p for the equilibrium in Example 19-n at 1000 K from the value at 25°C (1.4×10^{-13}), assuming $\Delta H°$ and $\Delta S°$ are the same at the two temperatures.

$$COCl_2(g) \rightleftharpoons CO(g) + Cl_2(g)$$

Solution.

$$\Delta G° = \Delta H° - T\Delta S° = -(2.303)(RT) \log K_p$$

$$\Delta H° = \Delta H_f° \text{(products)} - \Delta H_f° \text{(reactants)}$$

$$= (-110.5 + 0) - (-223.0) = +112.5 \text{ kJ}$$

$$\Delta S° = S° \text{(products)} - S° \text{(reactants)}$$

$$= (197.9 + 223.0) - (289.1) = +131.8 \text{ J/K}$$

Since $-2.303RT \log K_p = \Delta G^\circ = \Delta H^\circ - T\Delta S^\circ$

$$\log K_p = \frac{\Delta H^\circ - T\Delta S^\circ}{-2.303RT} = \frac{-\Delta H^\circ}{2.303RT} + \frac{\Delta S^\circ}{2.303R}$$

Substituting,

$$\log K_p = \frac{-112{,}500}{(2.303)(8.314)(1000)} + \frac{131.8}{(2.303)(8.314)}$$

$$= -5.88 + 6.88 = 1.00$$

$$K_p = 1.0 \times 10^1$$

19-D. Electrochemical Cells and Free Energy

Just as equilibrium constants for gaseous reactions can be related to standard free energy changes, equilibrium constants for reactions in solution can also be related to standard frcc energy changes, that is,

$$\Delta G^\circ = -2.303RT \log K \tag{1}$$

In Section 18-B-4, it was pointed out that the standard potential of an electrochemical cell was related to the equilibrium constant for the cell reaction by the equation:

$$\mathscr{E}^\circ_{\text{cell}} = 2.303 \frac{RT}{nF} \log K \tag{2}$$

where n = number of moles of electrons transferred in the balanced chemical equation; $\mathscr{E}^\circ_{\text{cell}}$ is the difference in the standard reduction potentials of the two half-reactions making up the cell, and F is the Faraday constant expressed as 96,487 J/(volt-equiv). By combining Equations (1) and (2)

$$\Delta G^\circ = -nF\mathscr{E}^\circ_{\text{cell}}$$

For reactions in solution, K is usually expressed in terms of concentrations (molarity) of reactants and products. As with standard reduction potentials, standard state conditions are: all solids are as pure substances in their most stable form; all gases are at 1 atm pressure, and all solutes are at a concentration of 1 molar. The solvent in dilute solutions is considered the same as a pure liquid, hence it is in its standard state.

For conditions other than standard conditions,

$$\Delta G = -nF\mathscr{E}_{\text{cell}} \tag{3}$$

where ε_{cell} is given by the equation:

$$\mathscr{E}_{cell} = \mathscr{E}^{\circ}_{cell} - \frac{RT}{nF}(2.303)\log Q \tag{4}$$

where Q is as defined in Section 19-C. Combining (3) and (4) above, we obtain

$$\Delta G = -nF\mathscr{E}^{\circ}_{cell} + (2.303)RT\log Q$$

Example 19-p. Determine ΔG at 25°C for the reaction in aqueous solution:

$$2\,Ag^{+}(0.10\,M) + Sn(s) \rightleftharpoons 2\,Ag(s) + Sn^{2+}(0.50\,M) \quad \mathscr{E}^{\circ}_{cell} = +0.94 \text{ volts}$$

Solution.

$$\Delta G = -nF\mathscr{E}^{\circ}_{cell} + 2.303RT\log\frac{[Sn^{2+}]}{[Ag^{+}]^2}$$

$$\Delta G = -(2)(96{,}487)(0.94) + (2.303)(8.314)(298)\log\frac{0.50}{(0.10)^2}$$

$$= -181.4\text{ kJ} + [5.706\log(50)]\text{ kJ}$$

$$= -181.4\text{ kJ} + 9.7\text{ kJ} = -171.7\text{ kJ} = -1.7 \times 10^2\text{ kJ}$$

General Problems

19-1. Calculate ΔH° for the reaction at 25°C

$$CH_4(g) + 2\,O_2(g) \longrightarrow CO_2(g) + 2\,H_2O(l)$$

19-2. The heat of combustion of C_2H_2 at constant volume (ΔE) and 20°C is -1301.6 kJ/mole. Calculate the heat of combustion at constant pressure (ΔH) of 1 atm and 20°C.

19-3. Calculate the work done in isothermally expanding 1.0 mole of gas from 5.0 liters to 15.0 liters at 27°C.

19-4. Using standard entropy values in Table 19-2, calculate ΔS° for the following reactions:

(a) $\frac{1}{2}H_2(g, 1\text{ atm}) + \frac{1}{2}Cl_2(g, 1\text{ atm}) \rightarrow HCl(g, 1\text{ atm})$
(b) $C(s) + O_2(g, 1\text{ atm}) \rightarrow CO_2(g, 1\text{ atm})$
(c) $Zn(s) + H_2SO_4(aq, 1\,M) \rightarrow ZnSO_4(aq, 1\,M) + H_2(g, 1\text{ atm})$
(d) $Na(s) + \frac{1}{2}Cl_2(g, 1\text{ atm}) \rightarrow NaCl(s)$

19-5. Calculate ΔH° and ΔE° for the following reaction at 25°C:

$$C_2H_6(g, 1 \text{ atm}) + 3\tfrac{1}{2}\,O_2(g, 1 \text{ atm}) \longrightarrow 2\,CO_2(g, 1 \text{ atm}) + 3\,H_2O(l)$$

19-6. Calculate ΔE° and ΔH° for the following reactions at 25°C and 1.0 atm pressure. ΔH_f° for $(NH_4)_2C_2O_4$ and NH_4HCO_3 are -1117.8 and -849.8 kJ/mole, respectively.
(a) $(NH_4)_2C_2O_4(s) \rightarrow 2\,NH_3(g) + CO_2(g) + CO(g) + H_2O(l)$
(b) $NH_4HCO_3(s) \rightarrow NH_3(g) + CO_2(g) + H_2O(l)$

19-7. Calculate ΔS_{total} for the reaction $C(s) + O_2(g) \rightarrow CO_2(g)$ at 25°C. Can this reaction occur spontaneously?

19-8. Use data from Example 19-o in this exercise. $COCl_2$ is pumped into a cylinder at 1000 K and allowed to come to equilibrium with its dissociation products. At equilibrium the partial pressure of $COCl_2$ is 5.0 atm. Calculate the partial pressures of CO and Cl_2.

19-9. Calculate ΔG° and K_p at 25°C for the equilibrium

$$PCl_5(g) \rightleftharpoons PCl_3(g) + Cl_2(g)$$

Will the reaction occur spontaneously at standard conditions?

19-10. Calculate K_p at 600 K for the equilibrium in Problem 19-9. Heats of formation: $PCl_3 = -306.3$ kJ/mole, $PCl_5 = -399.2$ kJ/mole. *Hint*: Calculate ΔS° from ΔG° above.

19-11. The heat of vaporization of Hg at its boiling point 356.6°C is 59.26 kJ/mole. Calculate ΔH, ΔE, ΔS, and ΔG for the vaporization of 1 mole of Hg at the B.P. and 1 atm pressure.

19-12. Heats of formation of $HCl(g)$, $CO_2(g)$, and $NaCl(s)$ arc respectively -92.3, -393.5, and -411.3 kJ/mole. Calculate ΔG° at 298 K for parts (a), (b), and (d) in Problem 19-4.

19-13. Determine ΔG° for the following reactions:
(a) $CH_4(g) + 2\,O_2(g) \rightarrow CO_2(g) + 2\,H_2O(l)$
(b) $2\,P(s) + 3\,Br_2(l) \rightarrow 2\,PBr_3(g)$
(c) $C(s) + O_2(g) \rightarrow CO_2(g)$
(d) $C_2H_6(g) + 3\tfrac{1}{2}\,O_2(g) \rightarrow 2\,CO_2(g) + 3\,H_2O(l)$
Indicate whether these reactions can occur spontaneously as written under standard conditions.

19-14. Determine the boiling point of Hg from the following data: heat of vaporization = 59.26 kJ/mole. At the boiling point and standard pressure the entropy change is 94.1 J/(mole K).

19-15. The heat of fusion of ethyl alcohol, C_2H_5OH, at its melting point of -114°C is 4.79 kJ/mole. If one mole of liquid ethyl alcohol is converted to solid at $-114°C$, calculate ΔS. Is ΔS positive or negative and how do you determine this? Discuss the value of ΔG for the change.

19-16. Using data in Table 19-1 and Table 19-2, calculate $\Delta H°$, $\Delta S°$, and $\Delta G°$ for the following reaction at 25°C.

$$CO(g) + \tfrac{1}{2} O_2(g) \longrightarrow CO_2(g)$$

What do you conclude regarding the spontaneity of the reaction under standard conditions?

19-17. (a) Calculate ΔS_{total} for the reaction $CO(g) + \frac{1}{2}O_2(g) \rightarrow CO_2(g)$ at 25°C. (b) Using the discussion presented in Section 19-B-2, calculate $\Delta G°$ from ΔS_{total}. Compare this $\Delta G°$ with that obtained from tabulated free energy of formation values as in Problem 19-16.

19-18. Calculate ΔG at 298 K for the reactions under the conditions indicated in the following equations:

(a) $CO(g, 0.60\ atm) + Cl_2(g, 0.10\ atm) \rightleftharpoons COCl_2(g, 3.0\ atm)$

(b) $CH_4(g, 0.30\ atm) + Cl_2(g, 0.30\ atm) \rightleftharpoons CH_3Cl(g, 0.90\ atm) + HCl(g, 1.0\ atm)$

(c) $CH_3Cl(g, 0.20\ atm) + H_2O(g, 0.010\ atm) \rightleftharpoons CH_3OH(l) + HCl(g, 0.40\ atm)$

(d) $N_2(g, 40.0\ atm) + 3\ H_2(g, 50.0\ atm) \rightleftharpoons 2\ NH_3(g, 1.0\ atm)$

(e) $H_2(g, 5.0\ atm) + I_2(g, 0.10\ atm) \rightleftharpoons 2\ HI(g, 0.40\ atm)$

(f) $2\ NOCl(g, 0.20\ atm) + CO(g, 0.30\ atm) \rightleftharpoons 2\ NO(g, 0.40\ atm) + COCl_2(g, 1.0\ atm)$

(g) $CO(g, 40.0\ atm) + H_2O(g, 30.0\ atm) \rightleftharpoons CO_2(g, 3.0\ atm) + H_2(g, 1.0\ atm)$

19-19. Calculate K_p at 298 K for the equilibria represented by the following equations:

(a) $Br_2(l) + Cl_2(g) \rightleftharpoons 2\ BrCl(g)$

(b) $2\ CO(g) + O_2(g) \rightleftharpoons 2\ CO_2(g)$

(c) $2\ HI(g) + \frac{1}{2} O_2(g) \rightleftharpoons H_2O(g) + I_2(g)$

(d) $C_2H_4(g) + H_2(g) \rightleftharpoons C_2H_6(g)$

(e) $CH_4(g) + Cl_2(g) \rightleftharpoons CH_3Cl(g) + HCl(g)$

(f) $2\ Na(s) + Cl_2(g) \rightleftharpoons 2\ NaCl(s)$

(g) $2\ NOCl(g) + CO(g) \rightleftharpoons 2\ NO(g) + COCl_2(g)$

(h) $H_2(g) + I_2(g) \rightleftharpoons 2\ HI(g)$

(i) $H_2(g) + F_2(g) \rightleftharpoons 2\ HF(g)$

19-20. Using data in Table 18-1, determine $\Delta G°$ for each of the following reactions in aqueous solution as shown by the equations. (*Note*: The standard state for a solvent is taken as the pure liquid; however, in dilute solutions the solvent is assumed to be in its standard state. This assumption is made here and a term for the "concentration of water" does not appear in the expression for either Q or K.)

(a) $Zn(s) + 2\,Cr^{3+} \rightleftharpoons Zn^{2+} + 2\,Cr^{2+}$

(b) $Sn^{2+} + 2\,Fe^{3+} \rightleftharpoons Sn^{4+} + 2\,Fe^{2+}$

(c) $Cr_2O_7^{2-} + 14\,H^+ + 6\,Fe^{2+} \rightleftharpoons 6\,Fe^{3+} + 2\,Cr^{3+} + 7\,H_2O$

(d) $3\,H_2S + 2\,H^+ + 2\,NO_3^- \rightleftharpoons 3\,S + 2\,NO + 4\,H_2O(l)$

(e) $Cu + 4\,H^+ + 2\,NO_3^- \rightleftharpoons Cu^{2+} + 2\,NO_2 + 2\,H_2O(l)$

19-21. Using data in Table 18-1, determine ΔG at 25°C for each of the following reactions:

(a) $Zn(s) + 2\,Cr^{3+}(1.0\,M) \rightleftharpoons Zn^{2+}(0.0010\,M) + 2\,Cr^{2+}(0.010\,M)$

(b) $Sn^{2+}(1.0\,M) + 2\,Fe^{3+}(1.0\,M) \rightleftharpoons Sn^{4+}(0.0010\,M) + 2\,Fe^{2+}(0.010\,M)$

(c) $Sn^{2+}(0.0010\,M) + 2\,Fe^{3+}(0.010\,M) \rightleftharpoons Sn^{4+}(1.0\,M) + 2\,Fe^{2+}(1.0\,M)$

(d) $Cr_2O_7^{2-}(1.0\,M) + 14\,H^+(0.10\,M) + 6\,Fe^{2+}(1.0\,M) \rightleftharpoons$
$6\,Fe^{3+}(1.0\,M) + 2\,Cr^{3+}(1.0\,M) + 7\,H_2O(l)$

(e) $Cu(s) + 4\,H^+(0.010\,M) + 2\,NO_3^-(0.010\,M) \rightleftharpoons$
$Cu^{2+}(0.10\,M) + 2\,NO_2(g, 1.0\text{ atm}) + 2\,H_2O(l)$

19-22. ΔG at 298 K for the reaction and conditions shown by the equation

$$Co(s) + 2\,Fe^{3+}(0.10\,M) \longrightarrow Co^{2+}(0.20\,M) + 2\,Fe^{2+}(0.010\,M)$$

was found experimentally to be -216.5 kJ. With the help of Table 18-1, determine $\varepsilon°$ for the half-reaction $Co^{2+} + 2\,e^- \rightleftharpoons Co$.

Chapter 20

Nuclear Changes

20-A. Radioactivity

Radioactivity is the spontaneous emission of high energy radiation due to decay or disintegration of the nuclei of atoms. Atoms of atomic number greater than 82 and a few of lower atomic number are naturally radioactive (for example, U, Th, Pa) and emit alpha, beta and/or gamma radiation. Many isotopes of the lighter elements have been produced by nuclear reactions, and some of these also exhibit the property of radioactivity. In addition to the decay processes mentioned, lighter isotopes may also decay by positron emission or K-electron capture.

Radioactivity is a property of the nucleus and the rate of disintegration is independent of external effects such as temperature and pressure. Each radioactive isotope has a characteristic rate of decay that is expressed as a **half-life**. The half-life is the time required for one-half of any given amount of the element to decompose. For example, the half-life of $^{226}_{88}Ra$* is 1620 years. Starting with 1 g of Ra today, after 1620 years, $\frac{1}{2}$ g would remain; in two half-lives (3240 years) only 1/4 g would remain undecomposed; in three half-lives (4860 years) only 1/8 of the original amount remains, and so on. This process is analogous to dividing the original amount by 2 a number of times equal to the number of half-lives elapsed. Therefore, the amount of isotope left after three-half-lives is

$$\frac{\text{Original amount}}{2 \times 2 \times 2} = \frac{\text{Original amount}}{2^3} = \frac{\text{Original amount}}{8}$$

* See Section 20-C for an explanation of symbol notation.

In this case 1/8 g would be left after 4860 years. In a more general statement, the amount **remaining** after n half-lives = original amount/$(2)^n$.

Example 20-a. The half-life of $^{222}_{86}Rn$ gas is 4 days. Starting with 10 mg of the isotope, what amount would decay in 16 days and what amount would remain?

Solution. Sixteen days represent four (16/4) half-lives. Starting with 10 mg, 5 mg would decay in one half-life period (4 days).

$5\text{ mg} + 2\tfrac{1}{2}\text{ mg} = 7\tfrac{1}{2}$ mg would decay in two half-life periods (8 days)

$5\text{ mg} + 2\tfrac{1}{2}\text{ mg} + 1\tfrac{1}{4}\text{ mg} = 8\tfrac{3}{4}$ mg would decay in three half-life periods (12 days)

$5\text{ mg} + 2\tfrac{1}{2}\text{ mg} + 1\tfrac{1}{4}\text{ mg} + \tfrac{5}{8}\text{ mg} = 9\tfrac{3}{8}$ mg would decay in four half-life periods (16 days)

The amount remaining after 16 days would be $10 - 9\tfrac{3}{8} = \tfrac{5}{8}$ mg; or, using the formula above,

$$\text{Amount remaining} = \frac{10\text{ mg}}{(2)^4} = \frac{10\text{ mg}}{16} = \frac{5}{8}\text{ mg}$$

Example 20-b. A certain isotope has a half-life of 15 min. In what period of time would the activity be reduced to approximately 10% of the original?

Solution.

In one half-life, or 15 min, activity will be reduced to 50%.
In two half-lives, or 30 min, activity will be reduced to 25%.
In three half-lives, or 45 min, activity will be reduced to $12\tfrac{1}{2}$%.
In four half-lives, or 60 min, activity will be reduced to $6\tfrac{1}{4}$%.

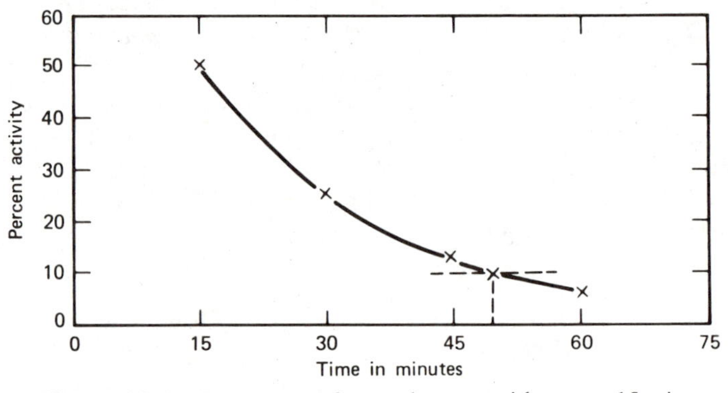

Figure 20-1 Decay curve for a substance with $t_{1/2} = 15$ min.

Thus the activity would be reduced to 10% in approximately 50 min. A more precise estimation of the time can be obtained by plotting the data as shown in Figure 20-1. Reading from the graph, the time is about 49.5 min.

Problems (Radioactivity)

20-1. The half-life of $^{214}_{82}Pb$ is 27 min. Starting with 128 mg of this isotope, what quantity would be left at the end of 1 hr and 48 min?

20-2. The $^{238}_{92}U$ isotope, which occurs naturally, has a half-life of 4.5×10^9 yr. In what period of time would the activity be reduced to 1/32 of the original?

20-3. The half-life of $^{14}_{6}C$ is 5730 yr. Plot a decay curve for this isotope similar to the one shown in Figure 20-1. Estimate the time to reduce the original activity to 15%.

20-4. A piece of charcoal found in a cave from a wood fire of long ago shows an activity of 20.0% of freshly prepared charcoal. Using the graph from Problem 20-3, estimate the age of the charcoal sample.

20-5. 4.00 grams of $^{218}_{83}Bi$ decay to 0.250 g in 80.0 min. What is the half-life of this isotope?

20-6. Explain what changes in mass and charge take place in the nucleus on the emission of (a) an alpha particle, (b) an electron, (c) gamma radiation, and (d) a positron.

20-7. A radioactive isotope has a half-life of 0.75 hr. Starting with 3.2×10^{21} atoms, how many atoms will be left at the end of 4.5 hrs?

20-8. Determine the half-life of an element, 87.5% of which decays in 6.0 years.

20-9. The half-life of $^{35}_{16}S$ is 87 days. Starting with 8.0×10^5 atoms of ^{35}S, how many atoms would **decompose** in 435 days?

20-10. What is the half-life of a radioactive element 7/8 of which decomposes in 12 sec?

20-11. The half-life of francium is 21 min. Starting with 8.0×10^{18} atoms of francium, how many atoms would **decompose** in 84 min?

20-12. The half-life of a radioactive element is 35 sec. In what period of time would the activity of the sample be reduced to 1/64 of the original activity?

20-13. In one of the radioactive disintegration series, $^{238}_{92}U$ ultimately forms $^{206}_{82}Pb$. Knowing that the emission of an alpha particle results in a decrease of two units in atomic number and a loss of four units in mass number and that the emission of a β^- particle results in an increase of one in atomic number and no change in mass number, (a) how many α particles must have been emitted per atom of $^{206}_{82}Pb$ formed, (b) how many β^- particles?

20-14. $^{226}_{88}Ra$ decays according to the equation $^{226}_{88}Ra \rightarrow ^{222}_{86}Rn + ^{4}_{2}He$. Starting with 1 mole of Ra, (a) what volume of He (at STP) will be produced by this reaction when the decay is 100% complete? (b) If this He gas is collected in a 2.00 liter container, what will the pressure in the container be at 0°C? (c) How long a time will it actually take for 100% of the sample to decompose?

20-15. $^{239}_{94}Pu$ is an α emitter with a half-life of 2.70×10^4 yr. Starting with 4.8 mg of $^{239}_{94}Pu$, how many Pu atoms will have decayed after 1.08×10^5 yr?

20-16. $^{212}_{83}Bi$ is an α emitter with a half-life of 60.0 min. Starting with 21.2 g of $^{212}_{83}Bi$, what period of time will be required to form 2.10 liters of He gas measured at STP?

20-B. Radioactive Decay as a First-Order Rate Process

Since the decay of a radioactive element is a first-order process, precise calculations may be made from the rate equation for first-order reactions

$$2.303 \log \frac{A}{A_0} = -kt \qquad (1)$$

where A_0 is the amount of substance at the start, A is the amount present at time t, and k is the rate constant. At the end of one half-life period, $t_{1/2}$, the amount remaining is one-half the original amount. That is, when $t = t_{1/2}$, $A = A_0/2$. By substitution of the quantities into Equation (1), it can then be shown that

$$t_{1/2} = \frac{0.693}{k} \qquad (2)$$

Example 20-c. The half-life of $^{90}_{38}Sr$ is 28 yr. In what period of time would the activity of a given sample be reduced to 1/10 of the original activity?

Solution. This is analogous to asking the question "When will $A/A_0 = 1/10$?" First evaluate k from Equation (2).

$$k = \frac{0.693}{28 \text{ yr}} = 0.025/\text{yr}$$

Substitute for A/A_0 and k in Equation (1) above and solve for t.

$$2.303 \log \frac{1}{10} = -\frac{0.025\,t}{\text{yr}}$$

$$t = \frac{2.303(-1)}{-0.025/\text{yr}} = 92 \text{ yr}$$

Example 20-d. 5.0% of a certain radioactive element decays in 21.0 min. Determine its half-life.

Solution. The fraction remaining after 21.0 min is 0.950 or $A/A_0 = 0.950$. Substituting into (1)

$$2.303 \log 0.950 = -k(21.0 \text{ min})$$

$$k = \frac{(2.303)(-0.0223)}{-21.0 \text{ min}} = 0.00245/\text{min}$$

$$t_{1/2} = \frac{0.693}{k} = \frac{0.693}{0.00245/\text{min}} = 283 \text{ min}$$

Problems (Using First-Order Rate Equations)

20-17. If 65.5% of a radioactive isotope decays in 38.0 min, what is its half-life?

20-18. A certain isotope has a half-life of 15.0 min. In what period of time would the activity be reduced to 10.0% of the original? Compare this value with the result obtained by graphical means in Example 20-b.

20-19. Work Problem 20-3 using the first-order rate expression. Then compare the answer with that obtained graphically.

20-20. $^{131}_{53}I$ has a half-life of 8.0 days. Starting with 9.1×10^{20} atoms, how many would be left at the end of 72 days?

20-21. A certain isotope has a half-life of 1.0×10^5 yr. What fraction of the original sample would be left at the end of 1.0×10^6 yr?

20-22. A certain radioactive element X has a half-life of 1.00 yr. Starting with 3.25 g of X (atomic weight 233), how many disintegrations will occur in 20.0 sec?

20-23. 1.50 moles of a certain radioactive element emits 3.09×10^{10} β^- particles per second. Assuming each beta particle corresponds to the decay of one atom of the radioactive element, what is the half-life of the element?

20-24. A sample of a radioactive substance initially registers 675 counts per sec, but 5.50 hr later the activity is 137 counts per sec. Determine the half-life of this substance.

20-25. A particular mineral of uranium was found to contain 0.604 g of $^{206}_{82}Pb$ for every 1.340 g of $^{238}_{92}U$. If all the $^{206}_{82}Pb$ present was formed by radioactive decay of $^{238}_{92}U$, what is the age of the mineral? Note that numbers derived from such a source are usually taken as indicative of the age of the earth. The half-life of $^{238}_{92}U$ is 4.51×10^9 yr; the half-lives of the elements between U and Pb are short in comparison with $^{238}_{92}U$.

20-26. The procedure for determining the age of carbon-containing objects was worked out by Professor Willard F. Libby while he was at the University of Chicago. This method is based on the fact that $^{14}_{6}C$, a β^- emitter with a half-life of 5730 yr, is produced by the interaction of cosmic rays with nitrogen of the atmosphere. It is presumed to have achieved a steady-state concentration so that it decomposes at the same rate it is produced. Hence a constant ratio of $^{14}_{6}C$ to $^{12}_{6}C$ exists in the CO_2 of the atmosphere. Living organisms are composed of carbon with essentially the same ratio of ^{14}C to ^{12}C as atmospheric CO_2. On death, no further carbon is taken into the organism, and from this time on, the ratio of ^{14}C to ^{12}C decreases because of the decay of ^{14}C. (a) A sample of wood charcoal obtained from an Indian fishing site (now known as the Marmes Site) near the confluence of the Snake and Palouse rivers in the state of Washington was found to have a ^{14}C to ^{12}C ratio of 27.5% of present-day charcoal. Assuming the ^{14}C to ^{12}C ratio in the atmosphere has been constant over the intervening years, how long ago was the site occupied? (b) A site on the northwest coast of the state of Washington was inhabited until the early twentieth century by ancestors of the present-day Makah Indian tribe. This village, now called the Ozette Site, was once covered by a mud slide that covered dwellings and protected them and their contents from damage by the elements of nature. Recently, woven baskets, fishing gear, and other artifacts have been uncovered by Professor R. D. Daugherty and his coworkers. Samples of the baskets show a ^{14}C to ^{12}C ratio, which is 94.1% of that of recently grown, similar materials. What are the approximate ages of the artifacts?

20-27. A dating procedure for water has also been proposed by Professor Libby based on the β^- activity of $^{3}_{1}H$. Ordinary water contains tritium or hydrogen of mass 3 that is formed in the earth's atmosphere from cosmic rays. It decomposes by beta emission at a rate equal to that at which it is formed, and hence water exposed to the atmosphere will contain a constant mole percentage of tritium. Wine that has been bottled will not have its tritium replaced after it decays, and hence, the older the wine, the lower the tritium content. A certain sample was found to have 44.8% of the tritium content of present-day water. Assuming tritium is lost only by β^- decay, what is the age of the wine? $t_{1/2}$ for $^{3}_{1}H = 12.26$ yr.

20-28. A sample of uranium ore is found to contain 5.25 g of $^{238}_{92}U$. On heating, helium gas is driven off. If all of the helium originally came from the decay of $^{238}_{92}U$ to $^{206}_{82}Pb$ during the 2.50×10^9 yr of existence of the rock, what volume of He was collected at 25°C and 1.00 atm pressure? The half-life of $^{238}_{92}U$ is 4.51×10^9 yr; the half-lives of the elements between U and Pb are short in comparison with $^{238}_{92}U$. (See Problem 20-13.)

20-29. $^{90}_{38}Sr$ is one of the fission products of the atomic bomb that is present in radioactive fallout. If a given sample of soil has 3.71×10^{-11} g of $^{90}_{38}Sr$, how long will it take for the activity to decrease to 1/10 of its present amount? The half-life of $^{90}_{38}Sr$ is 28.0 yr.

20-C. Nuclear Equations

In writing nuclear equations it is customary to show the mass number and the atomic number of the atoms or particles involved. The mass number is shown as a superscript to the left of the symbol and the atomic number as a subscript to the left, for example, $^{238}_{92}U$. For bombarding and ejected particles, the following representations may be used:

alpha particle	$^{4}_{2}He$	or (α)
proton	$^{1}_{1}H$	or (p)
deuteron	$^{2}_{1}H$	(or $^{2}_{1}D$ or d)
tritium	$^{3}_{1}H$	(or $^{3}_{1}T$ or t)
beta particle (negative electron)	$^{0}_{-1}e$	or (β^- or e^-)
positron	$^{0}_{+1}e$	or (β^+ or e^+)
neutron	$^{1}_{0}n$	or (n)
gamma ray	$^{0}_{0}\gamma$	

In a balanced nuclear equation, the sum of the superscripts (mass number) must be the same on both sides of the equation; likewise the sum of the subscripts (atomic number) must balance on the two sides of the equation. Examples are

1. $^{14}_{7}N + ^{4}_{2}He \rightarrow ^{17}_{8}O + ^{1}_{1}H$
2. $^{14}_{7}N + ^{1}_{0}n \rightarrow ^{14}_{6}C + ^{1}_{1}H$
3. $^{68}_{30}Zn + ^{1}_{0}n \rightarrow ^{65}_{28}Ni + ^{4}_{2}He$
4. $^{55}_{25}Mn + ^{2}_{1}H \rightarrow ^{55}_{26}Fe + 2^{1}_{0}n$
5. $^{35}_{17}Cl + ^{1}_{0}n \rightarrow ^{35}_{16}S + ^{1}_{1}H$
6. $^{23}_{11}Na + ^{2}_{1}H \rightarrow ^{24}_{11}Na + ^{1}_{1}H$
7. $^{6}_{3}Li + ^{1}_{0}n \rightarrow ^{4}_{2}He + ^{3}_{1}H$

(*Note*: In all these equations both the subscripts and the superscripts balance on the two sides of the equations.)

A somewhat shorter notation may be employed to show nuclear processes in which the bombarding particles (the **in** particles) and the ejected particles (the **out** particles) are shown in parenthesis between the target atom and the product atom; thus reactions 1 through 4 above may be shown as:

1. ${}^{14}_{7}N(\alpha, p){}^{17}_{8}O$
2. ${}^{14}_{7}N(n, p){}^{14}_{6}C$
3. ${}^{68}_{30}Zn(n, \alpha){}^{65}_{28}Ni$
4. ${}^{55}_{25}Mn(d, 2n){}^{55}_{26}Fe$

Problems (Nuclear Equations)

20-30. Complete the following:

(a) ${}^{11}_{5}B + ? \rightarrow {}^{14}_{7}N + {}^{1}_{0}n$

(b) ${}^{9}_{4}Be + {}^{4}_{2}He \rightarrow {}^{12}_{6}C + ?$

(c) ${}^{30}_{15}P \rightarrow {}^{30}_{14}Si + ?$

(d) ${}^{7}_{3}Li + {}^{1}_{1}H \rightarrow 2?$

(e) ${}^{235}_{92}U + {}^{1}_{0}n \rightarrow {}^{94}_{38}Sr + {}^{139}_{54}Xe + ?$

20-31. Write balanced equations for the following nuclear processes:

(a) ${}^{39}_{19}K(p, n)$

(b) ${}^{31}_{15}P(n, \gamma)$

(c) ${}^{191}_{77}Ir(\alpha, n)$

(d) ${}^{24}_{12}Mg(\alpha, n)$

(e) ${}^{242}_{96}Cm(\alpha, 2n)$

20-32. From your textbook select nuclear reactions that represent the following: (a) fission, (b) fusion, (c) artificial radioactivity, (d) natural radioactivity, (e) proton bombardment, and (f) K-electron capture.

20-33. Complete the following equations:

(a) ____ $+ {}^{4}_{2}He \rightarrow {}^{30}_{14}Si + {}^{1}_{1}H$

(b) ${}^{9}_{4}Be +$ ____ $\rightarrow {}^{2}_{1}H + {}^{8}_{4}Be$

(c) ${}^{27}_{13}Al + {}^{4}_{2}He \rightarrow {}^{1}_{0}n +$ ____

(d) ${}^{85}_{37}Rb + {}^{1}_{0}n \rightarrow 2\,{}^{1}_{0}n +$ ____

(e) ${}^{246}_{96}Cm + {}^{12}_{6}C \rightarrow {}^{254}_{102}No + 4$____

20-34. Write balanced equations for the following:

(a) ${}^{253}_{99}Es(\alpha, n)$

(b) ${}^{241}_{95}Am(\alpha, 2n)$

(c) ${}^{65}_{29}Cu(n, 2n)$

(d) ${}^{54}_{26}Fe(\alpha, 2p)$

(e) ${}^{250}_{98}Cf({}^{11}_{5}B, 4n)$

20-35. When the plutonium isotope $^{239}_{94}Pu$ is irradiated with neutrons in a pile, the nuclide begins to undergo a series of neutron absorptions and negative electron emissions. (a) In one experiment, a product nuclide was found that could only have resulted by the absorption of 15 neutrons and emission of 6 beta particles. Characterize the final nuclide. (b) In a second experiment employing $^{238}_{92}U$, a nuclide was found that resulted from the absorption of 15 neutrons and the emission of 7 beta particles. Characterize the final nuclide.

20-36. (a) The strontium isotope $^{93}_{38}Sr$ emits 3 successive beta particles. What is the product nucleus? Write equations for the three successive reactions. (b) The oxygen isotope $^{13}_{8}O$ emits 2 successive positrons. What is the final product nucleus? Write equations for the 2 successive reactions. (c) The isotope $^{254}_{100}Fm$ emits 6 successive α particles and then 2 beta particles. What is the product nucleus at the end of this sequence? Write equations for the successive reactions.

20-D. Units of Radioactivity

The **curie** is the standard unit of radioactivity and is defined as that amount of a radioactive substance that produces 3.70×10^{10} disintegrations per second. Smaller units are the **millicurie** (3.70×10^{7} dis/sec) and the **microcurie** (3.70×10^{4} dis/sec).

Example 20-e. Calculate the mass of radium that has a radioactivity of one microcurie. $t_{1/2}$ for Ra = 1622 yr (5.12×10^{10} sec).

Solution. Since the decay of a radioactive element is a first-order process,

$$\frac{-d[A]}{dt} = k[A]$$ where $[A]$ is the number of atoms of the disintegrating substance (see Section 13-A).

In this case

$$\frac{-d[A]}{dt} = 1 \text{ microcurie} = 3.70 \times 10^{4} \text{ atoms/sec}$$

$$t_{1/2} = \frac{0.693}{k}$$

Hence

$$k = \frac{0.693}{t_{1/2}}$$

$$k = \frac{0.693}{5.12 \times 10^{10} \text{ sec}} = 1.35 \times 10^{-11}/\text{sec}$$

Substituting in the first-order rate equation,

$$3.70 \times 10^4 \text{ atoms/sec} = (1.35 \times 10^{-11}/\text{sec})[A]$$

or

$$[A] = \frac{3.70 \times 10^4 \text{ atoms/}\cancel{\text{sec}}}{1.35 \times 10^{-11}/\cancel{\text{sec}}} = 2.74 \times 10^{15} \text{ atoms}$$

This is the number of Ra atoms present. To convert this to mass of Ra:

$$\text{Mass of Ra} = 2.74 \times 10^{15} \cancel{\text{atoms}} \times \frac{1 \cancel{\text{mole}}}{6.02 \times 10^{23} \cancel{\text{atoms}}} \times \frac{226 \text{ g Ra}}{1 \cancel{\text{mole}}}$$

$$= 1.03 \times 10^{-6} \text{ g Ra}$$

Radiation is harmful to humans because it can cause ionization and ultimately destruction of the molecules that make up our living cells. There are two ways of expressing this effect on body tissues. The **rad** is defined as the absorption of 1.0×10^{-5} J of energy per gram of tissue. Some types of radiation are more harmful than others. The **rem** is defined as the number of rads absorbed multiplied by a correction factor describing the relative harmfulness of the type of radiation, that is,

$$\text{Number of rems} = a \times \text{Number of rads}$$

where $a = 1$ for β^- and γ radiation and $a = 10$ for α radiation

Example 20-f. A 57 kg human (~125 lbs.) is exposed to α radiation equivalent to the absorption of 1.3 J. (a) How many rads of radiation have been absorbed? (b) What radiation dosage, in rems, has been received?

Solution.

$$\text{(a) Rads absorbed} = \frac{1.3 \text{ J}}{57 \text{ kg}} \times \frac{1 \text{ kg}}{10^3 \text{ g}} \times \frac{1 \text{ rad}}{1.0 \times 10^{-5} \text{ J/g}}$$

$$= \frac{1.3 \cancel{\text{J}}}{57 \cancel{\text{kg}}} \times \frac{1 \cancel{\text{kg}}}{10^3 \cancel{\text{g}}} \times \frac{1 \text{ rad} \times \cancel{\text{g}}}{1.0 \times 10^{-5} \cancel{\text{J}}} = 2.3 \text{ rads}$$

$$\text{(b) Rems absorbed} = a \times \text{Number of rads}$$

$$= 10 \times 2.3 = 23 \text{ rems}$$

Problems (Units of Radioactivity)

20-37. The half-life of $^{214}_{83}Bi$ is 19.7 min. Calculate the mass of this radioisotope that will give an activity of 2.5 millicurie.

20-38. What mass of $^{238}_{92}U$ has an activity of 3.8 microcuries? $t_{1/2} = 4.5 \times 10^9$ yr.

20-39. The half-life of $^{239}_{94}Pu$ is 2.4×10^4 yr. (a) Determine the mass of Pu that has an activity of 1.3 curies. (b) Determine the rate of decay in disintegrations per second per milligram of Pu.

20-40. Radioactive iodine in the form of potassium iodide is used to treat disorders in the thyroid gland of human beings. The iodine isotope used is $^{131}_{53}I$, which has a half-life of 8.0 days. Assuming a patient requires an initial radiation level of 2.0 millicuries, what mass of KI, containing $^{131}_{53}I$, is required?

20-41. Suppose a 1.8×10^2 lb man is exposed to β^- radiation equivalent to the absorption of 0.78 J. (a) How many rads of radiation have been absorbed? (b) What radiation dosage, in rems, has been received?

20-E. Mass-Energy Relations

The very large amount of energy liberated to the surroundings in a nuclear change is a result of a loss in mass during the process. Mass is converted into **its energy equivalent** as given by the Einstein relation

$$E = Mc^2$$

$$\text{Energy liberated (joules)} = \text{Mass decrease (kilograms)} \times [\text{Velocity of light (m/sec)}]^2$$

Although the SI system would dictate the expression of energy released in terms of *joules* (kg m²/sec²), nuclear chemists still prefer the use of other units. Hence, since various energy units are employed in describing these transformations, the student must know the definitions of these units and the relationships among them.

An **atomic mass unit, amu**, is defined as 1/12 the mass of the $^{12}_{6}C$ atom, whose isotopic mass is exactly 12 *amu*. As shown in Chapter 3, 1 amu is equivalent to 1.661×10^{-24} g.

Energies may be expressed in **electron volts**. An electron volt is defined as the energy required to raise one electron through a potential difference of one volt. This is a very small unit; hence, one million electron volts, Mev, is commonly used as a unit of energy for nuclear processes. Other energy units used are calories and joules.

$$1 \text{ Mev} = 1.602 \times 10^{-13} \text{ joules} \quad \textit{or} \quad 1 \text{ joule} = 6.24 \times 10^{12} \text{ Mev}$$

Since 4.184 joules = 1 cal,

$$1 \text{ Mev} = 3.829 \times 10^{-14} \text{ cal} \qquad \textit{or} \qquad 1 \text{ cal} = 2.612 \times 10^{13} \text{ Mev}$$

Now let us return to the Einstein equation and determine the relationship between mass and energy in joules, Mev, and calories.

$$\text{Energy (joules)} = 1\ \text{kg} \times (2.998 \times 10^8\ \text{m/sec})^2 = 8.988 \times 10^{16}\ \text{joules.}$$

Thus the **energy equivalent of mass** is 8.988×10^{16} J/kg

or

$$\frac{8.988 \times 10^{16}\ \text{J}}{1\ \cancel{\text{kg}}} \times \frac{1\ \cancel{\text{kg}}}{10^3\ \text{g}} = 8.988 \times 10^{13}\ \text{J/g}$$

or

$$\frac{8.988 \times 10^{13}\ \cancel{\text{J}}}{1\ \text{g}} \times \frac{1\ \text{cal}}{4.184\ \cancel{\text{J}}} = 2.148 \times 10^{13}\ \text{cal/g}$$

or

$$\frac{8.988 \times 10^{13}\ \cancel{\text{J}}}{1\ \text{g}} \times \frac{6.24 \times 10^{12}\ \text{Mev}}{1\ \cancel{\text{J}}} = 5.61 \times 10^{26}\ \text{Mev/g}$$

Since 1.661×10^{-24} g = 1 amu, the **energy equivalent of mass** is also

$$5.61 \times 10^{26}\ \frac{\text{Mev}}{\cancel{\text{g}}} \times 1.66 \times 10^{-24}\ \frac{\cancel{\text{g}}}{\text{amu}} = 931\ \text{Mev/amu}$$

Example 20-g. Consider the following nuclear fusion reaction:

$$^{7.01601}_{3}{}^{*}\text{Li} + {}^{1.00782}_{1}\text{H} \longrightarrow 2\ {}^{4.00265}_{2}\text{He}$$

Determine (a) the mass lost in amu per 2 atoms of He formed, (b) the corresponding energy release in Mev and calories, and (c) the energy release per 2 moles of He formed.

* To conserve space and as a matter of convenience, the exact isotopic mass has been used in place of the mass number in stating the problems in this chapter. For example, the $^{7}_{3}$Li isotope is written $^{7.01601}_{3}$Li to indicate that the isotope of mass number 7 has an exact isotopic mass of 7.01601.

Solution.

(a) Mass of reactants: 7.01601 + 1.00782 = 8.02383 amu

Mass of product: 2 × 4.00265 = 8.00530 amu

Mass loss = 0.01853 amu

(b) Since 1 amu = 931 Mev, the energy release per 2 atoms of He formed

$$= \frac{0.01853 \text{ amu}}{(2 \text{ atoms He})}\left(931 \frac{\text{Mev}}{\text{amu}}\right) = 17.25 \text{ Mev/(2 atoms He)}$$

or, expressed in calories, the energy release

$$= \frac{17.25 \text{ Mev}}{(2 \text{ atoms He})} \times \frac{3.829 \times 10^{-14} \text{ cal}}{1 \text{ Mev}} = 6.605 \times 10^{-13} \text{ cal/(2 atoms He)}$$

(c) For 1 mole of Li reacting with 1 mole of H to form 2 moles of He,

$$\text{Energy release} = \frac{17.25 \text{ Mev}}{(2 \text{ atoms He})}\left(6.02 \times 10^{23} \frac{\text{atoms He}}{\text{mole He}}\right)$$

$$= 1.04 \times 10^{25} \text{ Mev/(2 moles He)}$$

or

$$\text{Energy release} = \frac{1.04 \times 10^{25} \text{ Mev}}{2 \text{ moles He}} \times \frac{1 \text{ J}}{6.24 \times 10^{12} \text{ Mev}}$$

$$= 1.67 \times 10^{12} \text{ J/(2 moles He)}$$

20-F. Binding Energy

The mass of an atom is always less than the sum of the masses of its constituent particles. Thus the helium atom may be regarded as being derived from 2 hydrogen atoms and 2 neutrons:

$$\text{Mass of } 2\,{}^{1.007825}_{1}\text{H} = 2.01565 \text{ amu}$$

$$\text{Mass of } 2\,{}^{1.008675}_{0}\text{n} = 2.01735 \text{ amu}$$

$$\text{Mass of constituent particles} = 4.03300 \text{ amu}$$

$$\text{Mass of } {}^{4.00265}_{2}\text{He} = 4.00265 \text{ amu}$$

$$\text{Mass difference} = 0.03035 \text{ amu}$$

$$\text{Energy equivalent} = (0.03035 \text{ amu})\left(931 \frac{\text{Mev}}{\text{amu}}\right) = 28.3 \text{ Mev}$$

This energy is termed the **binding energy** of the nucleus, that is, the energy that holds it together. Nuclear particles, protons, and neutrons, are termed

nucleons. The binding energy per nucleon for He, which has 4 nucleons, is 28.3/4 = 7.1 Mev. The binding energy is related to stability and the most stable of naturally occurring elements are those with high binding energies.

Example 20-h. Compute the binding energy per nucleon for the isotope

$$^{55.9349}_{26}\text{Fe}$$

Solution. The iron atom may be regarded as composed of 26 H atoms and 30 neutrons.

$$\begin{aligned}
26\ ^{1.00782}_{1}\text{H have mass } 26 \times 1.00782 &= 26.2033 \text{ amu} \\
30\ ^{1.00868}_{0}\text{n have mass } 30 \times 1.00868 &= \underline{30.2604 \text{ amu}} \\
\text{Mass of constituent particles} &= 56.4637 \text{ amu} \\
\text{Mass of } ^{56}_{26}\text{Fe isotope} &= \underline{55.9349 \text{ amu}} \\
\text{Mass difference} &= 0.5288 \text{ amu}
\end{aligned}$$

$$\text{Binding energy} = (0.529 \cancel{\text{amu}})\left(931 \frac{\text{Mev}}{\cancel{\text{amu}}}\right) = 492 \text{ Mev}$$

$$\begin{aligned}
\text{Binding energy per nucleon} &= 492 \text{ Mev/56 nucleons} \\
&= 8.8 \text{ Mev/nucleon}
\end{aligned}$$

Problems

20-42. The conversion of 4.50 g of matter to energy by the Einstein equation yields how many joules?

20-43. Calculate the energy equivalent, in Mev, of 5.25 mg of matter.

20-44. For the nuclear reaction

$$^{2.01410}_{1}\text{D} + {}^{3.01605}_{1}\text{T} \longrightarrow {}^{4.00265}_{2}\text{He} + {}^{1.00868}_{0}\text{n},$$

determine (a) the mass loss in amu per He atom formed, (b) the energy released in Mev and joules in (a), and (c) the energy released, in Mev, per mole of He formed.

20-45. For each of the following nuclear reactions, calculate (a) the mass loss in amu per atom of bombarded substance, (b) the energy released, in Mev and J, per atom, and (c) the energy released, in Mev and J, when 1 mole of each species reacts.

(a) $^{11.00931}_{5}\text{B} + {}^{4.00265}_{2}\text{He} \rightarrow {}^{14.00310}_{7}\text{N} + {}^{1.00868}_{0}\text{n}$

(b) $^{9.01220}_{4}\text{Be} + {}^{4.00265}_{2}\text{He} \rightarrow {}^{12.00000}_{6}\text{C} + {}^{1.00868}_{0}\text{n}$

(c) $^{6.01515}_{3}\text{Li} + {}^{1.00868}_{0}\text{n} \rightarrow {}^{4.00265}_{2}\text{He} + {}^{3.01605}_{1}\text{H}$

20-46. Calculate the binding energy per nucleon for each of the following isotopes (follow the procedure in Example 20-h).

(a) ${}^{11.00931}_{5}B$ (c) ${}^{136.9056}_{56}Ba$ (e) ${}^{87.9056}_{38}Sr$
(b) ${}^{34.96885}_{17}Cl$ (d) ${}^{39.96238}_{18}Ar$ (f) ${}^{235.044}_{92}U$

20-47. The release of 2.8×10^{13} J in a nuclear reaction corresponds to the conversion of how many grams of matter to energy?

20-48. The release of 8.32×10^{30} Mev of energy corresponds to the conversion of how many grams of matter to energy?

20-49. For each of the following nuclear reactions, calculate (a) the mass loss or gain in amu per atom of bombarded substance, (b) the energy released or absorbed per atom, and (c) the energy released or absorbed when 1 mole of each species reacts.

(1) ${}^{10.01294}_{5}B + {}^{4.00265}_{2}He \rightarrow {}^{13.00335}_{6}C + {}^{1.007825}_{1}H$
(2) ${}^{12.00000}_{6}C + {}^{2.01410}_{1}H \rightarrow {}^{10.01294}_{5}B + {}^{4.00265}_{2}He$
(3) ${}^{238.0508}_{92}U + {}^{4.00265}_{2}He \rightarrow {}^{240.0540}_{94}Pu + 2\,{}^{1.00868}_{0}n$
(4) ${}^{53.93962}_{26}Fe + {}^{4.00265}_{2}He \rightarrow {}^{55.93494}_{26}Fe + 2\,{}^{1.007825}_{1}H$

20-50. Calculate the binding energy per nucleon for each of the following isotopes:

(a) ${}^{3.01603}_{2}He$ (d) ${}^{126.90447}_{53}I$
(b) ${}^{61.92834}_{28}Ni$ (e) ${}^{193.9628}_{78}Pt$
(c) ${}^{138.90614}_{57}La$ (f) ${}^{239.0522}_{94}Pu$

Appendix I

Mathematical Review

Without exception, students of chemistry will encounter mathematical problems that must be solved. Generally, in a first-year course, these problems will involve simple arithmetic and very elementary algebra. In the following discussion we assume that students have a knowledge of simple mathematics and need only to be reminded of the operations involved in the solution of equations.

An algebraic statement is one representing an equality. Generally it contains several terms and usually at least one term whose numerical value is unknown. The equality

$$\frac{2X}{3} - 6 = 10$$

is such an algebraic statement. We are usually interested in solving for the unknown term X, that is, getting X on one side of the equality sign and the remainder of the terms on the other.

A. Solution of Simple Equations

To solve the most commonly encountered equations, a few simple rules are employed, having to do with the transfer of quantities from one side of the equation to the other.

(a) First, the addition or subtraction of a number to **both** sides of the equation does not destroy the equality. In the example just given, the technique of transferring the term -6 to the right-hand side involves merely a change of sign, so that the equation now reads

$$\frac{2X}{3} = 10 + 6$$

This is obviously analogous to adding $+6$ to both sides of an equation and then, since $+6$ and -6 exactly cancel each other, the equation takes the form

$$\frac{2X}{3} - \cancel{6} + \cancel{6} = 10 + 6 \quad or \quad \frac{2X}{3} = 16$$

(b) Second, the multiplication or division of **both** sides of an equation by the same number does not alter the equality. In the above equality, let us next eliminate fractions by multiplying both sides of the equation by 3. In this event

$$\frac{2X \cdot \cancel{3}}{\cancel{3}} = 16 \cdot 3$$

Since the number 3 now appears in both the numerator and denominator on the left-hand side of the equation, they cancel; and the equation assumes the form

$$2X = 16 \cdot 3 \qquad or \qquad 2X = 48$$

The final step in solving for the unknown (X) is the division of both sides of the equation by 2, which yields the relation

$$\frac{\cancel{2}X}{\cancel{2}} = \frac{48}{2}$$

Again the number 2 appears in both the numerator and denominator of the left-hand term and therefore they cancel each other, yielding the relationship

$$X = \tfrac{48}{2} \quad or \quad X = 24$$

Thus $X = 24$ is the solution to the original equation. To check this, we merely substitute the number 24 for X in the original equation and determine whether the equality holds.

Check.

$$\frac{2X}{3} - 6 = 10$$

Substituting 24 for X

$$\frac{2 \cdot 24}{3} - 6 = 10$$

$$\frac{48}{3} - 6 = 10$$

Dividing 48 by 3

$$16 - 6 = 10$$

and

$$10 = 10$$

Therefore, we may conclude that 24 is a solution to the original equation.

B. Sign Conventions

Sign conventions are of great importance in solving algebraic equations. In such operations, the use of negative numbers is commonly encountered. The addition of two numbers of like sign requires that the sum be of the same sign, for example, $-5 - 3 = -8$. The addition of numbers of opposite sign is carried out by subtracting the smaller from the larger and assigning the sign of the larger to the difference, for example, $-5X + 3X = -2X$. The multiplication of numbers of the same sign requires the product to have a positive sign, for example, $(-5)(-3) = +15$; whereas the product has a negative sign if the original numbers are of opposite sign, for example, $(-5)(3) = -15$. A similar rule applies to division; the quotient of either two positive or two negative numbers is positive, whereas the quotient of a positive and a negative number is negative.

C. Solution of First-Degree Equations

First-degree equations are those in which the exponent of the unknown quantity is equal to one. These equations are easily solved using the principles outlined above; for example, solve the equation $(5X + 3 = X - 5)$ for X.

Solution.

$$5X + 3 = X - 5$$
$$5X - X = -5 - 3$$
$$4X = -8$$
$$X = -2$$

Similarly, solve the equation

$$\frac{6X - 3}{5} = \frac{6X}{4} \quad \text{for } X.$$

Solution.

$$\frac{6X - 3}{5} = \frac{6X}{4}$$
$$4(6X - 3) = (6X)5$$
$$24X - 12 = 30X$$
$$-12 = 30X - 24X$$
$$-12 = 6X$$
$$X = -2$$

D. Solution of Quadratic Equations

Occasionally a student will need to solve for an unknown in a simple quadratic such as $9X^2 - 25 = 0$. A quadratic equation is a second-degree equation, that is, one where the largest exponent of the unknown is 2. The quadratic equation above is a very simple one that is easily solved in the following manner.

Solution.

$$9X^2 - 25 = 0$$
$$9X^2 = 25$$
$$X^2 = \frac{25}{9}$$
$$X = \sqrt{\frac{25}{9}}$$
$$X = \pm\frac{5}{3}$$

An alternative method of solution of quadratic equations that are not so simple is shown by the solution to the general quadratic

$$aX^2 + bX + c = 0$$

where a and b are the coefficients of X^2 and X in the quadratic, and c is a constant. The solution to this equation is given by the relation

$$X = \frac{-b \pm \sqrt{b^2 - 4ac}}{2a}$$

In the example $9X^2 - 25 = 0$, $a = 9$, $b = 0$, and $c = -25$. The solution to the equation is

$$X = \frac{-0 \pm \sqrt{0^2 - (4)(9)(-25)}}{(2)(9)} = \frac{\pm\sqrt{(36)(25)}}{18} = \frac{\pm\sqrt{900}}{18}$$

$$X = \pm\tfrac{30}{18} = \pm\tfrac{5}{3}$$

Problems (Appendix I)

Solve the following algebraic equations for the numerical value of X.

I-1. $3X - 2 = 10$

I-2. $6X + 4 = 2X + 8$

I-3. $\dfrac{2X}{6} - 7 = 5$

I-4. $\dfrac{3X + 5}{4} = 20$

I-5. $\dfrac{4X}{5} = \dfrac{6X - 18}{3}$

I-6. $\dfrac{8X - 4}{14} = \dfrac{3X + 8}{35}$

I-7. $(4 - X)(3) = 6$

I-8. $\dfrac{(X - 6)(4)}{3} = 4$

I-9. $\dfrac{(3X - 20)(5)}{-4} = \dfrac{(2X + 3)(3)}{2}$

I-10. $X^2 - 24 = 12$

I-11. $5X^2 - 7 = 18$

I-12. $2X^2 - 3X = 54$

I-13. $3X^2 - 5X = 20$

I-14. $6X^2 - 7X - 10 = 0$

Appendix II
Exponential Numbers

A. Expressing Exponential Numbers as Decimal Numbers

In science, very large numbers and very small numbers are expressed exponentially, usually as the product of a number and a power of 10. Two such numbers are the number of atoms in one mole of atoms, 6.02×10^{23}, and the diameter of a hydrogen atom, 1.04×10^{-8} cm.

Consider first the meaning of a "power of 10." A power of 10 is the number of times a quantity must be multiplied by 10 to produce the equivalent decimal number. In the examples below, this principle is illustrated.

$$1 \times 10^3 = 1 \times 10 \times 10 \times 10 = 1000$$

$$5 \times 10^5 = 5 \times 10 \times 10 \times 10 \times 10 \times 10 = 500{,}000$$

$$5 \times 10^0 = 5$$

(10 raised to the 0 power means that 5 is multiplied by 10 zero times; hence $10^0 = 1$. Note that this operation is not the same as multiplication by zero.)

The power of 10 tells the number of places the decimal point must be moved to produce the desired number. If the exponent is positive, the decimal point is moved to the right exactly that number of places. If the exponent is negative, the decimal point is moved to the left that number of places in order to express the number as a decimal number.

$$2 \times 10^{-4} = 0.0002$$

$$7.6 \times 10^{+2} = 760$$

B. Expressing Decimal Numbers as Exponentials

To express a number as a power of 10, the number of places the decimal point must be moved indicates the power of 10 involved. Thus the number 21,000,000 is usually written as 2.1×10^7, but can also be written as 21×10^6 or 210×10^5, or 0.21×10^8, or 0.021×10^9, etc. Note that if the decimal point is moved to the right, the power of 10 is decreased by 1 for each decimal place moved and if the decimal point is moved to the left, the power of 10 is increased by 1 for each decimal place.

It is customary to express a given number in exponential form so that the first number to the left of the decimal point is between 1 and 10; thus 20,000,000 is usually expressed as 2×10^7 (where 2 is between 1 and 10) rather than 20×10^6, 0.2×10^8, and so on.

To express numbers less than unity as powers of 10, the decimal point must be moved to the right and multiplied by negative powers of 10. These exponents must correspond to the number of decimal places moved. Consider the following series of numbers that are converted to powers of 10.

$$0.1 = 1 \times 10^{-1}$$

$$0.03 = 3 \times 10^{-2}$$

$$0.0000018 = 1.8 \times 10^{-6}$$

C. Addition of Numbers with Exponents

Numbers consisting of exponentials may be added or subtracted as ordinary numbers provided the exponents are the same for all. In the event that this is not the case, the numbers must be converted to a series having the same exponents; for example, consider the following series of numbers.

$$2.47 \times 10^4$$
$$1.580 \times 10^5$$
$$7.26 \times 10^4$$
$$8.0 \times 10^3$$

To add this column of figures, a common exponent must be selected for all, and in this instance it would be simplest if all were converted to the fourth power of 10. The numbers then become

$$
\begin{array}{lr}
 & 2.47 \times 10^4 \\
 & 15.80 \times 10^4 \\
 & 7.26 \times 10^4 \\
 & 0.80 \times 10^4 \\
\hline
\textbf{Answer} & 26.33 \times 10^4
\end{array}
$$

Thus the sum is 26.33×10^4 or 2.633×10^5.

D. Subtraction of Numbers with Exponents

Subtraction is accomplished in a similar manner. For example, the difference between 4.88×10^{-2} and 0.081×10^{-1} is easily obtained by converting 0.081×10^{-1} to 0.81×10^{-2} and then subtracting 0.81 from 4.88.

$$
\begin{array}{lr}
 & 4.88 \times 10^{-2} \\
 & -0.81 \times 10^{-2} \\
\hline
\textbf{Answer} & 4.07 \times 10^{-2}
\end{array}
$$

Note that in addition and subtraction of numbers with exponents, the value of the exponent does not change unless the decimal point of the number is subsequently moved.

E. Multiplication of Numbers with Exponents

In multiplication of two numbers with exponents, the numbers are multiplied together as usual and the exponents of 10 are added algebraically to give a new exponent of 10 for the product.

(a) $3.3 \times 10^3 \times 2.0 \times 10^4 = (3.3 \times 2.0) \times (10^3 \times 10^4) = 6.6 \times 10^7$

(b) $8.0 \times 10^3 \times 5.0 \times 10^{-2} = (8.0 \times 5.0) \times (10^3 \times 10^{-2}) = 40 \times 10^1$
$= 4.0 \times 10^2$

(c) $1.6 \times 10^{-5} \times 4.0 \times 10^{-2} = (1.6 \times 4.0) \times (10^{-5} \times 10^{-2}) = 6.4 \times 10^{-7}$

F. Division of Numbers with Exponents

In division of numbers with exponents, the ordinary numbers are handled in the usual manner, but the exponent of 10 in the denominator is subtracted algebraically from the exponent of 10 in the numerator.

(a) $\dfrac{6 \times 10^4}{3 \times 10^1} = \left(\dfrac{6}{3}\right) \times \left(\dfrac{10^4}{10^1}\right) = 2 \times 10^3$

(b) $\dfrac{7.48 \times 10^4}{2.14 \times 10^{-3}} = 3.50 \times 10^7$

(c) $\dfrac{1.0 \times 10^{-5}}{4.0 \times 10^{-7}} = \dfrac{10 \times 10^{-6}}{4.0 \times 10^{-7}} = 2.5 \times 10^1$

It should be observed that the process of division by a power of 10 is analogous to multiplication by the negative of that power of 10. The ratio

(d) $\dfrac{5 \times 10^9}{10^4} = 5 \times 10^9 \times 10^{-4} = 5 \times 10^5$

and the ratio

(e) $\dfrac{6 \times 10^8}{3 \times 10^{-3}} = 2 \times 10^8 \times 10^3 = 2 \times 10^{11}$

G. Extracting Square Roots and Cube Roots

In extracting the square root of a number, the numbers without exponents are handled in the usual fashion and the exponent of 10 is divided by 2. In order not to have fractional powers of 10 for the answer, the number is adjusted so that the power of 10 is divisible by two an integral number of times prior to taking the square root.

(a) $\sqrt{4 \times 10^2} = \sqrt{4} \times \sqrt{10^2} = \pm 2 \times 10^1$
(b) $\sqrt{4 \times 10^{-2}} = \sqrt{4} \times \sqrt{10^{-2}} = \pm 2 \times 10^{-1}$
(c) $\sqrt{36 \times 10^8} = \sqrt{36} \times \sqrt{10^8} = \pm 6.0 \times 10^4$
(d) $\sqrt{6.4 \times 10^{-9}} = \sqrt{64 \times 10^{-10}} = \sqrt{64} \times \sqrt{10^{-10}} = \pm 8.0 \times 10^{-5}$

Similarly, to obtain the cube root of a number, the decimal number is handled in the usual way and the exponent of 10 is divided by 3. Again, in order to avoid fractional exponents, it is necessary to adjust the number so that the exponent of 10 is divisible by 3 an integral number of times.

(e) $\sqrt[3]{0.27 \times 10^{-4}} = \sqrt[3]{27 \times 10^{-6}} = \sqrt[3]{27} \times \sqrt[3]{10^{-6}} = 3.0 \times 10^{-2}$

(f) $\sqrt[3]{1.25 \times 10^{23}} = \sqrt[3]{125 \times 10^{21}} = \sqrt[3]{125} \times \sqrt[3]{10^{21}} = 5.00 \times 10^{7}$

Problems

II-1. Express the following numbers in exponential form:
(a) 35,000 (d) 0.00014
(b) 6,000,000 (e) 0.0077
(c) 235,000 (f) 0.0345

II-2. Express in decimal form:
(a) 7.5×10^{3} (d) 2.24×10^{3}
(b) 3.2×10^{-4} (e) 0.005×10^{4}
(c) 0.82×10^{-6} (f) 89×10^{0}

II-3. Solve for X in the following:
(a) $X = (32.00 \times 10^{4}) + (1.8 \times 10^{3}) + (0.150 \times 10^{5})$
(b) $X = (1.80 \times 10^{-5}) - (6.00 \times 10^{-7})$
(c) $X = (3.0 \times 10^{4})(5.0 \times 10^{8})$
(d) $X = (2 \times 10^{3})(4 \times 10^{-5})$
(e) $X = (1.5 \times 10^{-3})(4 \times 10^{-6})(8 \times 10^{8})$
(f) $X = \dfrac{1.8 \times 10^{5}}{3.0 \times 10^{2}}$
(g) $X = \dfrac{(4 \times 10^{-3})(6 \times 10^{4})}{(2 \times 10^{-4})(3 \times 10^{-8})}$
(h) $X^2 = 36 \times 10^{8}$
(i) $X^2 = 2.5 \times 10^{-7}$
(j) $X^3 = 2.7 \times 10^{-8}$
(k) $X^2 = \dfrac{(2 \times 10^{2})^{2}(4 \times 10^{-3})^{3}}{(1 \times 10^{-5})(8 \times 10^{2})^{2}}$

Appendix III

Logarithms

The logarithm of a number is the exponent that must be given 10 to produce the number; for example, the logarithm of 100 is 2 since this is the exponent that must be given 10 in order to produce 100.

$$10^2 = 100 \qquad \text{Therefore } \log 100 = \log 10^2 = 2$$

$$10^{-2} = 1/10^2 = 1/100 = 0.01 \qquad \text{Therefore } \log 0.01 = -2$$

$$10^6 = 1{,}000{,}000 \qquad \text{Therefore } \log 1{,}000{,}000 = \log 10^6 = 6$$

The logarithms of numbers that are not integral powers of 10 are not obvious. It is therefore necessary to resort to tables of computed values of logarithms to obtain the required information.

Logarithms of Numbers*

No.	0	1	2	3	4	5	6	7	8	9
1	0.000	0.041	0.079	0.114	0.146	0.176	0.204	0.230	0.255	0.279
2	0.301	0.322	0.342	0.362	0.380	0.398	0.415	0.431	0.447	0.462
3	0.477	0.491	0.505	0.519	0.532	0.544	0.556	0.568	0.580	0.591
4	0.602	0.613	0.623	0.634	0.644	0.653	0.663	0.672	0.681	0.690
5	0.699	0.708	0.716	0.724	0.732	0.740	0.748	0.756	0.763	0.771
6	0.778	0.785	0.792	0.799	0.806	0.813	0.820	0.826	0.833	0.839
7	0.845	0.851	0.857	0.863	0.869	0.875	0.881	0.887	0.892	0.898
8	0.903	0.909	0.914	0.919	0.924	0.929	0.935	0.940	0.945	0.949
9	0.954	0.959	0.964	0.969	0.973	0.978	0.982	0.987	0.991	0.996

* This table is for illustrative purposes only. A four-place table of logarithms is found in Appendix X.

To find the logarithm of a number, for example, the number 8.5, we look down the first column to the number 8 and then across to the column heading 5. The number recorded at this point is 0.929 and this is the logarithm of 8.5. As a matter of interest, from the definition of a logarithm we see that $10^{0.929} = 8.5$.

To find the logarithm of a number not lying between 1 and 10, such as the number 4900, the number should be converted to a decimal number with only a single digit to the left of the decimal point multiplied by the appropriate power of 10, that is, $4900 = 4.9 \times 10^3$. Since the logarithm of a product of two numbers is equal to the sum of the logarithms of the individual numbers, the logarithm of 4.9×10^3 becomes equal to the log of 4.9 plus the log of 10^3. In summary,

$$\begin{aligned} \log(4900) &= \log(4.9 \times 10^3) \\ &= \log 4.9 + \log 10^3 \\ &= 0.690 + 3 \\ &= 3.690 \end{aligned}$$

Similarly, the log of 0.00068 may be found

$$\begin{aligned} \log(0.00068) &= \log(6.8 \times 10^{-4}) \\ &= \log 6.8 + \log 10^{-4} \\ &= 0.833 - 4 \\ &= -3.167 \end{aligned}$$

Sooner or later, every student will be faced with the problem of finding a number whose logarithm is known; for example, find the number whose logarithm is -4, or stated algebraically, $\log X = -4$. Since this is an integral logarithm, the number X equals 10^{-4} or 0.0001. For a number where the logarithm is not integral, such as $\log X = 5.800$, we must again resort to a table of logarithms and proceed as follows. The logarithm is first written as the sum of two numbers, one of which is integral (called the "characteristic") and the other a decimal fraction (called the "mantissa"). The mantissa always has a positive value, whereas the characteristic, which locates the decimal point, may be either positive or negative.

$$\begin{aligned} \log X &= 5.800 \\ \log X &= 5 + 0.800 \end{aligned}$$

The number whose logarithm equals 5 is 10^5, and from the table, the number whose logarithm equals 0.800 is 6.3 or

$$5 = \log 10^5$$

and

$$0.800 = \log 6.3$$

On substitution,

$$\log X = \log 10^5 + \log 6.3$$
$$\log X = \log (6.3 \times 10^5)$$
$$X = 6.3 \times 10^5$$

6.3×10^5 is called the "antilog" of 5.800.

An example of finding the number whose logarithm is negative is the following:

$$\log X = -7.510$$
$$\log X = 0.490 - 8$$

(converted to this form so the mantissa has a positive value.)

$$\log X = \log 3.1 + \log 10^{-8}$$
$$\log X = \log (3.1 \times 10^{-8})$$
$$X = 3.1 \times 10^{-8}$$

3.1×10^{-8} is called the antilog of -7.510.

A. Multiplication

As indicated above, the logarithm of the product of two numbers is the **sum** of the logarithms of the numbers. To carry out multiplication, we must (1) obtain the logarithms of the numbers to be multiplied, (2) obtain the sum of the logarithms, and (3) obtain the antilog of the sum.

This process is illustrated by the following examples.

(a) Determine X if $X = (3)(50)(100)$

$$\log X = \log (3)(50)(100)$$
$$\log 3 = 0.477$$
$$\log 50 = 1.699$$
$$\log 100 = \underline{2.000}$$
$$\log X = 4.176$$
$$X = \text{antilog of } 4.176 = 1.5 \times 10^4$$

(b) $Y = (1.5 \times 10^{-2})(6.2 \times 10^{6})$. Determine the product.

$$\log Y = \log(1.5 \times 10^{-2}) + \log(6.2 \times 10^{6})$$

$$\log 1.5 \times 10^{-2} = \log 1.5 + \log 10^{-2} = 0.176 + (-2.000) = -1.824$$

$$\log 6.2 \times 10^{6} = \log 6.2 + \log 10^{6} = 0.792 + 6.000 = 6.792$$

$$\log Y = 4.968$$

$$Y = \text{antilog of } 4.968 = 9.3 \times 10^{4}$$

B. Division

To carry out the operation of division, the logarithm of the divisor is **subtracted** from the logarithm of the dividend. The antilog of this difference is determined to obtain the quotient.

Determine X if

$$X = \frac{(4.7 \times 10^{-6})}{(9.4 \times 10^{4})}$$

$$\log X = \log(4.7 \times 10^{-6}) - \log(9.4 \times 10^{4})$$

$$\log(4.7 \times 10^{-6}) = \log 4.7 + \log 10^{-6} = 0.672 + (-6.000) = -5.328$$

$$-[\log(9.4 \times 10^{4})] = -[\log 9.4 + \log 10^{4}] = -[0.973 + 4.000] = -4.973$$

$$\log X = -10.301$$

In order to obtain a positive mantissa, this number (-10.301) may be expressed as ($0.699 - 11$).

$$X = \text{antilog of } (0.699 - 11) = 5.0 \times 10^{-11}$$

C. Roots

To extract the nth root of a number, simply divide the logarithm of the number by n, and determine the antilog. For extracting square roots, $n = 2$, and for extracting cube roots, $n = 3$.

$$\sqrt{8100} = X$$

$$\log \sqrt{8100} = \frac{\log 8100}{2} = \frac{3.909}{2} = 1.954$$

$$X = \text{antilog of } 1.954 = 9.0 \times 10^{1}$$

Similarly

$$\sqrt[3]{5800} = X$$

$$\log \sqrt[3]{5800} = \frac{\log 5800}{3} = \frac{3.763}{3} = 1.254$$

$$X = \text{antilog of } 1.254 = 18$$

D. Powers

To raise a number to a power, obtain the logarithm, multiply the logarithm of the number by the power, and determine the antilog.

$$(210)^3 = X$$
$$\log (210)^3 = 3 \log 210 = (3)(2.322) = 6.966$$
$$X = \text{antilog of } 6.966 = 9.2 \times 10^6$$

Problems

III-1. Obtain the logarithm of the following numbers. (Use log tables in Appendix X.)
(a) 3.00×10^4 (c) 6.022×10^{23} (e) 0.004325
(b) 1.700×10^3 (d) 1.600×10^{-19} (f) 24.27

III-2. Find the number (antilog) corresponding to the following logarithm values.
(a) 4.0000 (c) 1.800 (e) −3.4660
(b) −8.7700 (d) 23.780 (f) −0.0540

III-3. Perform the following operations by logarithms.
(a) $(1.4 \times 10^4)(6.3 \times 10^{-2})$
(b) $(4.5 \times 10^{-7})(3.2 \times 10^6)$
(c) $\dfrac{3.7 \times 10^{14}}{8.0 \times 10^{-4}}$
(d) $\dfrac{8.8 \times 10^{-8}}{22 \times 10^5}$
(e) $\sqrt[4]{9.600 \times 10^3}$
(f) $(2.600 \times 10^{-7})^3$

Appendix IV
Significant Figures

A quantitative measurement of some property requires the placing of numerical values on that property and also a statement of the units in which the measurement is made. The number of digits used to designate the amount is referred to as the number of significant figures. Some very simple rules should be followed in their use.

1. Significant Figures

The significant figures in a number are those digits that give meaningful but not misleading information. Only the last digit contains an uncertainty. For example, the volume of a liquid may be recorded as 675 ml. This number has three significant figures, the last of which is uncertain. The uncertainty in the last digit depends on the accuracy of the measurement. If the measurement of volume is accurate to 1 ml, the volume may be specified as 675 ± 1 ml; that is, the volume is not less than 674 ml or more than 676 ml.

Consider another example in which the number of significant figures in a measurement depends on the degree of refinement of the devices used in making the measurement. In determining atmospheric pressure by measuring the height of a mercury column supported by the atmosphere, the number of significant figures depends primarily on the rulings on the scale; see Fig. IV-1.

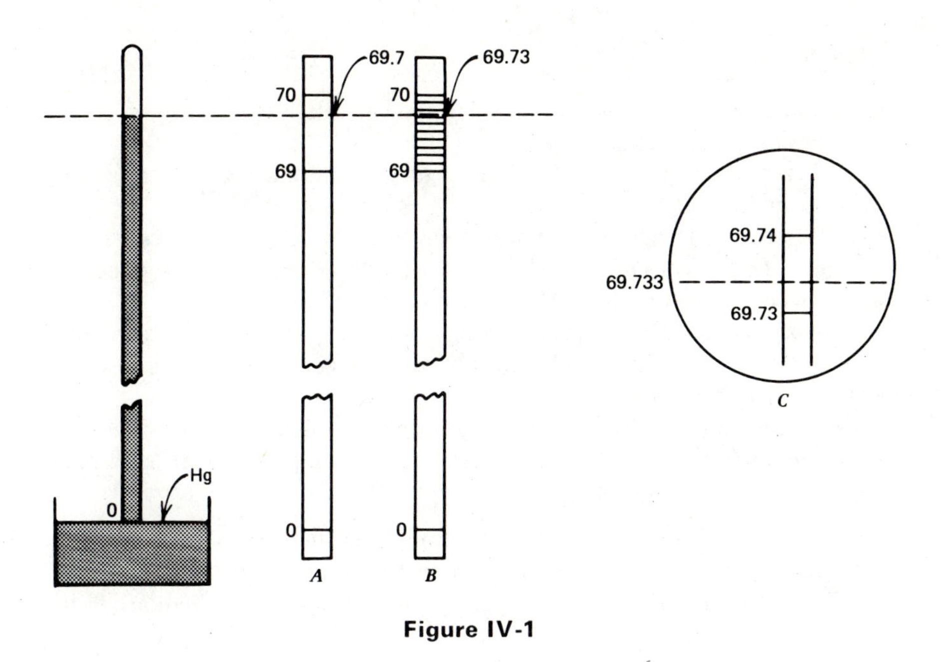

Figure IV-1

The height of the mercury column is fixed and if measured with scale *A*, it is noted that the upper part of the meniscus is approximately 7/10 of the distance between 69 and 70. Hence the height of the column is reported as 69.7 cm. On scale B, with more rulings, it is estimated that the meniscus falls 3/10 of the distance between 69.7 and 69.8. Therefore this is recorded as 69.73 cm. If an instrument called a cathetometer (a device employing a telescope for measuring differences in heights accurately) were used, an additional significant figure could be obtained and the height recorded as 69.733 cm (scale *C*).

The number 69.7 has three significant figures.
The number 69.73 has four significant figures.
The number 69.733 has five significant figures.

Similarly, the number 0.00697 has three significant figures, and the number 697 has three significant figures. However, the number 3970 may have three or four significant figures. The number of significant figures is decided by the rules outlined in the following section.

2. Zeros

Zeros must be considered as a special case in relation to their position with respect to the decimal point.

(a) A zero between two significant figures is also a significant figure. Thus the number 107.8 consists of four significant figures.

(b) A zero to the right of a digit beyond the decimal point is included in the count of significant figures. The number 10.780 consists of five significant figures, whereas the number 10.78 consists of only four.

(c) A zero is not significant if it occurs merely to fix the decimal point; for example, the zeros in 0.054 are not considered significant figures; in 5400 they may or may not be significant.

(d) To indicate the proper number of significant figures for the numbers 0.054 and 5400, the number is best represented as a power of 10 with one digit to the left of the decimal.

> 0.054 becomes 5.4×10^{-2} indicating only two significant figures. 5400 becomes 5.4×10^3 if there are two significant figures; 5.40×10^3 if there are three significant figures; and 5.400×10^3 if there are four significant figures.

3. Addition and Subtraction

In addition, no more digits are meaningful than the lowest digit common to all of the numbers to be summed. In the column of numbers below, the units digit is the lowest digit common to all numbers, and therefore their sum is rounded to the three significant figure number 535.

Lowest common digit (↓ units digit)

```
   28.2
  505
    1.732
  -------
  534.932 = 535
```

In subtraction a similar rule applies. For example, the difference between 505 and 1.7 is 503.3, but since the units digit is again the lowest common digit, the result is 503.

In obtaining the difference $0.505 - 0.455 = 0.050$, the remainder contains only two significant figures. Notice that it is possible in subtraction to have fewer significant figures in the answer than in either of the original numbers.

In this case, the 1/1000 digit is the smallest digit common to both three significant figure numbers, but the difference is a number containing only two significant figures. It is usually true that the difference of two numbers of nearly equal magnitude has fewer significant figures than either of the original numbers.

4. Multiplication and Division

In multiplication and division the same number of significant figures is retained as in the least accurate number. Occasionally the retention of one additional figure, but never more, is allowed if retaining only the prescribed number results in a loss of accuracy. A case where the number of digits retained is the same as in the least accurate number is

$$2.00 \times 102.14 = 204.28 = 204$$

The number 2.00 has only three significant figures, and hence the answer is rounded to the three significant figures 204.

An example where keeping an extra significant figure is permitted is

$$2.00 \times 62 = 124$$

The number 62 has only two significant figures but is known to 1 part in 62 or roughly 2%. Rounding the answer to 120 to retain only two significant figures results in a number uncertain to 10 parts in 120 or 8%. Under these conditions, retention of three significant figures is allowed.

Division is treated similarly to multiplication.

5. Rounding Off Unnecessary Figures

In rounding off unneeded figures and nonsignificant figures we consider the magnitude of the highest digit to be dropped.

(a) If this digit is 4 or less, the digit is dropped and the last digit retained is unchanged.

Thus, if 2.74 is to be rounded to two significant figures, it becomes 2.7. The number 2.748 is rounded to 2.7 and not to 2.8, and on rounding 0.56649 to three significant figures, it becomes 0.566, not 0.567.

(b) If the highest digit to be dropped is greater than 5, or is 5 followed by numbers other than zeros, the last digit retained is increased by one.

Thus in rounding 2.76 to two significant figures it becomes 2.8. In rounding 2.8502 to two significant figures it becomes 2.9.

(c) If the largest digit to be dropped is 5 followed only by zeros, the last digit retained is increased by one if it is odd and is left unchanged if it is even.

When rounded to two significant figures, 4.850 becomes 4.8 and 4.750 becomes 4.8. The number 4.8509 becomes 4.9, since the 5 is followed by digits other than zero.

6. Integral Numbers

Integral numbers are those representing an entity, a body, or a class that is not subdivided into parts. In such instances, the concept of significant figures is without meaning; for example, the formula O_2 represents a molecule of oxygen and consists of exactly two atoms per molecule. In the equation for the volume of a sphere, $V = \frac{4}{3}\pi r^3$, the ratio 4/3 is the ratio of integral numbers and, here also, the number of significant figures is meaningless. In a chemical equation such as $N_2 + 3\,H_2 \rightarrow 2\,NH_3$, the coefficients before the formulas tell that one molecule of N_2 reacts with three molecules of H_2 to form two molecules of NH_3. Similarly, the equation indicates that 1 mole of N_2 reacts with 3 moles of H_2 to form 2 moles of NH_3. These coefficients are integral numbers.

Problems (Significant Figures)

IV-1. Round each of the following numbers to three significant figures.

(a) 1785 (f) 4.8351
(b) 3.4624 (g) 3.4550
(c) 1.4888 (h) 0.00852
(d) 1.4848 (i) 3.920
(e) 4.8350 (j) 685.4999

IV-2. Obtain the sum of each of the following sets of numbers to the proper number of significant figures.

(a) 145 + 692 + 18.3 + 9.40
(b) 18.5 + 31.3 + 0.88 + 2.75
(c) 38.775 + 3.0075 + 1.8473 + 22.8361
(d) 700 + 250 + 375
(e) 0.002 + 0.630 + 0.06

IV-3. For each of the following pairs of numbers, obtain the difference, and express the answer to the correct number of significant figures.

(a) 38.50 − 16.49 (c) 0.6001 − 0.6 (e) 37.4 − 22
(b) 0.737 − 0.73 (d) 39.95 − 0.015

IV-4. Obtain the quotient for each of the following ratios and round to the correct number of significant figures.

(a) 36.36/4.04 (c) 351.4/9.547 (e) 99/72
(b) 0.7406/0.141 (d) 270.4/270 (f) 99/15

IV-5. Round the products of the numbers listed below to the proper number of significant figures.

(a) 88×2.01 (e) 0.0860×104.0
(b) 49×75 (f) 210×1.25
(c) 200×1.85 (g) 0.0057×250.4
(d) 27.00×0.0751

Appendix V

Experimental Error

The experimental determination of any quantity, such as the measurement of the pressure of the atmosphere, is subject to error because it is impossible to carry out such measurements with absolute certainty. The extent of the error in determination of the barometric pressure is a function of the quality of the instrument, the temperature, and the experience and capability of the observer.

The **accuracy** of a determination is a measure of the approach of the experimental value to the true or generally accepted value. This is usually described by means of the percent error, defined as

$$\text{Percent error} = \frac{\text{Experimental value} - \text{True value}}{\text{True value}} \times 100$$

The **precision** of a determination refers to the reproducibility of the measurement and is usually described in terms of the average or mean deviation. This is computed by determining first the average of the individual measurements in the usual manner. Then the absolute value of the difference between each individual determination and the average value is calculated. These differences are summed and divided by the number of values in order to get the average or mean deviation.

Measurements	Deviation from Mean
49.9	0.2
49.7	0.0
49.8	0.1
49.5	0.2
4)198.9	4)0.5
49.7	0.1

The average value is, therefore, 49.7, and the average deviation is ± 0.1 units. The number is generally reported as 49.7 ± 0.1 units.

Occasionally the deviation is expressed as the percentage deviation. This is simply obtained by dividing the average deviation by the average value of the quantity and multiplying by 100. In the above example, the percentage deviation is calculated as follows:

$$\text{Percentage deviation} = \frac{0.1}{49.7} \times 100 = 0.2\%$$

Appendix VI

Vapor Pressure of Water

Temperature (°C)	Pressure (torr)	Temperature (°C)	Pressure (torr)
0	4.6	24	22.4
2	5.3	25	23.8
4	6.1	26	25.2
6	7.0	27	26.7
8	8.0	28	28.3
10	9.2	29	30.0
12	10.5	30	31.8
15	12.8	35	42.1
16	13.6	40	55.3
17	14.5	50	92.5
18	15.5	60	149
19	16.5	70	234
20	17.5	80	355
21	18.6	90	526
22	19.8	100	760
23	21.1	110	1074

Appendix VII

Ionization Constants for Acids and Bases

Acid	Formula	K_a	Acid	Formula	K_a
Acetic	$HC_2H_3O_2$	1.8×10^{-5}	Hypobromous	HBrO	2.1×10^{-9}
Arsenic*	H_3AsO_4	5.6×10^{-3}	Hypochlorous	HClO	3.0×10^{-8}
		1.7×10^{-7}	Hypoiodous	HIO	2.3×10^{-11}
		4×10^{-12}	Hydrogen sulfide	H_2S	1.1×10^{-7}
Ascorbic	$H_2C_6H_6O_6$	7.9×10^{-5}			1×10^{-14}
		1.6×10^{-12}	Iodic	HIO_3	1.7×10^{-1}
Benzoic	$HC_7H_5O_2$	6.4×10^{-5}	Lactic	$HC_3H_5O_3$	1.4×10^{-4}
Bromoacetic	$HC_2H_2O_2Br$	2.0×10^{-3}	Nitrous	HNO_2	4.6×10^{-4}
Butyric	$HC_4H_7O_2$	1.5×10^{-5}	Oxalic	$H_2C_2O_4$	5.6×10^{-2}
Carbonic	H_2CO_3	4.3×10^{-7}			5.2×10^{-5}
		5.6×10^{-11}	Periodic	H_5IO_6	2.3×10^{-2}
Chloroacetic	$HC_2H_2O_2Cl$	1.4×10^{-3}	Phosphoric	H_3PO_4	7.5×10^{-3}
Chlorobenzoic	$HC_7H_4O_2Cl$	1.5×10^{-4}			6.2×10^{-8}
Formic	$HCHO_2$	1.8×10^{-4}			4.8×10^{-13}
Hydrocyanic	HCN	4.9×10^{-10}	Sulfurous	H_2SO_3	1.6×10^{-2}
Hydrofluoric	HF	6.6×10^{-4}			6.3×10^{-8}

Base	Formula	K_b	Base	Formula	K_b
Ammonia	NH_3	1.8×10^{-5}	Pyridine	C_5H_5N	1.6×10^{-9}
Hydrazine	N_2H_4	1.4×10^{-6}	Quinoline	C_9H_7N	6.3×10^{-10}
Hydroxylamine	NH_2OH	1.1×10^{-8}	Trimethylamine	C_3H_9N	6.3×10^{-5}

* (Successive ionization constants of polyprotic acids, shown in order.)

Appendix VIII
Solubility Product Constants

Compound	K_{sp}	Compound	K_{sp}
$Al(OH)_3$	2×10^{-32}	HgS	1.6×10^{-54}
$AgBr$	5.0×10^{-13}	Li_2CO_3	1.7×10^{-3}
$AgCl$	1.8×10^{-10}	$MgCO_3$	1×10^{-5}
AgI	8.3×10^{-17}	MgF_2	7.1×10^{-9}
$AgBrO_3$	4×10^{-5}	$Mg(OH)_2$	1.2×10^{-11}
$AgIO_3$	3.0×10^{-8}	$MgNH_4PO_4$	2.5×10^{-13}
Ag_3PO_4	1.3×10^{-20}	MnS	1.4×10^{-15}
Ag_2S	2×10^{-49}	NiS	1.4×10^{-24}
$BaCO_3$	8.1×10^{-9}	$PbCO_3$	3.3×10^{-14}
BaC_2O_4	2.3×10^{-8}	$PbBr_2$	9×10^{-6}
$BaCrO_4$	2.4×10^{-10}	$PbCrO_4$	1.8×10^{-14}
$BaSO_4$	1.1×10^{-10}	$PbCl_2$	1.6×10^{-5}
BiI_3	8×10^{-19}	PbI_2	7.1×10^{-9}
Bi_2S_3	3×10^{-96}	$PbSO_4$	1.6×10^{-8}
$Be(OH)_2$	2.0×10^{-18}	PbS	8×10^{-28}
$CaCO_3$	4.8×10^{-9}	Sb_2S_3	1.7×10^{-93}
CaC_2O_4	2.6×10^{-9}	$SrCO_3$	1.6×10^{-9}
CaF_2	4.0×10^{-11}	$SrSO_4$	2.8×10^{-7}
$CaSO_4$	2.0×10^{-4}	$TlBr$	3.4×10^{-6}
CdS	1.6×10^{-28}	$TlBrO_3$	8.5×10^{-5}
$Cd_3(PO_4)_2$	2.2×10^{-33}	TlI	4.0×10^{-8}
CuS	9×10^{-36}	Tl_2S	5×10^{-22}
$Fe(OH)_3$	4×10^{-38}	$TlCl$	1.7×10^{-4}
FeS	3.7×10^{-19}	$Zn(OH)_2$	3×10^{-17}
Hg_2Cl_2	1.3×10^{-18}	ZnS	1.2×10^{-23}
$HgCl_2$	9.5×10^{-3}		

Appendix IX

Dissociation Constants for a Few Complex Ions

Complex ion	K_d
$Ag(CN)_2^-$	1.4×10^{-20}
$Ag(NH_3)_2^+$	6.2×10^{-8}
$Al(OH)_4^-$	1×10^{-34}
$Cd(NH_3)_4^{2+}$	2.8×10^{-7}
$Co(NH_3)_6^{3+}$	2.2×10^{-34}
$Cu(CN)_2^-$	1×10^{-16}
$Cu(NH_3)_4^{2+}$	4.9×10^{-14}
$Fe(CN)_6^{4-}$	1×10^{-35}
$Ni(NH_3)_6^{2+}$	2×10^{-9}
$Zn(NH_3)_4^{2+}$	3×10^{-10}
$Zn(OH)_4^{2-}$	3×10^{-16}

Appendix X

Table of Logarithms

Natural Numbers	0	1	2	3	4	5	6	7	8	9	Proportional parts								
											1	2	3	4	5	6	7	8	9
10	0000	0043	0086	0128	0170	0212	0253	0294	0334	0374	4	8	12	17	21	25	29	33	37
11	0414	0453	0492	0531	0569	0607	0645	0682	0719	0755	4	8	11	15	19	23	26	30	34
12	0792	0828	0864	0899	0934	0969	1004	1038	1072	1106	3	7	10	14	17	21	24	28	31
13	1139	1173	1206	1239	1271	1303	1335	1367	1399	1430	3	6	10	13	16	19	23	26	29
14	1461	1492	1523	1553	1584	1614	1644	1673	1703	1732	3	6	9	12	15	18	21	24	27
15	1761	1790	1818	1847	1875	1903	1931	1959	1987	2014	3	6	8	11	14	17	20	22	25
16	2041	2068	2095	2122	2148	2175	2201	2227	2253	2279	3	5	8	11	13	16	18	21	24
17	2304	2330	2355	2380	2405	2430	2455	2480	2504	2529	2	5	7	10	12	15	17	20	22
18	2553	2577	2601	2625	2648	2672	2695	2718	2742	2765	2	5	7	9	12	14	16	19	21
19	2788	2810	2833	2856	2878	2900	2923	2945	2967	2989	2	4	7	9	11	13	16	18	20
20	3010	3032	3054	3075	3096	3118	3139	3160	3181	3201	2	4	6	8	11	13	15	17	19
21	3222	3243	3263	3284	3304	3324	3345	3365	3385	3404	2	4	6	8	10	12	14	16	18
22	3424	3444	3464	3483	3502	3522	3541	3560	3579	3598	2	4	6	8	10	12	14	15	17
23	3617	3636	3655	3674	3692	3711	3729	3747	3766	3784	2	4	6	7	9	11	13	15	17
24	3802	3820	3838	3856	3874	3892	3909	3927	3945	3962	2	4	5	7	9	11	12	14	16
25	3979	3997	4014	4031	4048	4065	4082	4099	4116	4133	2	3	5	7	9	10	12	14	15
26	4150	4166	4183	4200	4216	4232	4249	4265	4281	4298	2	3	5	7	8	10	11	13	15
27	4314	4330	4346	4362	4378	4393	4409	4425	4440	4456	2	3	5	6	8	9	11	13	14
28	4472	4487	4502	4518	4533	4548	4564	4579	4594	4609	2	3	5	6	8	9	11	12	14
29	4624	4639	4654	4669	4683	4698	4713	4728	4742	4757	1	3	4	6	7	9	10	12	13
30	4771	4786	4800	4814	4829	4843	4857	4871	4886	4900	1	3	4	6	7	9	10	11	13
31	4914	4928	4942	4955	4969	4983	4997	5011	5024	5038	1	3	4	6	7	8	10	11	12
32	5051	5065	5079	5092	5105	5119	5132	5145	5159	5172	1	3	4	5	7	8	9	11	12
33	5185	5198	5211	5224	5237	5250	5263	5276	5289	5302	1	3	4	5	6	8	9	10	12
34	5315	5328	5340	5353	5366	5378	5391	5403	5416	5428	1	3	4	5	6	8	9	10	11
35	5441	5453	5465	5478	5490	5502	5514	5527	5539	5551	1	2	4	5	6	7	9	10	11
36	5563	5575	5587	5599	5611	5623	5635	5647	5658	5670	1	2	4	5	6	7	8	10	11
37	5682	5694	5705	5717	5729	5740	5752	5763	5775	5786	1	2	3	5	6	7	8	9	10
38	5798	5809	5821	5832	5843	5855	5866	5877	5888	5899	1	2	3	5	6	7	8	9	10
39	5911	5922	5933	5944	5955	5966	5977	5988	5999	6010	1	2	3	4	5	7	8	9	10
40	6021	6031	6042	6053	6064	6075	6085	6096	6107	6117	1	2	3	4	5	6	8	9	10
41	6128	6138	6149	6160	6170	6180	6191	6201	6212	6222	1	2	3	4	5	6	7	8	9
42	6232	6243	6253	6263	6274	6284	6294	6304	6314	6325	1	2	3	4	5	6	7	8	9
43	6335	6345	6355	6365	6375	6385	6395	6405	6415	6425	1	2	3	4	5	6	7	8	9
44	6435	6444	6454	6464	6474	6484	6493	6503	6513	6422	1	2	3	4	5	6	7	8	9
45	6532	6542	6551	6561	6571	6580	6590	6599	6609	6618	1	2	3	4	5	6	7	8	9
46	6628	6637	6646	6656	6665	6675	6684	6693	6702	6712	1	2	3	4	5	6	7	7	8
47	6721	6730	6739	6749	6758	6767	6776	6785	6794	6803	1	2	3	4	5	5	6	7	8
48	6812	6821	6830	6839	6848	6857	6866	6875	6884	6893	1	2	3	4	4	5	6	7	8
49	6902	6911	6920	6928	6937	6946	6955	6964	6972	6981	1	2	3	4	4	5	6	7	8
50	6990	6998	7007	7016	7024	7033	7042	7050	7059	7067	1	2	3	3	4	5	6	7	8
51	7076	7084	7093	7101	7110	7118	7126	7135	7143	7152	1	2	3	3	4	5	6	7	8
52	7160	7168	7177	7185	7193	7202	7210	7218	7226	7235	1	2	2	3	4	5	6	7	7
53	7243	7251	7259	7267	7275	7284	7292	7300	7308	7316	1	2	2	3	4	5	6	6	7
54	7324	7332	7340	7348	7356	7364	7372	7380	7388	7396	1	2	2	3	4	5	6	6	7

Natural Numbers	0	1	2	3	4	5	6	7	8	9	Proportional parts								
											1	2	3	4	5	6	7	8	9
55	7404	7412	7419	7427	7435	7443	7451	7459	7466	7474	1	2	2	3	4	5	5	6	7
56	7482	7490	7497	7505	7513	7520	7528	7536	7543	7551	1	2	2	3	4	5	5	6	7
57	7559	7566	7574	7582	7589	7597	7604	7612	7619	7627	1	2	2	3	4	5	5	6	7
58	7634	7642	7649	7657	7664	7672	7679	7686	7694	7701	1	1	2	3	4	4	5	6	7
59	7709	7716	7723	7731	7738	7745	7752	7760	7767	7774	1	1	2	3	4	4	5	6	7
60	7782	7789	7796	7803	7810	7818	7825	7832	7839	7846	1	1	2	3	4	4	5	6	6
61	7853	7860	7868	7875	7882	7889	7896	7903	7910	7917	1	1	2	3	4	4	5	6	6
62	7924	7931	7938	7945	7952	7959	7966	7973	7980	7987	1	1	2	3	3	4	5	6	6
63	7993	8000	8007	8014	8021	8028	8035	8041	8048	8055	1	1	2	3	3	4	5	5	6
64	8062	8069	8075	8082	8089	8096	8102	8109	8116	8122	1	1	2	3	3	4	5	5	6
65	8129	8136	8142	8149	8156	8162	8169	8176	8182	8189	1	1	2	3	3	4	5	5	6
66	8195	8202	8209	8215	8222	8228	8235	8241	8248	8254	1	1	2	3	3	4	5	5	6
67	8261	8267	8274	8280	8287	8293	8299	8306	8312	8319	1	1	2	3	3	4	5	5	6
68	8325	8331	8338	8344	8351	8357	8363	8370	8376	8382	1	1	2	3	3	4	4	5	6
69	8388	8395	8401	8407	8414	8420	8426	8432	8439	8445	1	1	2	2	3	4	4	5	6
70	8451	8457	8463	8470	8476	8482	8488	8494	8500	8506	1	1	2	2	3	4	4	5	6
71	8513	8519	8525	8531	8537	8543	8549	8555	8561	8567	1	1	2	2	3	4	4	5	5
72	8573	8579	8585	8591	8597	8603	8609	8615	8621	8627	1	1	2	2	3	4	4	5	5
73	8633	8639	8645	8651	8657	8663	8669	8675	8681	8686	1	1	2	2	3	4	4	5	5
74	8692	8698	8704	8710	8716	8722	8727	8733	8739	8745	1	1	2	2	3	4	4	5	5
75	8751	8756	8762	8768	8774	8779	8785	8791	8797	8802	1	1	2	2	3	3	4	5	5
76	8808	8814	8820	8825	8831	8837	8842	8848	8854	8859	1	1	2	2	3	3	4	5	5
77	8865	8871	8876	8882	8887	8893	8899	8904	8910	8915	1	1	2	2	3	3	4	4	5
78	8921	8927	8932	8938	8943	8949	8954	8960	8965	8971	1	1	2	2	3	3	4	4	5
79	8976	8982	8987	8993	8998	9004	9009	9015	9020	9025	1	1	2	2	3	3	4	4	5
80	9031	9036	9042	9047	9053	9058	9063	9069	9074	9079	1	1	2	2	3	3	4	4	5
81	9085	9090	9096	9101	9106	9112	9117	9122	9128	9133	1	1	2	2	3	3	4	4	5
82	9138	9143	9149	9154	9159	9165	9170	9175	9180	9186	1	1	2	2	3	3	4	4	5
83	9191	9196	9201	9206	9212	9217	9222	9227	9232	9238	1	1	2	2	3	3	4	4	5
84	9243	9248	9253	9258	9263	9269	9274	9279	9284	9289	1	1	2	2	3	3	4	4	5
85	9294	9299	9304	9309	9315	9320	9325	9330	9335	9340	1	1	2	2	3	3	4	4	5
86	9345	9350	9355	9360	9365	9370	9375	9380	9385	9390	1	1	2	2	3	3	4	4	5
87	9395	9400	9405	9410	9415	9420	9425	9430	9435	9440	0	1	1	2	2	3	3	4	4
88	9445	9450	9455	9460	9465	9469	9474	9479	9484	9489	0	1	1	2	2	3	3	4	4
89	9494	9499	9504	9509	9513	9518	9523	9528	9533	9538	0	1	1	2	2	3	3	4	4
90	9542	9547	9552	9557	9562	9566	9571	9576	9581	9586	0	1	1	2	2	3	3	4	4
91	9590	9595	9600	9605	9609	9614	9619	9624	9628	9633	0	1	1	2	2	3	3	4	4
92	9638	9643	9647	9652	9657	9661	9666	9671	9675	9680	0	1	1	2	2	3	3	4	4
93	9685	9689	9694	9699	9703	9708	9713	9717	9722	9727	0	1	1	2	2	3	3	4	4
94	9731	9736	9741	9745	9750	9754	9759	9763	9768	9773	0	1	1	2	2	3	3	4	4
95	9777	9782	9786	9791	9795	9800	9805	9809	9814	9818	0	1	1	2	2	3	3	4	4
96	9823	9827	9832	9836	9841	9845	9850	9854	9859	9863	0	1	1	2	2	3	3	4	4
97	9868	9872	9877	9881	9886	9890	9894	9899	9903	9908	0	1	1	2	2	3	3	4	4
98	9912	9917	9921	9926	9930	9934	9939	9943	9948	9952	0	1	1	2	2	3	3	4	4
99	9956	9961	9965	9969	9974	9978	9983	9987	9991	9996	0	1	1	2	2	3	3	3	4

Appendix XI

Answers to Problems

Chapter 2

2-1. 5222.924 g

2-2. 3.311 g

2-3. 7287.09 m

2-4. 3.50×10^4 m^2

2-5. 3.0×10^1 Å

2-6. 1×10^3 liters

2-7. (a) 6.00×10^{-2} m
(b) 5.70×10^{-5} kg
(c) 2.76×10^{-2} liter
(d) 1.00×10^{-8} m^3
(e) 3.5 liters
(f) 6.7×10^3 g
(g) 9.2×10^{-3} m
(h) 7.53×10^3 cm^3
(i) 5.31×10^{-4} kg
(j) 2.2×10^5 cm

2-8. 5.8×10^{-1} μm
5.8×10^2 nm
5.8×10^3 Å

2-9. (a) 149°F
(b) 37.0°C
(c) 0 K
(d) −175°F
(e) −71°C
(f) −18°C
(g) 196 K
(h) 78 K
(i) 373 K
(j) 122°F

2-10. (a) 32°F, 0°C
(b) 257°F, 398 K
(c) 95°C, 368 K
(d) −5°C, 268 K
(e) −4°F, 253 K
(f) 45°C, 318 K
(g) −148°F, −100°C
(h) 95°F, 308 K
(i) 200°C, 473 K
(j) −11.4°C, 262 K
(k) −369°F, −223°C

2-11. −40°

2-12. 7.50 g/ml

2-13. 19.7 kg

2-14. 35.0 ml

2-15. 45.1%

2-16. 18.0 liters

2-17. 10.5 g/cm^3

2-18. 0.810 g/ml

2-19. 2.07 g/ml

2-20. 1.0×10^3 kg/m^3
2-21. 2.7×10^2 kg
2-22. 408 g
2-23. 4.00×10^{-2} liter
2-24. 2.00×10^2 ml
2-25. 21.5
2-26. 5.0×10^1 liters
2-27. 1.7×10^4 g
2-28. 1.39×10^2 kg
2-29. 2×10^1 μm
2-30. 1.1 m^2
2-31. 90°F
2-32. (a) +11.4°F or −11.4°C
(b) +5°F or −15°C
(c) 574°F or 574 K
2-33. 8×10^9 liters or 8%
2-34. 1.27×10^8 t/cm^3
2-35. 15 m
2-36. 7×10^9
2-37. 2.1×10^5 t
2-38. 1.2×10^3 liters
2-39. 2.4×10^6 yr
2-40. 36.2%
2-41. (a) 6.6×10^{24} kg
(b) 5.5 g/ml

Chapter 3

3-1. 72.0
3-2. (a) 125
(b) 128
(c) 23.0
(d) 187
(e) 287
(f) 98.0
(g) 122
(h) 294
(i) 158
3-3. 148
3-4. (a) 84.0
(b) 60.1
(c) 116
(d) 213
(e) 193
(f) 342
(g) 177
(h) 602
(i) 392
3-5. 27
3-6. 31.0
3-7. 0.500 mole
3-8. (a) 2.09 moles
(b) 3.26 moles
(c) 9.34 moles
(d) 2.93 moles
(e) 2.08 moles
(f) 1.82 moles
3-9. (a) 4.0×10^1 g
(b) 56 g
(c) 825 g
(d) 36.9 g
3-10. 1.96×10^{24} atoms
3-11. (a) 13 moles
(b) 7.8×10^{24} molecules
3-12. (a) 5.88 moles
(b) 3.03 moles
(c) 2.25 moles
(d) 2.82 moles
(e) 3.64 moles
(f) 2.50 moles
3-13. (a) 4.82×10^{24} atoms H
(b) 2.41×10^{24} atoms P
(c) 8.43×10^{24} atoms O
3-14. (a) 6.00 moles
(b) 3.61×10^{24} ions
(c) 1.20×10^{24} ions
3-15. (a) 1.674×10^{-24} g
(b) 3.156×10^{-23} g
(c) 3.953×10^{-22} g
3-16. (a) 256
(b) 32.0
(c) 4.25×10^{-22} g
3-17. (a) 8.55×10^{20} molecules
(b) 5.13×10^{21} atoms
3-18. 3.27×10^{-16} g
3-19. 172
3-20. 4.03×10^2 g
3-21. 18.9984
3-22. 10.81

3-23. 93.0% ^{39}K; 7.0% ^{41}K
3-24. (a) 15.999
(b) 31.998
3-25. (a) 47.0
(b) 122
(c) 116
(d) 56.1
(e) 3.10×10^2
(f) 137
3-26. 0.300 mole
3-27. 2.44 moles
3-28. 0.500 mole
3-29. (a) (i) 0.765 mole
(ii) 0.470 mole
(iii) 0.937 mole
(iv) 0.257 mole
(b) (i) 2.30 moles H
0.765 mole P
3.06 moles O
(ii) 0.470 mole Cu and S
1.88 moles O
(iii) 1.87 moles N
3.75 moles H
2.81 moles O
(iv) 0.514 mole Fe
0.771 mole C
2.31 moles O
3-30. 11.0 g
3-31. (a) 373 g
(b) 107 g
(c) 515 g
(d) 128 g
3-32. (a) 17.7 moles
(b) 2.16 moles
(c) 2.20 moles
(d) 2.99 moles
3-33. (a) 87 g
(b) 1.90×10^2 g
(c) 112 g
(d) 834 g
3-34. 0.20 mole
3-35. (a) 3.22 moles
(b) 1.94×10^{24} atoms
3-36. 0.630 mole each element
3-37. 7.48 moles
3-38. 5.00×10^{-2} mole
3-39. 2.44×10^3 g
3-40. 3.20 moles
3-41. 36.8 g
3-42. 4.68×10^{24} ions
3-43. 1.21×10^{23} atoms Mg
2.42×10^{23} atoms O
2.42×10^{23} atoms H
3-44. 3.01×10^{22} molecules
3-45. 8.4×10^{24} molecules
3-46. 5.11×10^4 lb Fe
3-47. (a) 0.32 g
(b) 1.8×10^{-3} mole
(c) 1.1×10^{21} molecules
3-48. (a) 3.00 moles
(b) 1.81×10^{24} molecules
(c) 9.00 moles O
(d) 5.42×10^{24} atoms O
3-49. (a) 0.200 mole
(b) 1.20 moles
(c) 14.4 g
(d) 1.20×10^{23} molecules
(e) 2.40×10^{24} atoms
3-50. 63.53
3-51. 95.99
3-52. 1.95×10^3 g
3-53. 60.5% ^{69}Ga
3-54. 6.94
3-55. 91.22
3-56. 57.50% ^{121}Sb
3-57. 236.14
3-58. 42% ^{191}Ir
3-59. 70.0% ^{205}Tl
3-60. 69.5% ^{63}Cu
3.61. 24.31

Chapter 4

4-1. (a) 32.9% K; 67.1% Br
(b) 60.4% Sb; 39.6% S
(c) 85.6% C; 14.4% H
4-2. (a) 44.9% K; 18.4% S; 36.8% O
(b) 24.6% Na; 19.9% Fe;
25.7% C; 29.9% N
(c) 36.1% Ce; 14.4% N; 49.5% O

4-3. 60.3% Mg; 39.7% O
4-4. (a) 10.0 g
(b) 6.44 g
(c) 15.8 g
(d) 9.55 g
4-5. 1.43 g
4-6. 35.4% Mg
4-7. 40.97% $BaCO_3$
4-8. PCl_3
4-9. HgI_2
4-10. (a) BrF_3
(b) BrF_3
4-11. (a) $AlCl_3$
(b) Al_2Cl_6
4-12. (a) C_2H_6O
(b) HNO_2
(c) $C_3H_6O_2$
(d) $C_6H_5O_2N$
(e) C_3H_7ON
4-13. (a) C_5H_7N
(b) $C_{10}H_{14}N_2$
4-14. 19.7% H_2O
4-15. 36.0% H_2O
4-16. $GdCl_3 \cdot 6H_2O$
4-17. $HClO_4 \cdot H_2O$
4-18. (a) XeF_6
(b) XeF_6
4-19. (a) 12.5% O
(b) 44.4% O
(c) 40.7% O
(d) 47.3% O
(e) 14.8% O
(f) 17.3% O
4-20. 92.7% Hg; 7.3% O
4-21. 79.8% Cu; 20.2% S
4-22. 0.813 kg O_2
4-23. 158 g
4-24. 35.6 g
4-25. $Ca_2P_2O_7$
4-26. (a) SiF_4
(b) SiF_4
4-27. C_6H_{12}
4-28. (a) CH
(b) NiO
(c) $NaSO_3$
(d) Sc_2O_3
(e) $TiOCl_2$
4-29. $Al_2C_6H_{18}$
4-30. (a) 55.9% H_2O
(b) 36.1% H_2O
(c) 29.0% H_2O
(d) 40.5% H_2O
(e) 44.8% H_2O
(f) 23.5% H_2O
4-31. $FeSO_4 \cdot 7H_2O$
4-32. $BeCl_2 \cdot 4H_2O$
4-33. $Na_2CO_3 \cdot 10H_2O$
4-34. $ZnCl_2 \cdot 4NH_3$
4-35. $Li_2B_4O_7 \cdot 6H_2O$
4-36. Ag_2O
4-37. $KClO_3$
4-38. $XeOF_2$
4-39. Fe_2O_3
4-40. 3.0×10^1% (wt) decomposed
4-41. 74.7% (wt) not decomposed
4-42. $[RuCl_2(NH_3)_5N_2]$
4-43. 4 molecules
4-44. $PtO_2[P(C_6H_5)_3]_2$
4-45. (a) $SiCl_3$
(b) Si_2Cl_6
4-46. Cu_2S
4-47. $XePtF_6$
4-48. $KClO_3$
4-49. $CoCl_2 \cdot 6H_2O$
4-50. (a) $3060
(b) $2920
(c) $2570
(d) $2810

Chapter 5

5-1. 2, 1, 2
5-2. 2, 1, 2, 1
5-3. 1, 2, 1, 1
5-4. 3, 1, 3, 1
5-5. 3, 2, 1, 6
5-6. 2, 3, 2, 3
5-7. 3, 2, 1, 6
5-8. 1, 1, 1, 2
5-9. 1, 3, 1, 3

5-10. 2, 2, 2, 1
5-11. 2, 13, 8, 10
5-12. 3, 4, 1, 12
5-13. 1, 3, 1, 3
5-14. 2, 5, 4, 2
5-15. 2, 6, 2, 3
5-16. 2, 2, 1
5-17. 2, 15, 12, 6
5-18. 1, 2, 1, 2
5-19. 1, 2, 1, 1, 1
5-20. 1, 3, 1, 2
5-21. 1, 3, 2, 3
5-22. 2, 1, 1, 2, 2
5-23. 2, 1, 1, 1, 1
5-24. 1, 5, 6, 4
5-25. 2, 3, 2, 2
5-26. 3, 1, 2, 1
5-27. 1, 4, 1, 4
5-28. 2, 2, 4, 1
5-29. 2, 1, 3, 3
5-30. 2, 2, 4, 1
5-31. 2, 2, 6, 2, 3
5-32. 1, 1, 1, 1
5-33. 1, 2, 1, 1
5-34. 1, 2, 1, 1, 1
5-35. 1, 3, 2
5-36. 1, 2, 1, 1
5-37. 1, 6, 2, 3
5-38. 3, 4, 1, 8
5-39. 1, 6, 2, 3
5-40. 3, 2, 1, 3
5-41. 1, 1, 3, 1, 1, 1
5-42. 1, 2, 2, 1, 1
5-43. 1, 1, 4, 1, 1, 1, 2
5-44. 1, 2, 2, 2
5-45. (a) 43
(b) 22
(c) 36
(d) 5
(e) 30
(f) 13
(g) 16
(h) 26
(i) 27
(j) 6
5-46. (a) 6
(b) 14
(c) 8
(d) 9
(e) 12
(f) 15
(g) 23
(h) 9
(i) 15
(j) 28
(k) 28
(l) 13
(m) 24
(n) 9
(o) 11
5-47. 2, 3, 2, 3
5-48. 5, 2, 6, 5, 2, 2, 8
5-49. 2, 3, 6, 3, 2, 2
5-50. 1, 2, 2, 1, 1
5-51. 6, 10, 3, 10, 5
5-52. 8, 5, 8, 5, 8, 12
5-53. 1, 4, 1, 1, 2
5-54. 10, 8, 2, 5, 1, 2, 8
5-55. 3, 4, 1, 3, 1, 2
5-56. 3, 6, 1, 3, 1, 3
5-57. 3, 8, 20, 12, 20, 16
5-58. 2, 3, 16, 4, 3, 11
5-59. 2, 11, 22, 2, 6, 11, 8
5-60. 3, 10, 28, 3, 9, 28, 5
5-61. 1, 15, 20, 2, 15, 3, 20, 30
5-62. 1, 1, 1, 1, 2
5-63. 4, 1, 11, 4, 1, 4
5-64. 8, 12, 5, 4, 8, 12, 5
5-65. 1, 1, 4, 1, 1, 2
5-66. 5, 1, 4, 5, 1, 1
5-67. 1, 3, 3, 1, 6, 6
5-68. 2, 5, 3, 1, 2, 10, 8
5-69. 1, 2, 4, 1, 2, 2
5-70. 2, 4, 1, 2, 1, 5
5-71. 1, 2, 1, 1, 4, 2
5-72. 1, 20, 28, 2, 5, 40
5-73. 1, 3, 2, 1, 6, 2, 8
5-74. 2, 1, 2, 1, 1
5-75. 5, 2, 4, 5, 1, 8
5-76. 3, 3, 6, 1

5-77. 4, 1, 10, 4, 4, 1, 3
5-78. 2, 10, 3, 2, 6, 8
5-79. 5, 2, 6, 5, 2, 3
5-80. 1, 4, 1, 1, 2
5-81. 5, 11, 1, 5, 22, 1
5-82. 1, 5, 2, 5, 1
5-83. 3, 2, 1, 3, 2, 2
5-84. 3, 7, 7, 3, 7, 2
5-85. 2, 4, 2, 1, 4, 2
5-86. 1, 2, 2, 1, 2, 2
5-87. 2, 5, 14, 2, 5, 7
5-88. 3, 8, 6, 1
5-89. 1, 3, 8, 2, 6, 7
5-90. 3, 6, 1, 5, 3
5-91. 1, 6, 2, 2, 4
5-92. 6, 1, 5, 3, 3
5-93. 2, 2, 1, 1, 2, 4
5-94. 7, 7, 1, 4, 6
5-95. 2, 3, 4, 3, 5, 2
5-96. 1, 4, 1, 1, 1, 1, 1, 2
5-97. 3, 4, 3, 4, 1, 4
5-98. 1, 1, 4, 1, 1, 1, 2
5-99. 7, 1, 3, 1, 6, 2
5-100. 5, 9, 3, 1, 5, 8
5-101. 3, 4, 28, 6, 9, 28, 4
5-102. 1, 10, 1, 10, 3, 5
5-103. 2, 15, 1, 4, 11

Chapter 6

6-1. 1/3
6-2. 86.4 ml
6-3. 9.00×10^2 ft^3
6-4. 1.51 liters
6-5. 11.4 g/liter
6-6. 199°C
6-7. 328 ml
6-8. 428 liters
6-9. 4.0×10^1°C
6-10. 2.00×10^2 ml
6-11. 243 ml
6-12. 3.39 liters
6-13. 0.667 atm
6-14. 8.91 liters
6-15. 2.00×10^2 K
6-16. 65.7 atm
6-17. 0.45 m^3
6-18. 2.47×10^3 K
6-19. 3.0 atm
6-20. 135 ml
6-21. 0.40 atm O_2; 0.20 atm N_2; 0.60 atm total
6-22. 27.4 liters
6-23. 0.396 liter
6-24. 50.6 atm A; 61.2 atm B; 38.1 atm C
6-25. 362 torr CO_2; 241 torr H_2; 121 torr O_2
6-26. 79 kPa N_2; 19 kPa O_2
6-27. (a) H_2
(b) 5.0 times as fast
6-28. 9/4
6-29. 4.6 micromole/hr
6-30. 1.0×10^2
6-31. 48 cm from A end
6-32. 728 ml
6-33. 5.6 liters
6-34. (a) 62.5 ft^3
(b) 75.0 ft^3
6-35. 29.1
6-36. 6.02×10^{20} molecules
6-37. 51.7
6-38. 5.36 g
6-39. 3.01×10^{21} molecules
6-40. 1.88×10^{18} atoms O
6-41. 2.10 liters
6-42. (a) 0.125 mole
(b) 76
6-43. 403 K
6-44. 133
6-45. 7.04×10^{23} molecules
6-46. (a) 8.313×10^3 liter Pa mole^{-1} K^{-1}
(b) 62.4 liter torr mole^{-1} K^{-1}
6-47. 3.91 g
6-48. 246 atm
6-49 2.22×10^3 g
6-50. 1.06×10^2 atm
6-51. 4.50 liters
6-52. 0.318 liter

6-53. 16.8 liters
6-54. 2.20 g
6-55. (a) 0.35
(b) 0.25
(c) 0.23
(d) 0.16
(e) 0.18
(f) 0.71
(g) 0.20
6-56. 1.5×10^2
6-57. 1.05×10^{24} molecules
6-58. 0.438 liter
6-59. 50.4 liters
6-60. (a) 9.4 liters O_2
(b) 6.8 liters H_2O
6-61. 1.68×10^{-4} liter
6-62. 1.22×10^4 kPa
6-63. 42.0
6-64. (a) 1.50 g/liter
(b) 33.6
6-65. 2.0×10^3 kPa
6-66. 1.28 g/liter
6-67. 760 torr
6-68. 13 atm
6-69. 22.4 liters
6-70. 2.60×10^{17} molecules
6-71. (a) C_3H_4
(b) C_6H_8
6-72. 4.0×10^1 liters
6-73. 42.8 atm
6-74. 22.0 liters
6-75. (a) 20.17
(b) 20.17
(c) 20.18
6-76. (a) 1.29 g/liter
(b) 1.15 g/liter
6-77. (a) 32.1
(b) N_2H_4
6-78. 6.02×10^{23}

Chapter 7

7-1. 5.00×10^2 kg
7-2. (a) 0.518 mole
(b) 0.345 mole
(c) 16.6 g
(d) 511 g
7-3. (a) 112 moles
(b) 3.58×10^3 g
(c) 135 moles
(d) 1.95 g
7-4. (a) 18.9 moles
(b) 15.1 moles
(c) 10.0 moles
(d) 8.00 moles
(e) 3.20×10^2 g
(f) 352 g
(g) 1.20 moles
(h) 31.2 g
7-5. (a) 17.5 moles
(b) 298 g
(c) 7.00×10^2 g
(d) 20.4 g
7-6. (a) 5.00 moles
(b) 1.60×10^2 g
(c) 2.50 moles
(d) 245 g
7-7. (a) 76.4 g
(b) 25.0 g
(c) 36.9 g
7-8. (a) 27.7 g
(b) 7.13 g
(c) 12.6 g
(d) 20.2 g
7-9. (a) 35.3 g
(b) 6.04 g
(c) 31.6 g
7-10. (a) 56.2 kg
(b) 175 kg
(c) 206 kg
(d) 75.0 kg
7-11. 4.25 liters
7-12. 87.1 g
7-13. (a) 175 liters
(b) 50.4 liters
7-14. (a) 11.2 liters
(b) 5.60 liters
(c) 8.05 liters

7-15. 1.21×10^3 liters
7-16. (a) 2, 5, 4, 6
(b) 28.0 liters
(c) 22.4 liters
7-17. (a) 1, 6, 4
(b) 105 liters
(c) 1.40×10^2 liters
7-18. (a) 71.0 liters
(b) 142 liters
(c) 50.7 liters
7-19. (a) NH_3 in excess
(b) 2.40×10^2 g NH_4NO_3
(c) 53.0 g NH_3
7-20. 23.8 g
7-21. (a) No
(b) 3.63×10^6 kg
7-22. 55.6%
7-23. (a) 2, 25, 16, 18
(b) 99.1%
7-24. 84.5%
7-25. (a) 2.70×10^2 g
(b) 3.30×10^2 g
(c) 3.00 moles
(d) 189 g
7-26. (a) 2.00 moles
(b) 22.0 g
(c) 0.500 mole of each
(d) 1.00 kg products
(e) 4.20×10^2 g
(f) 2.50 moles
7-27. (a) 144 kg
(b) 92.7 kg
7-28. 304 kg
7-29. (a) 186 g
(b) 289 g
7-30. 104 g
7-31. 0.407 mole
7-32. (a) $(NH_4)_3PO_4$ in excess
(b) 0.5001 mole $Ba_3(PO_4)_2$
7-33. (a) 83.8 g Fe
(b) 2.04 g H_2
7-34. 1.43×10^3 g
7-35. 5.58%
7-36. 33.0%
7-37. 34.5%
7-38. 5.00 kg $Ca_3(PO_4)_2$;
2.91 kg SiO_2; 0.484 kg C
7-39. 4.50×10^2 kg
7-40. (a) Pb_3O_4
(b) $Pb_3O_4 + 4H_2 \rightarrow 3Pb + 4H_2O$
7-41. (a) Fe_2O_3
(b) 70.00% Fe
(c) 5.416 g H_2O
7-42. 75.2% MgO; 24.8% Mg_3N_2
7-43. 35.4533
7-44. 6.940
7-45. 2.62 liters
7-46. (a) 2.28 kg
(b) 718 liters
(c) 718 liters
(d) 64.7 g
(e) 1.13 kg
7-47. (a) 1.96×10^3 liters
(b) 1.25×10^3 liters
7-48. (a) 179 g
(b) 125 liters
(c) 651 m^3
7-49. (a) 70.0 liters
(b) 1.00 liter
(c) 1.00×10^2 g
(d) 20.5 liters
7-50. 3.60×10^3 liters
7-51. 3.31 g NaBr; 1.58 g H_3PO_4
7-52. (a) 57.6 g H_2O
(b) 3.3 g C_3H_8
7-53. 0.893 kg
7-54. (a) 0.833 mole
(b) 54.0 g $KMnO_4$; 95.8 g KOH
(c) 7.00 liters
7-55. 3.36×10^5 liters
7-56. 71.8%
7-57. 51.9% Zn; 48.1% Cd
7-58. XeF_4
7-59. (a) 5.9 atm
(b) 0.15 mole O_2; 0.30 mole H_2O
(c) 4.4 atm

Chapter 8

8-1. (a) 1 (b) 4 (c) 1
(d) 4 (e) 2 (f) 1

8-2. 6, 8, 12

8-3. (a) 0.186 nm
(b) 0.428 nm
(c) 23.6 cm^3

8-4. (a) 0.409 nm
(b) 10.5 g/cm^3

8-5. (a) 0.495 nm
(b) 1.87×10^{-22} cm^3
(c) 56.3 cm^3
(d) 1.52 g/cm^3

8-6. (a) 1
(b) 6.80 g/cm^3; 17.5 cm^3

8-7. (a) 4
(b) 1.04×10^{-22} cm^3
(c) 15.7 cm^3
(d) 7.32 g/cm^3

8-8. (a) 4
(b) 0.2814 nm
(c) 0.3980 nm
(d) 1.783×10^{-22} cm^3

8-9. (a) 0.267 nm
(b) 0.534 nm
(c) 0.378 nm
(d) 1.52×10^{-22} cm^3

8-10. (a) 0.366 nm
(b) 0.423 nm
(c) 6
(d) 7.57×10^{-23} cm^3

8-11. (a) 0.654 nm
(b) 2.87 g/cm^3

8-12. (a) 0.456 nm
(b) 57.1 cm^3

8-13. (a) 0.4302 nm
(b) 0.3726 nm

8-14. (a) 0.3022 nm; 0.3181 nm
(b) 4, 2

8-15. 0.263 nm

8-16. (a) 0.124 nm
(b) 12

8-17. 0.325 nm; 0.337 nm

8-18. (a) 1.68×10^{-22} cm^3
(b) 0.390 nm
(c) 12
(d) 6.05 g/cm^3

8-19. (a) 0.5924 nm
(b) 2.079×10^{-22} cm^3
(c) 0.4189 nm
(d) 31.30 cm^3

8-20. 6.020×10^{23}

8-21. 0.515 nm

8-22. 4, 4

8-23. 6.024×10^{23}

8-24. (a) 0.524; 0.680; 0.740
(b) face-centered

8-25. Answers in order without units
(a) 0.3267; 14.85
(b) 0.600; 0.300; 4.12
(c) 0.4846; 0.3427; 48.47
(d) 0.4810; 3.348
(e) 0.46669; 0.33000; 43.28; 2.750
(f) 0.402; 0.284
(g) 0.3922; 0.2774; 25.70
(h) 0.4286; 0.3712; 4.49
(i) 0.4567; 0.3955
(j) 0.3834; 0.3320; 33.94
(k) 0.333; 0.288; 22.2; 9.91
(l) 0.2744; 7.10
(m) 0.390; 35.7

8-26. (a) 2%
(b) 0.2%
(c) 0.4%

Chapter 9

9-1. 1.80×10^2 g

9-2. 0.248 kg $C_{12}H_{22}O_{11}$;
0.752 kg H_2O

9-3. 69 g

9-4. NaCl = 0.0143; H_2O = 0.986

9-5. 0.119

9-6. 2.1×10^{-4}

9-7. 0.200 *M*

9-8. (a) 0.714 *M*
(b) 0.291 *M*
(c) 0.283 *M*
(d) 0.0834 *M*

9-9. 152.0 g
9-10. 5.26×10^{-3} M
9-11. 80.8 g
9-12. 0.141 M
9-13. (a) 1.59 g
(b) 5.56 g
(c) 0.912 g
(d) 1.56 g
9-14. 5.09 liters
9-15. 0.81 M
9-16. 1.20 liters
9-17. 0.333 M
9-18. (a) 0.200 mole
(b) 0.200 mole
(c) 0.800 M NaCl; 0.0667 M NaCl
9-19. 2.50 m
9-20. 1.60 kg
9-21. 506 g
9-22. 0.0147
9-23. 1.59 m
9-24. 62.0%
9-25. 84.0 g
9-26. 12.9 ml
9-27. 6.20 g
9-28. 202 ml
9-29. 0.690 liter
9-30. 0.717 g
9-31. (a) 0.160 N
(b) 0.0802 M
9-32. 0.742 g
9-33. 25.4 g
9-34. 0.125 N
9-35. (a) 26.1 g
(b) 0.415 mole
9-36. 0.100 M
9-37. 0.500 M
9-38. 52.6 g
9-39. 15.2 g
9-40. 9.09%
9-41. 3.20 m
9-42. Urea = 0.0602; H_2O = 0.940
9-43. 36 g
9-44. 0.0788 eq
9-45. 0.120 N
9-46. 5.06 g
9-47. 0.133 N
9-48. (a) 0.200 M
(b) 0.600 N
9-49. 35.7 g
9-50. 0.514 M
9-51. 2.68 M NaOH; 2.68 N NaOH
9-52. (a) 104 g
(b) 1.06 moles
9-53. (a) 2.35 g
(b) 15.0 g
(c) 4.08 g
(d) 29.2 g
(e) 39.0 g
9-54. (a) 444 ml
(b) 1.00×10^2 ml
(c) 133 ml
(d) 4.0×10^1 ml
9-55. (a) 4.56 g
(b) 0.600 N
(c) 1.47 g
9-56. 453 ml
9-57. 1.98 g
9-58. 4.90 g
9-59. (a) 15.99%
(b) 0.02081
9-60. (a) 0.02002
(b) 1.128 M
9-61. (a) 20.0%
(b) 2.33 M
(c) 2.55 m
(d) 0.0440
9-62. 1.09 g/ml
9-63. (a) 14.2 ml
(b) 34.1 ml
(c) 2.71 ml
9-64. (a) 2.22 m; 0.0385
(b) 4.80 m; 0.0793
(c) 11.6 m; 0.172
9-65. 0.200 kg
9-66. (a) 0.766 m
(b) 12.1%
9-67. 46.4 g
9-68. (a) 2.91 m
(b) 2.80 M

(c) 0.0498
(d) 1.42 g/ml
(e) 32.0%

9-69. (a) 25.3%
(b) 3.97 *M*
(c) 4.54 *m*
(d) 0.0755

9-70. 3.24 *M*

9-71. 2.26×10^{23} molecules

9-72. (a) 19.0 g
(b) 4.00 *M*

Chapter 10

10-1. 1.20 *N*

10-2. 47 ml

10-3. 50.0 ml

10-4. (a) 0.300 *N*
(b) 0.150 *M*
(c) 0.368 g

10-5. 0.750 *N*

10-6. (a) 5.60%
(b) 5.71%
(c) 6.33%
(d) 7.71%

10-7. 122

10-8. 59.0

10-9. 69.7%

10-10. (a) 31.4 g
(b) 40.0 ml

10-11. 30.1 ml

10-12. 29.2%

10-13. (a) 63.5 g
(b) 158
(c) 37.7%

10-14. 0.200 *N*

10-15. 0.2253 *N*

10-16. 29.68 ml

10-17. 40.6 ml

10-18. 0.200 *N*

10-19. 0.505 *N*

10-20. 0.0741 *M*

10-21. 0.500 *N*

10-22. 88.9 ml

10-23. 0.189 liter

10-24. (a) 0.635 *N*
(b) 3.81%

10-25. (a) 7.94 *N*
(b) 14.1%

10-26. 75.0

10-27. (a) 5.00×10^{-3} eq
(b) 5.00×10^{-3} eq
(c) 0.200 *N*

10-28. (a) 5.489×10^{-3} eq
(b) 5.489×10^{-3} eq

10-29. 0.3751 *N*

10-30. (a) 4.00 meq
(b) 0.216 *N*

10-31. (a) 0.920 *N*
(b) 0.460 *M*

10-32. 0.100 g

10-33. 25.3%

10-34. 8.02%

Chapter 11

11-1. 23.00 torr

11-2. 21.76 torr

11-3. 75.0

11-4. 84.5 g

11-5. −1.86°C

11-6. 100.52°C

11-7. 101.3°C

11-8. −29.9°C

11-9. 52.5

11-10. −5.58°C

11-11. 1.40×10^2

11-12. 1.69 *m*

11-13. 3.73 atm

11-14. 8.08×10^4

11-15. 7.9 atm

11-16. 2.8×10^4

11-17. (a) −0.58°C
(b) 100.16°C

11-18. (a) −0.66°C
(b) 100.18°C

11-19. 0.765 *m*

11-20. 2.2%

11-21. −1.20°C

11-22. 3%

11-23. $-4.61°C$
11-24. 2 ions/formula unit
11-25. 1.20×10^2
11-26. 101.8°C
11-27. 787 g
11-28. 497 g
11-29. (a) 2.20 *m*
(b) 1.49 *M*
11-30. 0.226 *m*
11-31. $-13.5°C$
11-32. 138
11-33. 35.7
11-34. 69.6°C
11-35. 66.6
11-36. 3.40×10^2
11-37. $C_4H_6O_4$
11-38. 4.90°C/*m*
11-39. (a) 31.09 torr
(b) 33.6 atm
11-40. 119
11-41. (a) 2.83°C/*m*
(b) 10.0°C
11-42. $C_6H_5NO_2$
11-43. 29.8 atm
11-44. 6.83 atm
11-45. greater than 26 atm
11-46. (a) 17.0 torr; 100.91°C; −3.26°C; 42.1 atm
(b) 16.5 torr; 101.8°C; −6.51°C; 84.2 atm
(c) 17.2 torr; 100.58°C; −2.06°C; 26.7 atm

Chapter 12

12-1. 64.4 kJ
12-2. 301 kJ
12-3. 2.10×10^3 g
12-4. 37.0°C
12-5. 19.4 kcal
12-6. 72 g
12-7. −846.0 kJ
12-8. +414.3 kJ
12-9. −197.6 kJ
12-10. −108.8 kcal/mole
12-11. 6.58 kJ
12-12. −249.9 kJ/mole
12-13. 31°C
12-14. 25 g ice remain
12-15. 7.1×10^2 g ice
12-16. −178.3 kJ
12-17. −113.9 kJ
12-18. +532.6 kJ
12-19. (a) +53.3 kJ/mole
(b) +51.7 kJ/mole
(c) +53.4 kJ/mole
(d) +57.3 kJ/mole
12-20. −297 kJ/mole S
12-21. −128.7 kJ/mole
12-22. −68.6 kJ/2 moles HBr
12-23. 463 kJ/mole
12-24. (a) −2253 kJ
(b) −112 kJ/mole C_3H_8
(c) −12 kJ/mole ICl
(d) −311.4 kJ
(e) −46.1 kJ
(f) −36.4 kJ
(g) −226.0 kJ
12-25. (a) -3.50×10^2 kJ
(b) −267 kJ
(c) −314 kJ
12-26. +5.99 kJ/mole

Chapter 13

13-1. (a) 12.8 hr
(b) 0.143
13-2. (a) 0.891 = fraction converted
(b) 0.312 hr
13-3. (a) 3.93×10^{-2} min^{-1}
(b) 17.6 min
(c) 3.15×10^{-2} mole liter^{-1} min^{-1}
(d) 1.58×10^{-2} mole liter^{-1} min^{-1}
13-4. (a) 1.83×10^{-2} sec^{-1}
(b) 114 sec

13-5. (a) 0.648 hr^{-1}
(b) 0.125
(c) 8.10×10^{-2} mole $liter^{-1}$ hr^{-1}
(d) 3.1×10^{-2} mole $liter^{-1}$ hr^{-1}

13-6. (a) 0.156 min^{-1}
(b) 0.154
(c) 0.0623
(d) 1.84×10^{-2} mole $liter^{-1}$ min^{-1}

13-7. (a) Rate = $k[Co(NH_3)_5Cl^{2+}]$
(b) 1.3×10^{-4} min^{-1} for each

13-8. (a) Rate = $k[CO][NO_2]$
(b) 1; 1; 2
(c) 1.9 M^{-1} hr^{-1}
(d) 1.5×10^{-7} mole $liter^{-1}$ hr^{-1}

13-9. (a) Rate = $k[N_2O_5]$
(b) first order
(c) 0.27 mole $liter^{-1}$ sec^{-1}

13-10. (a) Rate = k
(b) zero order

13-11. (a) Rate = $k[HI]^2$
(b) second order
(c) 1.9×10^{-6} mole $liter^{-1}$ min^{-1}

13-12. $t_{1/2} = \dfrac{0.693}{k}$

13-14. 1.5×10^{-2} hr^{-1}

13-15. 5.16×10^{-6} mole $liter^{-1}$ min^{-1}

13-16. (a) 6.7×10^2 yr
(b) 6.24×10^{-3} g Ra

13-17. 1.38×10^4 yr

13-19. (a) Rate = $k[NO]^2[Cl_2]$
(b) 8.0×10^{-6} mole $liter^{-1}$ min^{-1}
(c) 6.4×10^{-5} mole $liter^{-1}$ min^{-1}

13-20. (a) 1.1×10^2 min
(b) 1.5×10^2 min
(c) 5.2×10^{-5} mole $liter^{-1}$ min^{-1}

Chapter 14

14-1. 4.0

14-2. 0.50

14-3. 0.42

14-4. 24

14-5. 4.0

14-6. 2.8 M

14-7. 1.62×10^{-2}

14-8. 0.4 M

14-9. 1.0×10^{-2} M

14-10. 4.7×10^{-2}

14-11. (a) 2.0 M
(b) 5.0×10^1 mole % N unreacted

14-12. 1.3 M

14-13. 0.248

14-14. (a) 3.06 M
(b) 55.0 mole % conversion

14-15. (a) 0.538%
(b) $[H_2] = [Br_2] = 5.38 \times 10^{-3}$ M; $[HBr] = 1.99$ M

14-16. 4.7

14-17. 4.8 cm Hg

14-18. 10 moles PCl_5

14-19. $p_{CO} = p_{H_2O} = 4.80$ atm

14-20. (a) 0.569
(b) $[COBr_2] = 0.440$ M; $[CO] = 0.310$ M; $[Br_2] = 0.810$ M

14-21. $[SO_3] = 0.72$ M; $[NO] = 0.42$ M; $[SO_2] = 0.18$ M; $[NO_2] = 0.33$ M

14-22. (a) $[H_2] = [I_2] = 1.20$ M; $[HI] = 0.10$ M
(b) 96%

14-23. (a) 4.0
(b) 0.42 mole CO_2 required

14-24. (a) $[N_2] = 0.87$ mole/liter; $[H_2] = 2.61$ mole/liter; $[NH_3] = 4.26$ mole/liter
(b) 1.17
(c) 6.64 moles NH_3 required

14-25. $[H_2] = [Br_2] = 1.35 \times 10^{-5}$ M; $[HBr] = 0.0200$ M

14-26. (a) 0.737
(b) 0.737 atm

14-27. (a) 0.3420
(b) 0.5637
14-28. 4.8
14-29. (a) 124
(b) 79.6 mole % CO;
20.4 mole % CO_2
14-30. 0.0926 mole $CaCO_3$
14-31. $p_{O_2} = p_{N_2} = 2$ atm
14-33. (a) 6.26×10^{-2}
(b) 156 atm
(c) $p_{H_2} = 74.0$ atm;
$p_{N_2} = 61.7$ atm;
$p_{NH_3} = 20.3$ atm
(d) 1.65×10^{-5}
14-34. $[PCl_3] = [Cl_2] = 7.27 \times 10^{-2}\ M$;
$[PCl_5] = 0.127\ M$
14-35. $[PCl_3] = 0.0444\ M$;
$[Cl_2] = 0.145\ M$;
$[PCl_5] = 0.155\ M$
14-36. $[HI] = 2.06 \times 10^{-2}\ M$;
$[H_2] = 5.22 \times 10^{-2}\ M$;
$[I_2] = 1.47 \times 10^{-2}\ M$

Chapter 15

15-1. $[H^+] = [A^-] = 4.5 \times 10^{-5}\ M$;
$[HA] = 0.050\ M$
15-2. 4.5×10^{-5}
15-3. $9.7 \times 10^{-4}\ M$
15-4. 1.1%
15-5. 3.8%
15-6. (a) $2.3 \times 10^{-5}\ M$
(b) $3.2 \times 10^{-6}\ M$
15-7. 3.9×10^{-4}
15-8. 4.9×10^{-10}
15-9. (a) $9.4 \times 10^{-2}\ M$
(b) $5.2 \times 10^{-5}\ M$
15-10. (a) $7.7 \times 10^{-2}\ M$
(b) $6.3 \times 10^{-8}\ M$
15-11.

	$[H^+]$	$[OH^-]$
(a)	$0.015\ M$;	$6.7 \times 10^{-13}\ M$
(b)	$0.0020\ M$;	$5.0 \times 10^{-12}\ M$
(c)	$3.50 \times 10^{-3}\ M$;	$2.86 \times 10^{-12}\ M$
(d)	$0.085\ M$;	$1.2 \times 10^{-13}\ M$
(e)	$9.5 \times 10^{-5}\ M$;	$1.1 \times 10^{-10}\ M$
(f)	$1.29 \times 10^{-11}\ M$;	$7.75 \times 10^{-4}\ M$
(g)	$2.0 \times 10^{-13}\ M$;	$0.050\ M$
(h)	$2.63 \times 10^{-13}\ M$;	$0.0380\ M$
(i)	$4.00 \times 10^{-14}\ M$;	$0.250\ M$

15-12. (a) 0.28
(b) 2.12
(c) 4.04
(d) 4.80
(e) 3.48
(f) 9.79
(g) 10.26
(h) 12.00
(i) 12.79
(j) 9.72
15-13. (a) pH = 11.08; pOH = 2.92
(b) pH = 11.48; pOH = 2.52
(c) pH = 2.37; pOH = 11.63
15-14. 0.20
15-15. 5.9×10^{-5}
15-16. 1.0×10^{-7}
15-17. 2.0×10^{-6}
15-18. 11.83
15-19. 1.0×10^{1}
15-20. 6.70
15-21. 2.28
15-22. 8.74
15-23. (a) $5.9 \times 10^{-3}\ M$
(b) $3.6 \times 10^{-4}\ M$
15-24. (a) $4.2 \times 10^{-6}\ M$
(b) $2.5 \times 10^{-5}\ M$
15-25. (a) $3.0 \times 10^{-3}\ M$
(b) $9.0 \times 10^{-5}\ M$
15-26. (a) $2.7 \times 10^{-3}\ M$
(b) $3.6 \times 10^{-5}\ M$
15-27. 4.97
15-28. $[OH^-] = 1.8 \times 10^{-5}\ M$;
$[NH_4^+] = [NH_3] = 0.50\ M$;
$[H^+] = 5.6 \times 10^{-10}\ M$
15-29. 8.56

15-30. (a) 4.92
(b) 4.57
15-31. (a) 9.26
(b) 0.82 mole HCl
15-32. 0.39 mole NaOH
15-33. $[H^+] = [A^-] = 3.0 \times 10^{-3}\ M$;
$[HA] = 0.20\ M$
15-34. $[OH^-] = [B^+] = 2.0 \times 10^{-2}\ M$;
$[BOH] = 0.98\ M$;
$[H^+] = 5.0 \times 10^{-13}\ M$
15-35. 5.3×10^{-4}
15-36. 5.3×10^{-3}
15-37. (a) 0.42%
(b) 1.3%
(c) 4.2%
15-38. 3.35
15-39. (a) $[OH^-] = [C_4H_9NH_3{}^+]$
$= 1.1 \times 10^{-2}\ M$;
$[C_4H_9NH_2] = 0.24\ M$
(b) 12.04
(c) (i) 2.2%
(ii) 2.0×10^1%
(iii) 5.0×10^1%
15-40. $0.27 \simeq 0.3$
15-41. 3.9×10^{-6}
15-42. (a)
15-43. (a) $1.1 \times 10^{-13}\ M$; 12.94
(b) $3.3 \times 10^{-12}\ M$; 11.48
(c) $1.3 \times 10^{-12}\ M$; 11.88
(d) $2.1 \times 10^{-11}\ M$; 10.68
(e) $1.7 \times 10^{-12}\ M$; 11.78
15-44. (a) $2.4 \times 10^{-12}\ M$; 11.61
(b) $4.3 \times 10^{-13}\ M$; 12.36
(c) $1.1 \times 10^{-12}\ M$; 11.95
(d) $5.0 \times 10^{-14}\ M$; 13.30
(e) $1.2 \times 10^{-12}\ M$; 11.90
15-45. 2.66
15-46. 11 g
15-47. 8.86
15-48. 1.3×10^{-6}
15-49. 2.3×10^{-8}
15-50. 1.7×10^{-10}
15-51. (a) 3.27
(b) $2 \times 10^{-11}\ M$
(c) $5.4 \times 10^{-4}\ M$
15-52. (a) 10.50
(b) $3.2 \times 10^{-4}\ M$
(c) $2.0 \times 10^{-12}\ M$
15-53. 3.6×10^{-7}
15-54. 4.74
15-55. 8.95
15-56. 1.7×10^{-5}
15-57. 1.0×10^{-3}
15-58. 0.56
15-59. (a) 10.36
(b) $2.3 \times 10^{-4}\ M$
(c) 0.41 mole HCl
15-60. 9.28
15-61. $K_1 = 8.1 \times 10^{-5}$;
$K_2 = 1.6 \times 10^{-12}$

Chapter 16

16-1. (a) $PbCO_3$
(b) $MgCO_3$
16-2. 1.7×10^{-4}
16-3. 1.3×10^{-20}
16-4. 3.9×10^{-28}
16-5. (a) 2.0×10^{-4} mole/liter
(b) 6.6×10^{-2} g/liter
16-6. (a) 1.2×10^{-3} mole/liter
(b) 0.55 g/liter
16-7. 6.9×10^{-7} mole/liter
16-8. No, trial product is 2.0×10^{-8}
16-9. Yes, trial product is 8.7×10^{-18}
16-10. Yes, trial product is 2.2×10^{-8}
16-11. (a) $6.3 \times 10^{-8}\ M$
(b) $2.0 \times 10^{-12}\ M$
16-12. (a) $3 \times 10^{-3}\ M$
(b) $1 \times 10^{-2}\ M$
16-13. $4 \times 10^{-3}\ M$
16-14. (a) 4×10^{-5} mole/liter
(b) 1.1 g/liter
16-15. 3.1×10^{-8} mole/liter
16-16. (a) 8.7×10^{-16} mole/liter
(b) 2.0×10^{-13} g AgI/liter
(c) 9.1×10^{-9} mole/liter
(in H_2O)
16-17. 3.2×10^{-3} mole/liter

16-18. 4.9×10^{-7} g
16-19. 2.2×10^{-3} mole/liter
16-20. (a) 2.3×10^{-4} mole/liter
(b) 3.7×10^{-3} mole/liter
16-21. (a) 1.0×10^{-5} mole/liter
(b) 1.0×10^{-2} mole/liter
16-22. Just under 1.8×10^{-2} M Pb^{2+}
16-23. Ag^+; just under 3.1×10^{-2} M Cl^-
16-24. (a) Just under 1×10^{-5} M $C_2O_4^{2-}$
(b) 1.6
16-25. (a) Approximately 1×10^{-20} M S^{2-}
(b) ~1.0
16-26. (a) 1.3×10^{-2} mole/liter
(b) 3.1 g/liter
16-27. (a) 4.0×10^{-4} mole/liter
(b) 9.6×10^{-2} g/liter
16-28. 7.1×10^{-7} mole/liters in H_2O
2.5×10^{-8} mole/liters in CH_3OH
16-29. 4.0×10^{-12} mole/liter
16-30. 1.6×10^{-5}
16-31. 4.0×10^{-12}
16-32. 6.9×10^{-9}
16-33. 3×10^{-11} mole/liter
16-34. 6.6×10^{-6} g
16-35. (a) 1.6×10^{-6} M
(b) 5.3×10^{-3} M
16-36. (a) 7.5×10^{-11} M
(b) 1.9×10^{-6} M
16-37. Yes, trial product is 9.1×10^{-13}
16-38. Yes, trial product is 4.9×10^{-5}
16-39. 8×10^{-4} mole/liter
16-40. 4.0×10^{-2} M
16-41. 2×10^{-6} mole/liter
16-42. No, trial product is 7×10^{-19}
16-43. (a) 4.0×10^{-7}
(b) 4.7×10^{-6} mole/liter
16-44. 8×10^{-5} mole/liter
16-45. 0.10 g PbI_2
16-46. 0.35 g $CaSO_4$
16-47. 2.2×10^{-13} M
16-48. Hg^{2+}, Ag^+, Bi^{3+}, Sb^{3+}, Cd^{2+}, Zn^{2+}, Tl^+
16-49. 5.0×10^{-8} M
16-50. (a) $[Pb^{2+}] = 3.2 \times 10^{-6}$ M
$[Sr^{2+}] = 5.6 \times 10^{-5}$ M
(b) 0.758 g $PbSO_4$
0.817 g $SrSO_4$
16-51. (a) 7.5×10^{-13} M
(b) 1.1×10^{-4} M
16-52. (a) 1.2×10^{-11}
(b) 5.3×10^{-8} mole/liter
16-53. (a) 2.0×10^{-32}
(b) 1.6×10^{-22} mole/liter
16-54. 2.57

Chapter 17

17-1. (a) 2.2×10^{-11}
(b) 8.54
17-2. 4.73
17-3. (a) 5.6×10^{-10}
(b) 1.1×10^{-5} M
(c) 9.1×10^{-10} M
(d) 4.95
(e) 1.1×10^{-5} M
17-4. 11.35
17-5. (a) 3.86
(b) 1.4×10^{-4} M
17-6. (a) 4.9×10^{-5}
(b) 2.2×10^{-3} M
17-7. 8.30
17-8. 4.98
17-9. 2.8×10^{-7}
17-10. 1.3×10^{-2}
17-11. 5.0×10^{-14}
17-12. 1.1×10^{-47}
17-13. $[Hg^{2+}] = 2.8 \times 10^{-4}$ M
$[Cl^-] = 1.1 \times 10^{-3}$ M
$[HgCl_4^{2-}] = 0.50$ M
17-14. (a) 1.4×10^{-2}%
(b) $[Hg^{2+}] = 1.4 \times 10^{-5}$ M
$[SCN^-] = 5.5 \times 10^{-5}$ M
$[Hg(SCN)_4^{2-}] = 0.100$ M
17-15. (a) $[Pb(OH)_3^-] = 7.7 \times 10^{-3}$ M
$[Pb^{2+}] = 1.0 \times 10^{-16}$ M
$[OH^-] = 1.0$ M
(b) 7.7×10^{-3} mole/liter

17-16. (a) 2.2×10^{30}
(b) 0.11 *M*
17-17. 8.70
17-18. 4.87
17-19. 2.8×10^{-5}
17-20. 6.2×10^{-6}
17-21. (a) 4.5×10^{-6}
(b) 2.0×10^{-5} *M*
17-22. 0.850
17-23. 9.8×10^{-15}
17-24. 6.0×10^{-4} *M*
17-25. (a) $[Fe^{3+}] = [F^-] = 2 \times 10^{-4}$ *M*
(b) 4×10^{-6}
17-26. 5.6×10^{-4} *M*
17-27. (a) 4.0×10^1 g/liter
(b) 0.21 M
17-28. 0.40; 5.0×10^{-30}
17-29. (a) $K_w = K_h K_d$
(b) 1×10^{-8}
17-30. (a) 4.4×10^{12}
(b) 1.25

Chapter 18

18-1. 2.90×10^4 coul
18-2. 33.9 g Co
18-3. (a) 0.5585 g Fe; 0.5870 g Ni; 0.3467 g Cr
(b) 0.3063 amp
18-4. 2.57×10^3 sec
18-5. (a) 3.07 liters
(b) 1.38×10^4 sec
18-6. 1.60×10^{-19} coul/e^-
18-7. 12.3 g Zn
18-8. (a) +0.59 volt
(b) $Ni + Cu^{2+} \rightleftharpoons Ni^{2+} + Cu$
(c) Ni to Cu
(d) 1×10^{20}
18-9. (a) No
(b) 3×10^{-92}
18-10. (a) +1.20 volt
(b) $Sn + Br_2 \rightleftharpoons Sn^{2+} + 2Br^-$
(c) Sn to Pt
(d) 5×10^{40}
18-11. (a) +0.713 volt
(b) +0.319 volt
(c) −0.352 volt
(d) −0.346 volt
18-12. +1.19 volts
18-13. (a) 1.5×10^{-9}
(b) 3×10^{28}
(c) 2×10^{16}
(d) 1×10^{21}
(e) 2×10^{40}
18-14. +0.05 volt
18-15. 9.07 g
18-16. 3.83 amp
18-17. (a) 29.7 g Ag; 8.74 g Cu; 8.07 g Ni; 4.12 g Sc; 54.2 g Au
(b) 2.65×10^4 coul
18-18. 41.7
18-19. 6.52 min
18-20. 6.12 liters H_2; 3.06 liters O_2
18-21. (a) +1.562 volts
(b) +0.51 volt
(c) +0.43 volt
(d) +1.52 volts
(e) +0.462 volt
(f) +0.53 volt
18-23. (a) Ni, Al
(b) Ag^+, Br_2, Cu^{2+}
18-24. (a), (c), (d)
18-25. +0.23 volt
18-26. (a) +1.41 volts
(b) +0.49 volt
(c) +0.48 volt
(d) +0.15 volt
(e) −0.41 volt
(f) +1.55 volts
18-27. (a) 1×10^{20}
(b) 4×10^{15}
(c) 3×10^{-36}
(d) 1.1×10^{-18}
(e) 9×10^{17}
(f) 2×10^{61}
18-28. (a) 3×10^8
(b) 1.4×10^{28}
(c) 4×10^{54}

(d) 3×10^{24}
(e) 1.3×10^{66}

18-29. (a) $+0.76$ volt
(b) $+0.91$ volt

18-30. (a) 1.5×10^{-10}
(b) 2×10^{-18}

Chapter 19

19-1. -890.3 kJ

19-2. -1305.3 kJ/mole

19-3. 2.7 kJ

19-4. (a) $+9.8 \text{ J K}^{-1}$
(b) $+3.1 \text{ J K}^{-1}$
(c) -136.7 J K^{-1}
(d) -90.1 J K^{-1}

19-5. $\Delta H^\circ = -1560.0$ kJ
$\Delta E^\circ = -1553.8$ kJ

19-6. (a) $\Delta H^\circ = +235.6$ kJ
$\Delta E^\circ = +225.7$ kJ
(b) $\Delta H^\circ = +124.3$ kJ
$\Delta E^\circ = +119.3$ kJ

19-7. $+1.323 \times 10^3 \text{ J K}^{-1}$

19-8. $p_{CO} = p_{Cl_2} = 7.1$ atm

19-9. $K_p = 1.8 \times 10^{-7}$; No

19-10. 3.0×10^1

19-11. $\Delta H = +59.26$ kJ/mole
$\Delta E = +54.03$ kJ/mole
$\Delta S = +94.12 \text{ J mole}^{-1} \text{ K}^{-1}$
$\Delta G = 0$

19-12. (a) -95.2 kJ
(b) -394.4 kJ
(d) -384.5 kJ

19-13. (a) -818.4 kJ
(b) -344.8 kJ
(c) -394.6 kJ
(d) -1467.9 kJ

19-14. 357°C

19-15. $-30.1 \text{ J mole}^{-1} \text{ K}^{-1}$

19-16. $\Delta H^\circ = -283.0$ kJ
$\Delta S^\circ = -86.6 \text{ J K}^{-1}$
$\Delta G^\circ = -257.4$ kJ
Spontaneous

19-17. (a) $+863 \text{ J K}^{-1}$
(b) -257 kJ

19-18. (a) -63.6 kJ
(b) -97.7 kJ
(c) $+38.8$ kJ
(d) -72 kJ
(e) -19.6 kJ
(f) -26.2 kJ
(g) -43.6 kJ

19-19. (a) 2.03
(b) 1.7×10^{90}
(c) 1.3×10^{37}
(d) 5×10^{17}
(e) 1.3×10^{18}
(f) 4×10^{134}
(g) 5.2×10^5
(h) 8.8×10^2
(i) 8×10^{94}

19-20. (a) -68 kJ
(b) -1.2×10^2 kJ
(c) -3.2×10^2 kJ
(d) -4.7×10^2 kJ
(e) -89 kJ

19-21. (a) -108 kJ
(b) -1.6×10^2 kJ
(c) -8×10^1 kJ
(d) -2.4×10^2 kJ
(e) -26 kJ

19-22. $E^\circ = -0.27$ volt

Chapter 20

20-1. 8.0 mg

20-2. 2.2×10^{10} yr

20-3. 1.6×10^4 yr

20-4. 1.3×10^4 yr

20-5. 20.0 min

20-6. (a) $-4, -2$
(b) $0, +1$
(c) $0, 0$
(d) $0, -1$

20-7. 5.0×10^{19} atoms

20-8. 2.0 yr

20-9. 7.8×10^5 atoms decayed

20-10. 4.0 sec
20-11. 7.5 × 10^{18} atoms decayed
20-12. 2.1 × 10^{2} sec
20-13. (a) 8 α
(b) 6 β^-
20-14. (a) 22.4 liters
(b) 11.2 atm
(c) infinite time
20-15. 1.1 × 10^{19} atoms decayed
20-16. 2.4 × 10^{2} min
20-17. 24.8 min
20-18. 49.8 min
20-19. 1.6 × 10^{4} yr
20-20. 1.8 × 10^{18} atoms remain
20-21. 1.0 × 10^{-3}
20-22. 3.69 × 10^{15} disintegrations
20-23. 2.03 × 10^{13} sec
20-24. 2.39 hr
20-25. 2.72 × 10^{9} yr
20-26. (a) 1.07 × 10^{4} yr
(b) 503 yr
20-27. 14.2 yr
20-28. 2.04 liters
20-29. 92.9 yr
20-30. (a) ${}^{4}_{2}He$
(b) ${}^{1}_{0}n$
(c) ${}^{0}_{-1}e$
(d) ${}^{4}_{2}He$
(e) $3\,{}^{1}_{0}n$
20-31. (a) ${}^{39}_{20}Ca$
(b) ${}^{32}_{15}P$
(c) ${}^{194}_{79}Au$
(d) ${}^{27}_{14}Si$
(e) ${}^{244}_{98}Cf$
20-33. (a) ${}^{27}_{13}Al$
(b) ${}^{1}_{1}H$
(c) ${}^{30}_{15}P$
(d) ${}^{84}_{37}Rb$
(e) ${}^{1}_{0}n$
20-34. (a) ${}^{256}_{101}Md$
(b) ${}^{243}_{97}Bk$
(c) ${}^{64}_{29}Cu$
(d) ${}^{56}_{26}Fe$
(e) ${}^{257}_{103}Lr$
20-35. (a) ${}^{254}_{100}Fm$
(b) ${}^{253}_{99}Es$
20-36. (a) ${}^{93}_{41}Nb$
(b) ${}^{13}_{6}C$
(c) ${}^{230}_{90}Th$
20-37. 5.6 × 10^{-11} g
20-38. 11 g
20-39. (a) 21 g
(b) 2.3 × 10^{6} dis/sec/mg
20-40. 2.1 × 10^{-8} g KI
20-41. (a) 0.95 rad
(b) 0.95 rem
20-42. 4.04 × 10^{14} J
20-43. 2.95 × 10^{24} Mev
20-44. (a) 0.01882 amu
(b) 17.5 Mev;
2.80 × 10^{-12} J
(c) 1.05 × 10^{25} Mev/mole
20-45. 1. (a) 1.8 × 10^{-4} amu
(b) 0.17 Mev; 2.7 × 10^{-14} J
(c) 1.0 × 10^{23} Mev/mole;
1.6 × 10^{10} J/mole
2. (a) 6.17 × 10^{-3} amu
(b) 5.74 Mev; 9.21 × 10^{-13} J
(c) 3.46 × 10^{24} Mev/mole;
5.54 × 10^{11} J/mole
3. (a) 5.13 × 10^{-3} amu
(b) 4.78 Mev; 7.65 × 10^{-13} J
(c) 2.88 × 10^{24} Mev/mole;
4.61 × 10^{11} J/mole
20-46. (a) 6.92 Mev/nucleon
(b) 8.51 Mev/nucleon
(c) 8.39 Mev/nucleon
(d) 8.60 Mev/nucleon
(e) 8.73 Mev/nucleon
(f) 7.57 Mev/nucleon
20-47. 0.312 g
20-48. 1.48 × 10^{4} g
20-49. 1. (a) 4.42 × 10^{-3} amu
(b) 4.11 Mev
(c) 3.96 × 10^{8} kJ/mole
2. (a) 1.49 × 10^{-3} amu
(b) 1.39 Mev
(c) 1.34 × 10^{8} kJ/mole

3. (a) 1.791×10^{-2} amu
(b) 16.7 Mev
(c) 1.62×10^{9} kJ/mole
4. (a) 8.32×10^{-3} amu
(b) 7.75 Mev
(c) 7.48×10^{8} kJ/mole

20-50. (a) 2.57 Mev/nucleon
(b) 8.80 Mev/nucleon
(c) 8.38 Mev/nucleon
(d) 8.45 Mev/nucleon
(e) 7.94 Mev/nucleon
(f) 7.56 Mev/nucleon

Appendix I

I-1. 4
I-2. 1
I-3. 36
I-4. 25
I-5. 5
I-6. $\frac{18}{17}$
I-7. 2
I-8. 9
I-9. $\frac{82}{27}$
I-10. ± 6
I-11. $\pm\sqrt{5}$
I-12. 6, $-\frac{9}{2}$
I-13. 3.55, -1.88
I-14. 2, $-\frac{5}{6}$

Appendix II

II-1. (a) 3.5×10^{4}
(b) 6×10^{6}
(c) 2.35×10^{5}
(d) 1.4×10^{-4}
(e) 7.7×10^{-3}
(f) 3.45×10^{-2}

II-2. (a) 7500
(b) 0.00032
(c) 0.00000082
(d) 2240
(e) 50
(f) 89

II-3. (a) 3.368×10^{5}
(b) 1.74×10^{-5}
(c) 1.5×10^{13}
(d) 8×10^{-2}
(e) 4.8
(f) 6.0×10^{2}
(g) 4×10^{13}
(h) $\pm 6.0 \times 10^{4}$
(i) $\pm 5.0 \times 10^{-4}$
(j) 3.0×10^{-3}
(k) $\pm 2 \times 10^{-2}$

Appendix III

III-1. (a) 4.4771
(b) 3.2304
(c) 23.7797
(d) -18.7959
(e) -2.3640
(f) 1.3850

III-2. (a) 1.000×10^{4}
(b) 1.698×10^{-9}
(c) 6.31×10^{1}
(d) 6.03×10^{23}
(e) 3.420×10^{-4}
(f) 0.8831

III-3. (a) 8.8×10^{2}
(b) 1.4
(c) 4.6×10^{17}
(d) 4.0×10^{-14}
(e) 9.898
(f) 1.758×10^{-20}

Appendix IV

IV-1. (a) 1780
(b) 3.46
(c) 1.49
(d) 1.48
(e) 4.84
(f) 4.84
(g) 3.46
(h) 0.00852
(i) 3.92
(j) 685

IV-2. (a) 865
(b) 53.4
(c) 66.466
(d) 1300
(e) 0.69

IV-3. (a) 22.01
(b) 0.01
(c) 0
(d) 39.94
(e) 15

IV-4. (a) 9.00
(b) 5.25
(c) 36.81
(d) 1.0
(e) 1.38
(f) 6.6

IV-5. (a) 177
(b) 3700
(c) 400
(d) 2.028
(e) 8.94
(f) 260
(g) 1.43

Index